POLYMER AND
COMPOSITE RHEOLOGY

PLASTICS ENGINEERING

Founding Editor

Donald E. Hudgin

Professor
Clemson University
Clemson, South Carolina

1. Plastics Waste: Recovery of Economic Value, *Jacob Leidner*
2. Polyester Molding Compounds, *Robert Burns*
3. Carbon Black-Polymer Composites: The Physics of Electrically Conducting Composites, *edited by Enid Keil Sichel*
4. The Strength and Stiffness of Polymers, *edited by Anagnostis E. Zachariades and Roger S. Porter*
5. Selecting Thermoplastics for Engineering Applications, *Charles P. Mac-Dermott*
6. Engineering with Rigid PVC: Processability and Applications, *edited by I. Luis Gomez*
7. Computer-Aided Design of Polymers and Composites, *D. H. Kaelble*
8. Engineering Thermoplastics: Properties and Applications, *edited by James M. Margolis*
9. Structural Foam: A Purchasing and Design Guide, *Bruce C. Wendle*
10. Plastics in Architecture: A Guide to Acrylic and Polycarbonate, *Ralph Montella*
11. Metal-Filled Polymers: Properties and Applications, *edited by Swapan K. Bhattacharya*
12. Plastics Technology Handbook, *Manas Chanda and Salil K. Roy*
13. Reaction Injection Molding Machinery and Processes, *F. Melvin Sweeney*
14. Practical Thermoforming: Principles and Applications, *John Florian*
15. Injection and Compression Molding Fundamentals, *edited by Avraam I. Isayev*
16. Polymer Mixing and Extrusion Technology, *Nicholas P. Cheremisinoff*
17. High Modulus Polymers: Approaches to Design and Development, *edited by Anagnostis E. Zachariades and Roger S. Porter*
18. Corrosion-Resistant Plastic Composites in Chemical Plant Design, *John H. Mallinson*
19. Handbook of Elastomers: New Developments and Technology, *edited by Anil K. Bhowmick and Howard L. Stephens*

Additional Volumes in Preparation

POLYMER AND COMPOSITE RHEOLOGY

Second Edition, Revised and Expanded

RAKESH K. GUPTA

West Virginia University
Morgantown, West Virginia

CRC Press
Taylor & Francis Group
Boca Raton London New York

CRC Press is an imprint of the
Taylor & Francis Group, an **informa** business

CRC Press
Taylor & Francis Group
6000 Broken Sound Parkway NW, Suite 300
Boca Raton, FL 33487-2742

First issued in paperback 2019

ISBN-13: 978-0-367-39848-4

Library of Congress Cataloging-in-Publication Data
A catalog record for this book is available from the Library of Congress.

Visit the Taylor & Francis Web site at
http://www.taylorandfrancis.com

and the CRC Press Web site at
http://www.crcpress.com

To my wife, Gunjan

BHARYA MOOLAM GRIHASTHASYA
BHARYA MOOLAM SUKHASYA CHA
BHARYA DHARMA FALA VAPTYAI
BHARYA SANTAN BRIDHYAYE

*A wife is the quintessence of a home. She is the source of happiness and of
everything that is good in life and the future of children.*

Skanda Puran (Hindu scripture)

Preface

Several outstanding books on rheology have appeared since *Polymer Rheology* was published in 1977. This book remains unique in the simple and straightforward manner in which it introduces the concepts of rheology and the observed flow behavior of polymers to the student as well as to the practicing engineer. The first edition emphasized general principles and their practical implications rather than theoretical constructs. This practical, industrial orientation has been maintained in the second edition, and the book has been updated to include developments of the past quarter century. The book has also been expanded to give essentially equal coverage to single-phase and multiphase systems. A knowledge of chemorheology is often essential for understanding the processing of (thermosetting) polymer-matrix composites; discussion of reactive polymers has been included for this reason. Few, if any, of the currently available rheology books treat as wide a range of topics as are covered in the second edition.

The specific aims of the second edition are: (i) to present the general behavior of polymer melts, polymer solutions, suspensions, emulsions, foams, granular powders, and polymer composites during flow, together with physical explanations of the observations; (ii) to provide information on the effects of the different factors that influence flow behavior, (iii) to describe and recommend methods of measuring, calculating, or theoretically estimating flow properties of polymers and polymeric composites; and (iv) to direct readers to the appropriate technical literature for further study and consideration of current research issues. All this is done using the minimum amount of mathematics—not an easy undertaking given the universally accepted complexity of the subject.

In pursuit of these goals and after an introductory chapter, standard techniques for making rheological measurements are presented in Chapter 2. Thereafter, each chapter begins with an explanation of the practical and theoretical importance of the topic being examined. This is followed by a presentation of typical data and how these data might be represented in graphical form and also by means of empirical equations. The body of each chapter considers the use of any specialized instruments, data reduction when employing the most relevant rheological techniques, and the influence of the various material, geometrical, and processing variables on the property of interest, and it provides physical explanations for the observations. There is discussion, with minimum mathematics, of available theoretical models and their ability to both predict observed behavior and quantitatively represent the data. Each chapter also elaborates on work in progress and research needs for the future. Finally, there is a listing of complete citations to the technical literature. The book concludes with a short chapter on the enigma of melt fracture, an annoying instability of rheological origin, that limits production rates during polymer processing operations.

This is a practical book aimed at both practicing engineers and graduate students. It is a storehouse of information but with an emphasis on science rather than on technology. No specific background has been assumed of the reader, and it is hoped that the book will be as useful to a chemist or an engineer who wants to learn about rheology owing to the requirements of a new assignment as it will be to a student engaged in advanced research. The choice of topics and the depth of coverage have been dictated as much by my own research interests as by my perception of the importance of the subject at hand. Time will tell whether these choices have been judicious ones or not.

I would like to thank Professor Hota V. S. GangaRao, my colleague at West Virginia University, for having introduced me to the fascinating field of polymer composites and also for involving me in research on a variety of topics related to the processing and use of composite materials. Professor Raj Chhabra of the Indian Institute of Technology at Kanpur read the entire manuscript, as did Dr. Deepak Doraiswamy of the DuPont Company (also an Adjunct Professor in the Department of Chemical Engineering at West Virginia University); both of them corrected errors, made suggestions for improvement, and directed me to appropriate work in the literature. Deepak, in addition, coauthored Chapter 10 on solid-in-liquid suspensions. I am grateful to them both for their constant help and encouragement. The book also benefited from the several suggestions of Professor Jan Mewis and Dr. Kurt Wissbrun, who helped with and read Chapter 8 on the rheology of liquid crystalline polymers. Parts of the book were written during 1998/99 while I was on sabbatical leave as a Visiting Research Scientist at DuPont's Washington Works in Parkersburg, West Virginia; I thank Dr. Robert Cook for making my stay there possible, and I thank all my DuPont associates for their hospitality. In the course of the three years that it took me to complete

the writing of this book. I badgered innumerable friends and professional col-
leagues for reprints, preprints, and thesis copies. Their courtesy in acceding to
my frantic requests made my work and the work of the ever-helpful interlibrary
loan personnel in our Evansdale Library so much easier. I also wish to thank
Ms. Linda Rogers, who typed all the equations for me and helped me incorporate
all the revisions in the manuscript.

I did not know the late Larry Nielsen personally even though I worked
briefly for the Monsanto Company. I feel honored to have been asked to revise
and expand *Polymer Rheology*. I have made a sincere effort to maintain the character of the first edition, and I have attempted to retain as much as possible of
the material that appeared in that edition. I hope that the readers will recognize
this book to be as much Nielsen's as mine.

Rakesh K. Gupta

Contents

Contents

1

Introduction to Polymer Rheology

I. RHEOLOGY

Rheology can be defined as the science of the flow and deformation of matter. For low-molecular-weight fluids, the study of rheology involves the measurement of viscosity. For such fluids, the viscosity depends primarily upon the temperature and hydrostatic pressure. However, the rheology of high-molecular-weight liquids, whether neat or filled, is much more complex because polymeric fluids show nonideal behavior. In addition to having complex shear viscosity behavior, polymeric liquids show elastic properties, such as unequal normal stresses in shear and a prominent tensile viscosity in extension. All these rheological properties depend upon the rate of deformation, the molecular weight and structure of the polymer, and the concentration of various additives and fillers, as well as upon the temperature. In addition, even at a constant rate of deformation, stresses are found to depend on time.

The subject of rheology is very important for both polymers and polymeric composites. This is true for two reasons. Firstly, flow is involved in the processing and fabrication of such materials in order to make useful objects. Thus, fluid rheology is relevant to polymer processing and determines stress levels in operations such as extrusion, calendering, fiber spinning, and film blowing. Similarly, rheology influences residual stresses, cycle times, and void content in composite processing operations such as bag molding, compression molding, and injection molding. Clearly, a quantitative description of polymer and composite rheology is essential for developing models of the various polymer processing operations; these models can be employed for process optimization and for predicting the onset of flow instabilities. In the use of polymers, though, it is generally the

1

mechanical properties that are important. Mechanical behavior, however, is influenced by rheological behavior, and this is the second important reason for studying polymer rheology. For example, molecular orientation has dramatic effects on the mechanical properties of molded objects, fibers, and films. For short-fiber composites, fiber orientation plays the role of molecular orientation in unfilled systems. The kind and degree of molecular or fiber orientation are largely determined by the rheological behavior of the polymer and the nature of the flow in the fabrication process.

Rheology is involved in many other aspects of polymer science. For example, many polymers are made from emulsions of monomers in stirred reaction vessels. The resulting latices flow through pipes and may end up as a paint that is applied to a surface by some process in which the rheological properties of the latex must be controlled carefully. Plastisols, which are a suspension of a polymer in a liquid, are fabricated into useful objects by processes such as rotational molding. Powdered polymers or granules must flow from bins and must perform properly in a fabrication process such as rotational molding and in powder coatings. The rheology of polymer powders is important also in the first sections of extruders and in injection molding machines before the polymer softens to a liquid. Note that at a less fundamental level, rheological measurements, both transient and steady state, can be employed for product characterization and quality control purposes. Such measurements are often also used for examining and understanding the interaction of the different constituents of a multicomponent or multiphase mixture and their influence on the flow and other properties of such materials.

II. MATERIAL FUNCTIONS IN VISCOMETRIC FLOWS

The flow field that is generated in most standard instruments used to measure rheological properties is a particular kind of shear flow called *viscometric flow*. All the motion in a viscometric flow is along one coordinate direction, say, x_1 in Figure 1.1, the velocity varies along a second coordinate direction, say, x_2, and the third direction is neutral. This is illustrated schematically in Figure 1.1, where a liquid is confined between two flat plates of area A separated by a distance D. When the upper plate is moved in the x_1 direction relative to the lower plate, the liquid is sheared with the amount of shear strain γ being defined as

$$\gamma = \tan \theta = \frac{\text{Amount of shear displacement } S}{\text{Distance between shearing surfaces } D} \tag{1.1}$$

The rate of shear strain will then be

$$\dot{\gamma} = \frac{\text{Relative velocity}}{\text{Distance } D} \tag{1.2}$$

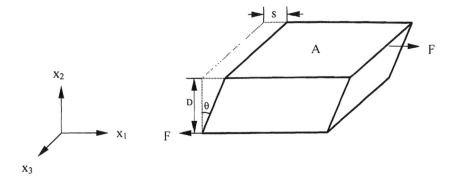

FIGURE 1.1 Schematic diagram for the measurement of shear viscosity.

which equals dv_1/dx_2 for small values of D. The rate of shear strain is commonly referred to as the *velocity gradient* or more simply as just the *shear rate*. It is evident that the shear rate will be independent of time if the relative velocity is constant.

A constant force F is typically required to move the top plate in Figure 1.1 at a constant velocity relative to the lower plate. This is a shear force, and on dividing this quantity by the area A of the shear face one gets the shear stress τ. For a Newtonian liquid, the shear stress is directly proportional to the shear rate, with the constant of proportionality being called the *dynamic viscosity* η. If the fluid is not Newtonian, a plot of the shear stress against the shear rate is not a straight line but a curve, such as the lower solid line shown in Figure 1.2. The liquid may be Newtonian at very low shear rates to give a limiting viscosity η_0 from the initial slope of the τ-versus-$\dot\gamma$ curve. When the τ-versus-$\dot\gamma$ curve is nonlinear, the viscosity (the adjective *dynamic* is commonly omitted) is no longer a constant, and it becomes a function of the shear rate. It is now defined as the slope of the secant line from the origin to the shear stress at the given value of the shear rate; that is,

$$\eta = \frac{\tau}{\dot\gamma} \tag{1.3}$$

Another kind of non-Newtonian behavior is Bingham plastic behavior. This is illustrated by the upper solid line in Figure 1.2. Here the τ-versus-$\dot\gamma$ curve is linear, but it does not pass through the origin; the value of the intercept on the shear stress axis is called the *yield stress*. Note that the slope of the shear stress curve at the chosen value of the shear rate is known as the consistency η_c.

The SI unit for viscosity is Pa-s, where Pa is the abbreviation for Pascal, which is Newtons per square meter. In the past, it was common to measure vis-

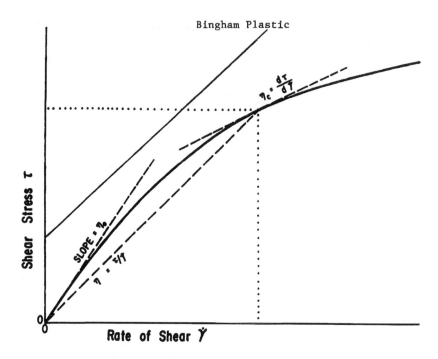

FIGURE 1.2 Measurement of viscosity from shear stress versus rate of shear curves.

cosity in the cgs unit of poise, which is the same as dyne second per cm² or g/cm-s. The viscosity of water is about 1 m Pa-s, or 1 centipoise, since 1 Pa-s equals 10 poise. Typical polymer melts have viscosities generally of the order of 10^2–10^3 Pa-s.

In general, it is found that the shear viscosity of polymers decreases as the shear rate increases. This behavior is known as *shear thinning*, and an important consequence of this result is that the well-known Hagen–Poiseuille equation does not hold for the flow of polymers through a circular tube. In particular, one finds that on doubling the flow rate through the tube, the necessary pressure drop is less than doubled.

For a fluid at rest, it is easy to show [1] that all the surface stresses that are present within the fluid are normal stresses and that at any location, these are not only equal in magnitude but also independent of the orientation of the surface. This single normal stress is compressive in character, and it is commonly called the *pressure* and denoted by the symbol p. From the definition of a fluid, the absence of motion of course implies the absence of shear stresses. Conversely, for a fluid in motion, it must necessarily be true that either shear stresses are present or the normal stresses are unequal. For the situation depicted in Figure 1.1, we have seen that flow arises due to the application of a shear stress. If the liquid being sheared is polymeric, one finds that the normal stresses along the three coordinate directions (see Fig. 1.3) become unequal; for Newtonian fluids, however, normal stresses remain equal to each other. Due to the fact that most liquids can be considered incompressible for all practical purposes, the application of equal normal stresses (or pressure) along the three coordinate directions does not lead to deformation or a change in volume; it is only unequal normal stresses that cause motion and deformation. As a consequence, one cannot determine the absolute value of the pressure from a measurement of fluid deformation, which is a measure of the change in distances between material points or material planes; one can only deduce normal stress differences.

It is common practice to define the first and second normal stress differences as follows:

First normal stress difference $N_1 = \sigma_{11} - \sigma_{22}$ $\hspace{2cm}$ (1.4)

Second normal stress difference $N_2 = \sigma_{22} - \sigma_{33}$ $\hspace{2cm}$ (1.5)

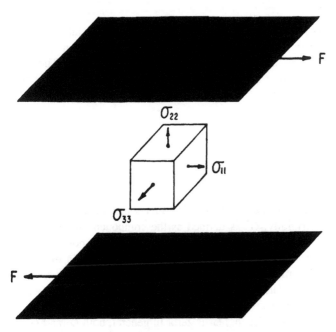

FIGURE 1.3   Schematic diagram showing the notation for normal stresses in a shear field.

While the first normal stress difference tends to force the shear plates in Figure 1.3 apart, the second normal stress difference tends to create bulges in the polymer at the edge of the plates either parallel or perpendicular to the direction of the applied force F. Indeed, it is the presence of the first normal stress difference in shear that is responsible for the swelling of an isothermal jet of polymer on emerging into air from a die or capillary. It is found that the normal stress differences are even functions of the shear rate, and it is customary to define the first and second normal stress coefficients, ψ_1 and ψ_2, as ratios of the respective stress differences with the square of the shear rate. At low shear rates, both these coefficients tend to attain constant values.

In cases where the relative velocity of the shearing plates in Figure 1.3 is not constant but varies in a sinusoidal manner so that the shear strain and the rate of shear strain are both cyclic and have a small amplitude, the shear stress is also sinusoidal. In these dynamic mechanical experiments, the stress is out of phase with the rate of strain unless the fluid is Newtonian. In this situation, a complex viscosity η^* can be measured. In general, the complex viscosity contains an elastic component in addition to a term similar to the ordinary steady state viscosity. The complex viscosity is defined by:

$$\eta^* = \eta' - i\eta'' \tag{1.6}$$

The dynamic viscosity η' is related to the steady state viscosity and is the part of the complex viscosity that measures the rate of energy dissipation. Similarly, the imaginary viscosity η'' measures the elasticity, or stored energy. These two viscosities are computed from the real and imaginary parts of the shear modulus using the following relations:

$$\eta'' = \frac{G'}{\omega} \tag{1.7}$$

$$\eta' = \frac{G''}{\omega} \tag{1.8}$$

where ω is the frequency of the oscillations in radians per second. The storage modulus, or dynamic rigidity, G', is defined as the component of the stress in phase with the strain divided by the strain amplitude; the loss modulus, G'', is defined as the component of the stress out of phase with the strain divided by the strain amplitude. The ratio of the loss modulus to the storage modulus, G''/G', is known as $\tan \delta$ and is an alternate measure of energy dissipation in viscoelastic materials. Note that for a perfectly elastic material, G'' is zero, while for Newtonian liquids (that are perfectly viscous) G' is zero. In general, both G' and G'' are functions of frequency.

The variation, with the rate of deformation, temperature and other variables, of the different material functions defined in this section is discussed in detail in later chapters for a large variety of polymeric fluids.

III. EXTENSIONAL FLOW

Besides viscometric flow, the other major category of flow of practical interest that can be generated in the laboratory is extensional flow. In mathematical terms, extensional flow may be represented in a rectangular Cartesian coordinate system x_i by the following set of equations for the three components of the velocity vector:

$$v_1 = \dot{\varepsilon}_1 x_1 \tag{1.9}$$

$$v_2 = \dot{\varepsilon}_2 x_2 \tag{1.10}$$

$$v_3 = \dot{\varepsilon}_3 x_3 \tag{1.11}$$

which also defines the stretch rates ε_i. In uniaxial extension at constant stretch rate (see Fig. 1.4), $\dot{\varepsilon}_1 = \dot{\varepsilon}$ and $\dot{\varepsilon}_2 = \dot{\varepsilon}_3 = -\dot{\varepsilon}/2$. The distance between material planes that are perpendicular to the flow direction increases with time in extensional flow. A material plane is a surface that always contains the same material points or particles. As material planes move apart, polymer molecules tend to

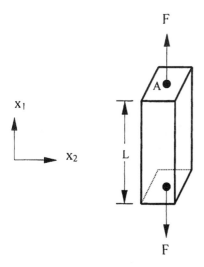

FIGURE 1.4 Schematic diagram for the measurement of extensional, or tensile, viscosity.

uncoil, and ultimately one can even have stretching of chemical bonds which results in chain scission. Stresses in the flow direction can, therefore, reach fairly large values. By integrating Eq. (1.9), one finds that, at constant stretch rate, the length of a strip of material increases exponentially:

$$\ln \frac{L}{L_0} = \dot{\varepsilon} t \tag{1.12}$$

in which L_0 is the initial length and L is the length at some later time t.

The only stress that one can measure experimentally in uniaxial extension is the tensile stress σ_E, which, with reference to Figure 1.4, is

$$\sigma_E = \frac{F}{A} \tag{1.13}$$

For constant-stretch-rate homogeneous deformation that begins from rest, one defines a tensile stress growth coefficient

$$\eta_E^+ (t, \dot{\varepsilon}) = \frac{\sigma_E}{\dot{\varepsilon}} \tag{1.14}$$

which has the dimensions of viscosity. The limiting value of η_E^+ as time approaches infinity is termed the *tensile* or *extensional* or *elongational viscosity*, η_E. In general, η_E is a function of the stretch rate, although in the limit of vanishingly low stretch rates [2]

$$\lim_{\varepsilon \to 0} \frac{\eta_E}{\eta_0} = 3 \tag{1.15}$$

where η_0 is the zero shear rate viscosity.

For Newtonian liquids, the extensional viscosity is three times the shear viscosity; but for polymeric liquids, the tensile viscosity may exceed the zero shear viscosity by a couple of orders of magnitude. Extensional viscosity is of great practical importance when polymers flow through channels or tubes in which the cross-sectional area is decreasing. Examples include flow through porous media, the spinning of fibers, and the filling of molds in injection molding.

IV. THE STRESS TENSOR

If one isolates a rectangular parallelepiped of material having infinitesimal dimensions such as shown in Figure 1.5, one finds that two kinds of forces act on the material element. These are body forces and surface forces. *Body forces* result from the action of external fields such as gravity; *surface forces* express the influence of material outside the parallelepiped but adjacent to a given surface.

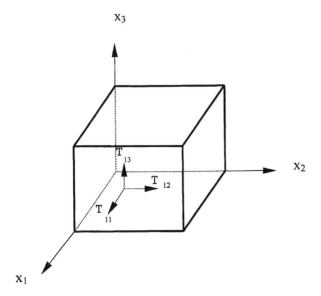

FIGURE 1.5 Diagram showing the notation used for components of the stress tensor.

Dividing the surface force by the area on which it acts yields the *stress vector*. Since the parallelepiped has six surfaces, there are six stress vectors.

Because each of the stress vectors can be resolved into three components parallel to the three coordinate axes, one has a total of 18 components. These are labeled T_{ij} or σ_{ij}, where the two subscripts help to identify a specific component. The first subscript, i, identifies the surface on which the stress acts; the surface, in turn, is identified by the direction of the outward-drawn normal. The second subscript, j, identifies the direction in which the stress component acts. Nine of the 18 components can be represented using a 3×3 matrix, called the *stress tensor*:

$$\begin{bmatrix} T_{11} & T_{12} & T_{13} \\ T_{21} & T_{22} & T_{23} \\ T_{31} & T_{32} & T_{33} \end{bmatrix}$$

T_{11}, for example, acts on a surface whose outward drawn normal points in the positive x_1 direction; the stress component itself also acts in the same direction. The other nine components of the stress are the same as these nine, but they act on opposite faces.

By means of a torque balance on a cubic element, it can be shown (see Ref. 1, for example) that T_{ij} equals T_{ji}. Thus, the stress tensor is symmetrical and only six of the nine components are independent quantities. Further, by examining the equilibrium of the tetrahedron shown in Figure 1.6, it can be demonstrated [1] that if the normal to the inclined surface is $\hat{n}$, the components of the surface stress $\underset{\sim}{f}$ acting on that surface, f_i, are (in matrix notation):

$$\begin{bmatrix} f_1 \\ f_2 \\ f_3 \end{bmatrix} = \begin{bmatrix} T_{11} & T_{12} & T_{13} \\ T_{12} & T_{22} & T_{23} \\ T_{13} & T_{23} & T_{33} \end{bmatrix} \begin{bmatrix} n_1 \\ n_2 \\ n_3 \end{bmatrix} \tag{1.16}$$

where n_i are the components of $\hat{n}$. Knowing the components of the stress tensor, therefore, allows one to obtain the stress vector acting on any plane described by the unit normal $\hat{n}$.

It is one of the major goals of rheologists to try to relate the stress tensor to the strain or the rate of strain. Recall that in Section II, it was shown that T_{12} is a unique function of the shear rate, while in Section III the net tensile stress was related to the stretch rate; the net tensile stress is nothing but $T_{11} - T_{22}$. These relationships between specific stress components and specific rate-of-strain components are known as *material functions*. The general equation relating the

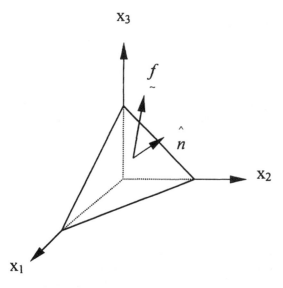

FIGURE 1.6 Equilibrium of a tetrahedron.

stress tensor to the three-dimensional strain or rate of strain tensor (the word *tensor* is simply another term for *matrix*) is known as a *constitutive equation*.

Due to fluid incompressibility, though, one cannot relate T_{ij} to deformation. This is because no amount of pushing, i.e., the application of hydrostatic pressure, can cause a change in volume. It is only when pressures are unequal that a strain, which can be understood as a change in the distance between neighboring particles, takes place. It is for this reason that it is usual to separate the stress tensor into two parts:

$$\begin{bmatrix} T_{11} & T_{12} & T_{13} \\ T_{12} & T_{22} & T_{23} \\ T_{13} & T_{23} & T_{33} \end{bmatrix} = \begin{bmatrix} -p & 0 & 0 \\ 0 & -p & 0 \\ 0 & 0 & -p \end{bmatrix} + \begin{bmatrix} \tau_{11} & \tau_{12} & \tau_{13} \\ \tau_{12} & \tau_{22} & \tau_{23} \\ \tau_{13} & \tau_{23} & \tau_{33} \end{bmatrix} \quad (1.17)$$

where p is the hydrostatic pressure whose presence causes no strain for incompressible materials and the τ_{ij} are components of the extra stress tensor whose presence causes strain to take place. Strain, therefore, is related to the extra stress $\underset{\sim}{\tau}$ rather than to the total stress $\underset{\sim}{T}$.

V. THE RATE OF STRAIN TENSOR

The shear rate defined in Eq. (1.2) can be generalized to three dimensions by defining the rate-of-strain tensor $\underset{\sim}{\dot{\gamma}}$ as

$$\dot{\gamma}_{ij} = \left(\frac{\partial v_i}{\partial x_j} + \frac{\partial v_j}{\partial x_i} \right) \quad (1.18)$$

and it is evident that for the flow situation in Figure 1.1, the shear rate is simply the 1,2 component of $\underset{\sim}{\dot{\gamma}}$. Furthermore, for a Newtonian liquid,

$$T_{ij} = \tau_{ij} = \eta \dot{\gamma}_{ij} \quad (1.19)$$

provided $i \neq j$, and

$$T_{ii} = -p + \eta \dot{\gamma}_{ii} \quad (1.20)$$

in which both i and j can be 1, 2, or 3.

Everything that we have defined so far has been with reference to a rectangular Cartesian coordinate system. In numerous practical situations, though, it is more appropriate to work in curvilinear coordinates. This parallel treatment may be found in more advanced books on rheology (see Ref. 3, for example).

Finally, we remark that the development of constitutive equations or, equivalently, the formulation of expressions similar to Eqs. (1.19) and (1.20) but applicable to polymeric fluids is merely a means to an end. In modeling polymer processing operations, one needs to solve boundary value problems involving the flow of polymeric fluids. The solution procedure involves inserting the constitutive equation into the momentum and energy balance equations and solving the resulting coupled differential equations subject to appropriate boundary conditions. In this regard, the no-slip boundary condition, which states that at a fluid–solid boundary the velocity of the fluid is the same as that of the solid, has been an article of faith for an extremely long time. Recent evidence [4], though, suggests that this boundary condition may not be universally valid. Indeed, slip at the solid–liquid boundary has been hypothesized to be the cause of polymer processing instabilities such as melt fracture; these instabilities tend to limit the rate of production during processes such as extrusion, fiber spinning, and film blowing [5].

REFERENCES

1. I. H. Shames. Mechanics of Fluids. 3rd ed. McGraw-Hill, New York, 1992.
2. K. Walters. Rheometry. Chapman and Hall, London, 1975.
3. R. B. Bird, R. C. Armstrong, and O. Hassager. Dynamics of Polymeric Liquids. 2nd ed. Vol. 1. Wiley, New York, 1987.
4. A. V. Ramamurthy. Wall slip in viscous fluids and influence of materials of construction. J. Rheol. 30:337–357 (1986).
5. C. J. S. Petrie and M. M. Denn. Instabilities in polymer processing. AIChE J. 22: 209–236 (1976).

2

Instruments for Shear Rheology

I. INTRODUCTION

A great variety of instruments have been used to measure the rheological properties of polymer solutions and polymer melts when these are subjected to shear
deformation [1–3]. These instruments range in sophistication from simple melt
indexers employed for quality control purposes to very complicated, research-
grade rheometers capable of measuring, as a function of temperature and rate of
deformation, both steady-state and transient values of the various material functions defined in the previous chapter. As might be expected, these instruments
span a wide price range [2,3], depending on the capability of the viscometer, and
they are manufactured and marketed by a large number of manufacturers all over
the world. In this chapter, we consider general-purpose, shear instruments that
one might expect to find in any laboratory devoted to rheological work; these
include capillary viscometers as well as rotational viscometers. We describe the
essential features of each type of instrument, present the basic equations used
for data analysis, and discuss equipment limitations and sources of error. A consideration of actual data, their possible molecular interpretation, and representation using constitutive equations is deferred to later chapters. Extensional viscometers are examined in Chapter 7, while specialized instruments are introduced
throughout this book as appropriate.

Since shear viscosity has been measured for a very long time, we might
expect shear rheometry to be a mature field. However, new instruments are continually being introduced. Some, such as the sliding-plate rheometer, involve new
concepts; others improve on existing instruments as far as the ease of measurement and data analysis or the range of measurement is concerned. Yet others

allow us to make measurements on complex materials or materials that might be reactive or temperature sensitive.

II. CAPILLARY VISCOMETERS

The standard instrument for the measurement of the steady shear viscosity of polymeric fluids at shear rates witnessed in typical polymer processing operations is the capillary viscometer, shown schematically in Figure 2.1. In this device, the polymer sample is forced by a piston or by pressure from a reservoir through a capillary; to obtain consistent results, the ratio of the reservoir diameter to the capillary diameter has to exceed 12. In commercial instruments, the capillary is usually vertical, and it can be detached from the viscometer so that we may make measurements with other capillaries having different entrance angles or different

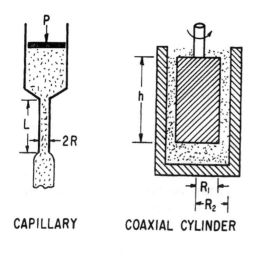

CAPILLARY   COAXIAL CYLINDER

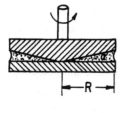

CONE AND PLATE

FIGURE 2.1   Schematic diagrams of three types of viscometers.

values of length L or diameter D. In actual use, we fill the cold reservoir, called the *barrel*, with a polymer solution or with pellets of the neat or filled polymer and then heat the system to the test temperature. Once thermal equilibrium is attained, extrusion is begun, and the volumetric flow rate Q of polymer coming from the capillary at a given pressure drop is recorded. If the plunger speed can be specified, as is commonly the case, it is not necessary actually to collect the extrudate. A load cell is employed to measure the force of extrusion.

The capillary rheometer has a number of advantages. First, the instrument is relatively easy to fill. This is an important consideration, since most polymer melts are too viscous to pour readily even at high temperatures. The test temperature and rate of shear are varied readily. The shear rates and flow geometry are similar to the conditions actually found in extrusion and in injection molding. In addition to the viscosity, some indication of polymer elasticity is found from the die swell of the extrudate. Finally, factors affecting the surface texture of the extrudate and the phenomenon of melt fracture can be studied. However, there are disadvantages too. The principal disadvantage is that the rate of shear is not constant but varies with radial position within the capillary. Another disadvantage is the necessity of making a number of corrections in order to get accurate viscosity values, especially with low-viscosity materials or when short capillaries are used.

In general, the velocity profile in the capillary is fully developed only in the region away from the inlet and the exit. At the capillary ends, there must be velocity rearrangements, and the pressure gradient varies with axial position and is larger than that in the fully developed region. If, however, the ratio of the capillary length to the capillary diameter is large, say, 50 or more, we may neglect end effects and assume that the pressure gradient in the entire capillary is $(p_1 - p_2)/L$, the difference between the inlet and exit pressures divided by the capillary length. By means of a force balance, then, the magnitude of the shear stress at the capillary wall, τ_w, is:

$$\tau_w = (p_1 - p_2)\frac{R}{2L} \tag{2.1}$$

where R is the capillary radius. To determine the shear viscosity, we need to divide the wall shear stress by the shear rate at the wall, $\dot{\gamma}_w$. For any non-Newtonian fluid, the expression for the wall shear rate is given as [4]:

$$\dot{\gamma}_w = \frac{4Q}{\pi R^3}\left(\frac{3}{4} + \frac{1}{4}\frac{d \ln Q}{d \ln \tau_w}\right) \tag{2.2}$$

To measure the viscosity as a function of shear rate, the flow rate is determined as a function of the pressure drop. This can be converted to flow rate as a function of the wall shear stress using Eq. (2.1), which then allows us to com-

pute the derivative appearing in Eq. (2.2) at any given value of the wall shear
stress. The ratio of this wall shear stress to the corresponding wall shear rate
calculated from Eq. (2.2) gives the shear viscosity at one shear rate—the one
corresponding to the chosen value of the shear stress at the wall. By repeating
this process at other values of the wall shear stress, we can trace out the entire
η-versus-$\dot\gamma$ curve or the flow curve.

If the non-Newtonian fluid obeys the power law, in which the shear stress
depends upon the shear rate to the nth power, Eq. (2.2) takes the form:

$$\dot\gamma_u = \frac{4Q}{\pi R^3}\left(\frac{3n + 1}{4n}\right) \tag{2.3}$$

where

$$n = \frac{d \log \tau}{d \log \dot\gamma} \tag{2.4}$$

In the special case of a Newtonian liquid, n equals unity and the wall shear
rate is simply $4Q/\pi R^3$. It is, therefore, seen that the quantity in brackets in Eq.
(2.2) corrects the rate of shear for non-Newtonian liquids [5]; this is known as
the *Rabinowitsch correction* and $4Q/\pi R^3$ is called the *apparent wall shear rate*.
Clearly, for a Newtonian liquid, the shear viscosity is:

$$\eta = \frac{\pi R^4(p_1 - p_2)}{8LQ} \tag{2.5}$$

The use of Eq. (2.5), which is the well-known Hagen–Poiseuille equation,
allows us to use Newtonian liquids to calibrate the viscometer and to check that
the experimental procedure is satisfactory. Note that the shear rate is always a
maximum at the wall but that the velocity is zero at the wall. The characteristics
of capillary flow are shown in the bottom section of Fig. 2.2. The solid lines are
for Newtonian liquids; the dashed lines are typical of polymeric non-Newtonian
fluids. Under conditions in which Q is the same for both Newtonian and non-
Newtonian fluids, the area under the velocity distribution curves should be the
same rather than as shown in Fig. 2.2, where conditions have been adjusted to
match the velocities at the center of the capillary.

A small-diameter capillary is preferred in order that one may make measurements using small-sized samples (working volumes of 25 ml are common).
Even for neat polymer melts, the practical lower limit is about 0.025 cm; otherwise, blockage can occur due to the presence of impurities in the polymer. Furthermore, the use of capillaries having large L/D values is desirable for theoretical
reasons. However, this is possible only for low-viscosity liquids for which the
pressure drop would otherwise be small. When working with molten polymers,
long capillaries cannot be used due to thermal degradation problems and because

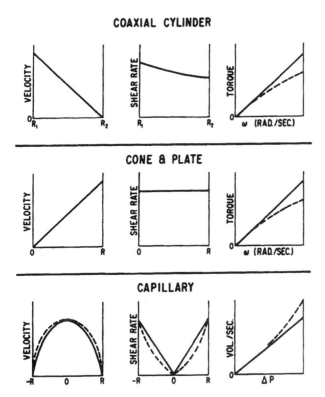

FIGURE 2.2 Characteristics of the three viscometers shown in Fig. 2.1. Solid lines refer to Newtonian fluids; broken lines are typical of non-Newtonian polymer melts.

the pressure drops are excessive; the latter fact causes problems with data analysis, since the viscosity can depend on pressure at high pressures.

If short capillaries are used, the variation in the pressure gradient along the tube axis cannot be neglected, and the effective length of the capillary becomes greater than the true length. This is taken into account through the Bagley [6,7] correction. The shear stress at the wall of Eq. (2.1) becomes

$$\tau_w = \frac{R(p_1 - p_2)}{2(L + eR)} = \frac{p_1 - p_2}{2(L/R + e)} = \frac{(p_1 - p_2) - p_0}{2L/R} \tag{2.6}$$

The Bagley correction factor e should be independent of capillary length, but in general it does vary somewhat with L/R because of the elasticity of polymer melts. The Bagley correction is determined by measuring the pressure drop at constant rate of shear for several capillary lengths and extrapolating to zero pressure drop, as shown in Figure 2.3. In Eq. (2.6), p_0 is the pressure drop corresponding to a capillary of zero length for a given rate of shear.

The upper limit on the shear rate in capillary viscometers is about 10^5-10^6 sec^{-1}. This comes about for a variety of reasons. As mentioned earlier, these instruments come in two basic designs, which differ in the method of melt extrusion. In the pressure-driven instrument, an inert gas is used to pump the liquid out of the reservoir, and we measure the volumetric flow rate corresponding to the applied pressure drop. For a capillary of given length and diameter, the maximum shear rate is set by the maximum pressure, but it can be increased further by increasing the capillary diameter. Increasing the diameter, though, causes a rather disproportionate increase in the volumetric flow rate, which can also lead to nonisothermal conditions due to viscous heating [8]. In a plunger-driven instrument, the extrusion velocity is fixed and the corresponding pressure drop is mea-

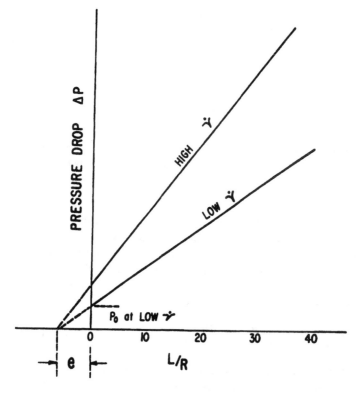

Figure 2.3 Bagley correction for capillary rheometers.

sured. Now the maximum shear rate is decided by the maximum speed of the plunger, but it can be increased further by reducing the capillary diameter. However, we cannot exceed the upper limit of the force transducer; for instruments designed to work with filled polymers, this force limit may be as high as 100 kN. In addition, data cannot be trusted once instabilities such as melt fracture set in. The lower limit of operation of capillary viscometers is about $1-10$ sec^{-1}. This is determined by the occurrence of drips in a pressure-driven instrument, while for a plunger-type viscometer the lower limit is set by the amount of friction between the plunger and the barrel. As the plunger speed decreases, friction can become a significant part of the total pressure drop and make the viscosity measurements unreliable. More details about capillary viscometry are available in References 1–4 and 9. The development of a capillary viscometer capable of operating at ambient pressures substantially above atmospheric has been described by Galvin et al. [10]. Commercial instruments are also available. For example, Polymer Laboratories (now part of Rheometrics Scientific) markets instruments that can make measurements at pressures as high as 200 Mpa.

Almost all present-day commercial instruments are plunger driven and are quite versatile. Some, such as the Rosand RH7 series viscometer, can operate at temperatures upto 500 C, can have twin capillaries that allow two simultaneous experiments, and have the facility of a nitrogen purge for work with materials that are sensitive to moisture and oxygen. In addition, some viscometers have pressure transducers mounted near the capillary entrance. This allows us to obtain a more accurate estimate of the pressure drop, since we do not have to account for pressure losses in the barrel. Often these viscometers are used as extruders in laboratory experiments, and it is common to employ them for spinning fibers and for determining the melt strength of polymers.

Although it is a fairly routine matter to use a capillary viscometer to measure the rheological properties of polymer solutions and polymer melts, suspensions and emulsions have to be treated with care. For both suspensions and emulsions, data analysis can be complicated due to (1) structural changes that occur during the process of measurement, and (2) apparent slip at the viscometer wall due to depletion of the dispersed phase in the vicinity of a solid surface. These effects are considered in later chapters. A note of caution also needs to be sounded on the practical side. Highly loaded suspensions are known to be shear thickening, and the sudden increase in the shear stress can make the viscometer seize, resulting in physical harm to the instrument and the investigator.

In closing this section, we note that the theory of Poiseuille flow through slits is similar to that for circular tubes [4], and slit rheometers are also commercially available [2,3]. An advantage of slit rheometry is that shear rates greater than 10^6 sec^{-1} can be attained and isothermal conditions still maintained. As shown by Lodge in Collyer and Clegg [3], this feature can be used to measure normal stress differences at shear rates higher than those accessible with rotational instruments. We discuss normal stress differences in detail in Chapter 5.

III. COAXIAL CYLINDER VISCOMETERS

The coaxial cylinder, or concentric cylinder, or Couette viscometer, is frequently used for measuring the steady shear viscosity and dynamic mechanical properties of low-viscosity liquids, polymer solutions, solid-in-liquid suspensions, and emulsions. However, it is used only occasionally at high temperatures or for high-viscosity melts. In this instrument, as shown in Figure 2.1, the fluid sample is placed in the annular space between two concentric cylinders. In one version of the apparatus, in order to measure the liquid viscosity, either one of the two cylinders is rotated at a constant speed while the torque acting on the other (stationary) cylinder is measured. In another version of the viscometer, one cylinder is held stationary while the other one is rotated under the influence of a constant torque; the speed of rotation is measured by some kind of a tachometer device. When it is desired to measure dynamic properties, the constant speed of rotation is replaced by oscillation at a fixed frequency and small amplitude.

A major advantage of the coaxial cylinder viscometer is the nearly constant shear rate throughout the entire volume of fluid being measured if the space between cylinders is small; this criterion is satisfied by ensuring that the ratio of the diameter of the inner cylinder to the diameter of the outer cylinder exceeds 0.97. This is an important factor with non-Newtonian polymeric fluids, in which the viscosity may be strongly dependent upon the rate of shear. Coaxial cylinder instruments are simple to use and are calibrated easily, and the corrections can be small. Most of these instruments are able to go to very low deformation rates, which is an advantage when working with blood and biological fluids, where one does not want structure alteration during the process of measurement. A major disadvantage of these instruments is the difficulty in filling them with a very viscous polymer melt. A less important, but annoying, disadvantage is the creeping of the polymer up the shaft of the inner cylinder because of normal stresses developed in the polymer by the rotation of the cylinder.

For Newtonian liquids, the important equations for coaxial cylinder viscometers (neglecting end corrections) are [11]:

$$\tau(r) = \frac{M}{2\pi r^2 h}; \quad R_1 \leq r \leq R_2 \tag{2.7}$$

$$\dot{\gamma}(r) = \frac{2\omega R_1^2 R_2^2}{r^2(R_2^2 - R_1^2)} \tag{2.8}$$

A torque M is produced by the rate of angular rotation ω in radians per second. As shown in Figure 2.1, the radii of the inner and outer cylinders are R_1 and R_2, respectively. The immersed length of cylinder is h, and r is any arbitrary radius between R_1 and R_2. The top part of Figure 2.2 illustrates the characteristics of coaxial cylinder instruments. The solid curves are for Newtonian liquids; the dashed line is typical of a non-Newtonian liquid. Note that Eq. (2.7), the relationship between the shear stress and the torque, is independent of the constitutive nature of the fluid. If the annular spacing is small compared to the diameter of the cylinders, the shear rate is nearly constant across the annular gap for all fluids; this shear rate is given by:

$$\dot{\gamma} = \frac{\omega R_1}{R_2 - R_1} \qquad (2.9)$$

The main correction that must be applied to data in order to get the correct viscosity is an end correction for a coaxial cylinder viscometer. The liquid at the bottom of the inner cylinder creates additional drag and gives rise to a torque in addition to the torque due to the liquid in the annular gap. Thus, the measured torque is equivalent to that of an inner cylinder with an apparent length greater than its true length. The simplest way to account for this additional length is to calibrate the viscometer with a liquid of known viscosity and to use the following equation:

$$\eta = \frac{kM}{\omega} \qquad (2.10)$$

The form of Eq. (2.10) follows from the fact that viscosity is the ratio of the shear stress to the shear rate, and these two quantities are given, respectively, by the right-hand sides of Eqs. (2.7) and (2.9). The instrument constant k takes into account any corrections as long as the volume of liquid in the viscometer is held constant. Of course, the end correction is negligible if the immersed length is a couple of orders of magnitude larger than the annular gap. The correction can also be minimized by shaping the bottom of the inner cylinder to resemble an inverted cone.

The maximum value of the shear rate achievable with coaxial cylinder viscometers is usually not an instrument limitation but depends on the nature of the fluid sample. For high viscosity liquids, viscous heating may become a problem; for low-viscosity liquids, the upper limit may be set by the occurrence of secondary flows or the transition to turbulent flow. Usually, though, we can make measurements at shear rates up to about 100 sec^{-1}. At the other end of the scale, one can often go down to shear rates as low as 10^{-2} sec^{-1}, especially with high viscosity liquids; the limiting factor here is the sensitivity of the torque transducer or the torsion bar or spring employed to measure the torque. The Contraves low-shear instruments, for example, are able to measure torque values as low as 10^{-9} Nm by replacing the torsion bar with a light-deflection compensating system. Here, the outer cylinder or cup is driven by a motor and the angular deflection of the stationary inner cylinder or bob is detected by a light beam reflected from a mirror attached to the bob axis and monitored by a photocell. In most coaxial cylinder viscometers, the sample size is about 1 ml and the maximum temperature is less than 80°C.

When the coaxial cylinder viscometer is used in the oscillatory mode, one cylinder is rigidly supported while the other one is given a small-amplitude angular dispacement of the form:

$$\phi = \phi_0 \sin(\omega t) \qquad (2.11)$$

Now the torque becomes time dependent, but the shear stress (and this is sinusoidal) is still related to the torque by Eq. (2.7). The sinusoidal strain is easily determined from the geometry and is given by:

$$\gamma(t) = \frac{R_2 \phi}{R_2 - R_1} \qquad (2.12)$$

A knowledge of the stress and strain as a function of time allows us to calculate the complex viscosity with the help of Eqs. (1.6–1.8) If, however, one of the cylinders is not rigidly held but is suspended from a torsion wire so that it too can oscillate, the theoretical analysis is somewhat involved, and the relevant results may be found in Walters [4].

A variation of the coaxial cylinder viscometer is the "Brookfield" viscometer—a cylindrical spindle rotating in a 600-ml container of fluid. While the Brookfield Company makes a variety of rotational viscometers, it is this simple, inexpensive device, commonly used for quality control purposes, that has become synonymous with the name of the company. The spindle is driven by a synchronous motor through a calibrated spring; the deflection of the spring is indicated by a pointer that displays a "dial" viscosity. When measurements are made on suspensions and pastes, the cylindrical spindle often cuts a hole in the material and the dial reading falls progressively. To prevent fluid channeling, we replace the cylindrical spindle with a T bar that is continuously raised and lowered in the fluid container. Again, the measured viscosity is not a fundamental quantity, but an empirical measure of the resistance to deformation of the fluid.

IV. CONE-AND-PLATE VISCOMETERS

The geometrical characteristics of a cone-and-plate rheometer are shown in the bottom section of Figure 2.1. A flat, circular plate and a linearly concentric cone are rotated relative to each other. The cone is normally truncated so that there is no physical contact between the two members. The liquid is in the space between the plate and cone. The plate radius R is typically a couple of centimeters, whereas the cone angle α is usually a few degrees. Either of the two members can be rotated or oscillated, and we measure the torque M needed to keep the other member stationary. We also measure the downward force N needed to hold the apex of the truncated cone at the center of the disk. From these measurements we deduce the various material functions defined in Chapter 1. Note that N equals zero for a Newtonian liquid.

A major advantage of a cone-and-plate viscometer is the constant shear rate throughout all the liquid. This comes about because, at a fixed radial position, the circumferential velocity varies linearly across the gap between the cone and the plate. The constancy of shear rate is especially important, since it makes data analysis extremely straightforward. A consequence of this is that the cone-and-plate viscometer is well suited to making time-dependent measurements, such as dynamic measurements or transient measurements following a step change in the rate of shear strain. If disposable fixtures are employed, measurements can be made readily on chemically reactive systems, allowing for the investigation of resin curing and the phenomenon of gelation. This instrument also has other advantages. The sample size is very small, so the instrument is valuable for evaluating the rheological behavior of experimental samples, where only small quantities of material may be available. The small sample size also results in less heat buildup at high rates of shear. Another advantage is the ease of loading the sample and the ease of cleaning the apparatus at the end of a test. The instrument has a number of disadvantages: Measurements can usually not be made at high shear rates. With increasing shear rate, there is a tendency for the development of secondary flows in the polymer and for the polymer to crawl out of the instrument under the influence of centrifugal forces and elastic instabilities. Even otherwise, the shape of the free surface at the edge of the sample can get distorted, leading to inaccuracies; this phenomenon can be aggravated by sample drying in the case of polymer solutions and suspensions. If this happens, coating the free surface with silicone oil or using a solvent trap can often provide relief. Accurate spacing of the cone and plate is required at all times, especially for the measurement of transient normal stresses. This makes the equipment more sophisticated and expensive, and it also demands a higher level of skill of the investigator.

For both Newtonian and non-Newtonian liquids, the basic equations for cone-and-plate rheometers are [4]:

$$\tau = \frac{3M}{2\pi R^3} \tag{2.13}$$

$$\dot{\gamma} = \frac{\omega}{\alpha} \tag{2.14}$$

$$N_1 = \frac{2N}{\pi R^2} \tag{2.15}$$

$$T_{\theta\theta} = -p_a + N_2 + (N_1 + 2N_2) \ln \frac{r}{R} \tag{2.16}$$

Here the cone angle is measured in radians, p_a is atmospheric pressure, and $T_{\theta\theta}$ is the total stress exerted by the fluid normal to the surface of the plate; the subscripts on T represent the θ direction in a spherical coordinate system.

The characteristics of the cone-and-plate viscometer are displayed in the middle portion of Figure 2.2. As before, the shear viscosity is obtained by dividing the shear stress by the shear rate. In principle, a knowledge of N_1 permits the determination of N_2 through the use of Eq. (2.16) provided that a pressure transducer is employed to measure $T_{\theta\theta}$ on the plate surface. This measurement, however, is not easy to make with any degree of accuracy. If the angular velocity is not constant but is some specified function of time—say, sinusoidal or a step function—the measured quantities M and N also depend on time but the preceding above equations remain unaltered. This makes it easy to compute dynamic mechanical properties or to measure stress growth and stress relaxation. Note that today's viscometers can be operated not just at particular values of the shear rate but also at specified values of the shear stress; the shear rate then becomes the dependent variable. The use of a constant-stress rheometer is one way of determining creep behavior and the value of the yield stress, if any, of a fluid at hand.

The cone-and-plate viscometer is the workhorse of the rheological testing laboratory, and it is manufactured by almost every company that makes rheometers. Depending on the capability of the torque transducer and the viscosity of the liquid sample, measurements can be made over a wide shear rate range, extending from 10^{-6} to 5,000 $\sec^{-1}$. As a general rule, high viscosities and low shear rates go hand in hand, and the low shear rate capability makes it possible to determine the zero-shear-rate viscosity of polymer melts. For oscillatory experiments, a frequency range of 10^{-2}–100 rad/sec is common. In the Rheometrics line of mechanical spectrometers, and these are some of the most sophisticated instruments available, the torque range is from 0.1 to 2000 gm-cm, and the normal force is measured in the range of 0.1–2000 g, employing a patented "force rebalance transducer" that prevents axial motion of the rotating members during the process of measurement. The sample is kept in an inert atmosphere in an oven that can be heated up to about 500°C or cooled down to about -150°C at heating and cooling rates as high as 60°C/min.

V. PARALLEL-PLATE VISCOMETERS

Even though the cone-and-plate viscometer is the preferred instrument when measurements are made at low shear rates, difficulties arise during measurements on filled polymers if the size of the particulates is comparable to the distance between the truncated cone and the surface of the plate. In such a situation, as shown in Figure 2.4a, one replaces the cone by a circular plate so that one has torsional flow between two identical, coaxial, parallel disks; the rest of the apparatus remains

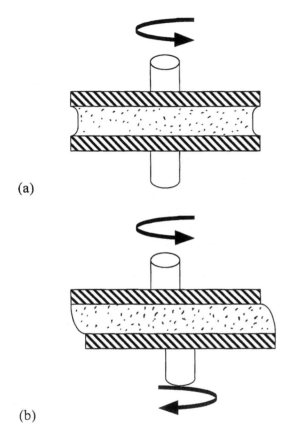

(a)

(b)

FIGURE 2.4.   Schematic diagrams of (a) the parallel-plate rheometer and (b) the orthogonal rheometer.

unchanged. One again measures the torque M and the normal force N tending to separate the two plates when the gap between them is maintained fixed. If one uses a large gap, an advantage of this arrangement is that fairly large and measurable torque readings are obtained during small-amplitude oscillatory flow. Another advantage is that sample preparation is easy, since disks of the appropriate size can be cut or punched out of a flat sheet of material. The major disadvantage of this instrument is that the shear strain and the shear rate vary with radial position, complicating the process of data analysis. In steady-shear flow, the shear rate and shear stress at the edge of the disks located at $r = R$ are given by [4]:

$$\dot{\gamma}(R) = \frac{\omega R}{h} \qquad\qquad (2.17)$$

$$\tau(R) = \frac{3M}{2\pi R^3}\left[1 + \frac{1}{3}\frac{d \ln M}{d \ln \dot{\gamma}(R)}\right] \tag{2.18}$$

$$N_1(R) - N_2(R) = \frac{2F}{\pi R^2}\left[1 + \frac{1}{2}\frac{d \ln N}{d \ln \dot{\gamma}(R)}\right] \tag{2.19}$$

in which h is the vertical gap between the two plates. It is clear that in order to get the shear stress or the normal stress differences at a single value of the shear rate, we have to measure the torque and normal force at several values of the shear rate and differentiate the data; data can be collected either by varying the gap spacing or, more easily, by changing the speed of rotation. Unfortunately, the two normal stress differences cannot be determined separately unless some very gross assumptions are made. However, if the first normal stress difference is measured using a cone-and-plate viscometer, the parallel-plate instrument can be utilized to compute the second normal stress difference. A variation of the experimental procedure already outlined [12] is to apply a known downward normal force to the upper plate and to measure the resulting value of h.

In small-amplitude sinusoidal deformation with an angular amplitude of ϕ_0 such that the entire sample is still within the linear viscoelastic region, a measurement of the torque amplitude M_0 and its phase difference δ with the angular deformation yields [2]:

$$G' = \frac{2M_0 h \cos \delta}{\pi R^4 \phi_0} \tag{2.20}$$

$$G'' = \frac{2M_0 h \sin \delta}{\pi R^4 \phi_0} \tag{2.20}$$

Instead of oscillating one plate relative to the other, the parallel-plate viscometer is sometimes operated in the "eccentric rotating disk" mode, as shown in Figure 2.4b. In this orthogonal rheometer, there is an offset between the axes of the two disks; as one disk rotates relative to the other, an eccentric oscillatory motion is set up in the polymer [13]. An analysis of this flow field [4] again yields G' and G".

Equipment limitations and sources of error for the parallel-plate viscometer are essentially the same as those for the cone-and-plate viscometer. When working with suspensions and emulsions, a common problem encountered with both types of instruments is the tendency of the sample to slip at the fixture walls [14]; roughening the fixture surfaces or using serrated surfaces is the normal method of attempting to overcome this difficulty.

VI. SLIDING-PLATE VISCOMETERS

Polymer processing operations are generally characterized by large deformations of short duration during which a steady state in the stress is usually not achieved.

Consequently, if the purpose of rheological testing is to examine polymer processability, most of the viscometers described in earlier sections of this chapter may be inadequate, since these are designed to measure steady-state stresses at relatively low rates of deformation. An instrument that is well suited to the investigation of large, transient deformations at high shear rates is the sliding-plate viscometer. This is shown in Fig. 2.5a, and it is very simple in concept. The fluid sample is placed between two horizontal parallel plates and sheared by moving one plate relative to the other. The shear stress is determined as the ratio of the measured shear force to the plate surface area; the shear rate is simply the plate velocity divided by the plate separation. The major difficulty with this experimental scheme is that stress measurements are influenced by edge effects, instrument friction, and fluctuations in plate spacing [3]. Dealy and coworkers [15,16] overcame these problems by measuring the shear stress locally with the use of a

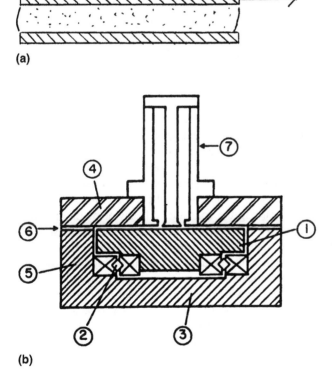

(a)

(b)

Figure 2.5. (a) Schematic diagram of the sliding-plate viscometer. (b) Cross section of the McGill University sliding-plate rheometer, showing the moving plate, 1, linear bearing, 2, back plate, 3, stationary plate, 4, side supports, 5, shims, 6, and the shear stress transducer, 7. (From Ref. 16.)

patented shear stress transducer; capacitance proximetry was employed to measure the deflection of a cantilever beam when the shear stress acted on the free end of the cantilever. The shear stress range of the transducer was 0.01–1 MPa.

The sliding-plate rheometer developed by Giacomin et al. [16] is shown in schematic form in Figure 2.5b. The sliding plate, mounted on linear bearings, is 9 inches long and 4 inches wide. The gap between the two plates is set using shims; the shear stress is monitored at the center of the stationary plate. The entire assembly is kept inside a forced-convection oven; the maximum service temperature is 350°C. The typical sample thickness is 1 mm, and shear rates as high as 250 sec^{-1} can be sustained for a total of 2 sec. The total strain limitation arises due to the fixed rheometer length. Besides the measurement of steady shear viscosity and dynamic mechanical properties, the sliding-plate viscometer can be used to study startup and shutdown of steady-shear flow, large-amplitude oscillatory flow, and situations in which the shear strain increases exponentially with time. Among other things, these results are useful for examining polymer entanglement effects and the tendency of anisotropic fillers to align themselves with the flow and for quantifying any slip between the polymer sample and viscometer walls. A commercial version of the instrument is manufactured by Interlaken Technology in Eden Prairie, MN.

VII.   OTHER VISCOMETERS

Even though one of the stated goals of rheometry is to generate information that can guide us in optimizing polymer processing operations, which typically involve complex flow fields, material response is usually characterized in simple flow situations using viscometers of the kind described in this chapter. The use of such viscometers ensures that, as far as possible, true material behavior is determined and the measurements are independent of the technique of measurement and any assumptions regarding the constitutive behavior of the material. In some instances, these data can be related directly to polymer processability, but in other cases we have to make use of mathematical models. Rheological data then form one kind of input to the models. In industrial practice, however, one often uses indexers or testers that tend to mimic a particular processing operation, and data from these devices provide direct information about polymer processability. It is, therefore, logical to ask if these testers can also be analyzed or calibrated to yield fundamental rheological information. Here we describe two common instruments for which the answer appears to be in the affirmative.

A.   Torque Rheometers

A torque rheometer is a batch mixer consisting of two irregularly shaped rollers or mixing elements contained in a heated mixing chamber shaped like a figure eight as shown in Figure 2.6 [17]. The mixer is filled with a polymer sample, which is then melted. The torque needed to turn the rollers in opposite directions in the melt is recorded as a function of time at a fixed speed of rotation. Also monitored are the sample temperature and the pressure inside the mixer. Even at a fixed speed of rotation, it is obvious that the deformation rate varies from position to position within the rheometer. Also, the flow is not purely a shear flow. Consequently, data cannot be analyzed unless drastic assumptions are made. If it is assumed that the polymer melt is a power-law fluid and that the mixing chamber is analogous to a concentric cylinder viscometer, it can be shown that under isothermal conditions [17]:

$$M = C(n)N^n \tag{2.22}$$

in which M is the measured torque, C is a constant that depends on the mixer geometry and the power-law index n, and N is the roller revolutions per minute (rpm). Thus, torque rheometer data can reveal the power-law index n. Also, since

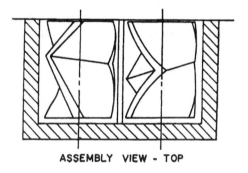

ASSEMBLY VIEW - TOP

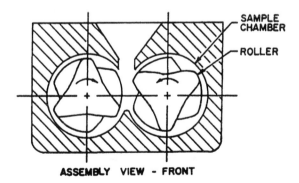

SAMPLE
CHAMBER

ROLLER

ASSEMBLY VIEW - FRONT

FIGURE 2.6   The Brabender torque rheometer. (From Ref. 17.)

viscosity is proportional to the torque at a fixed rpm, the temperature dependence of the viscosity can be obtained by plotting the logarithm of the torque as a function of the reciprocal of the absolute temperature.

Torque rheometers can be employed to study the processability of thermoplastics, thermosets, elastomers, and filled polymers. Data are useful for product development and quality control, in general, and in understanding the melting and degradation behavior of thermoplastics and the curing behavior of thermosets, in particular. The influence of additives can also be examined. Laboratory units have stainless steel mixing bowls of 50–150-ml capacity, and these accommodate removable mixing blades; cam, Banbury, and sigma blades are common. Measurements can be made at temperatures up to about 450°C. Torque rheometers are made commercially by the C.W. Brabender and Haake companies, among others. Additional details are available in the review by Chung [18].

B. Squeezing Flow Viscometer

As shown in Figure 2.7, this is an extremely simple instrument. It consists of two parallel, coaxial disks that are made to approach each other either at constant force or at constant velocity; the liquid sample is contained in the gap between the plates, and it is squeezed out of the gap in the process of testing. One measures the time dependence of the applied force F in constant velocity squeezing or the plate separation $2H$ as a function of time in constant-force squeezing. Instruments of this general nature have been around for a very long time, and the parallel-

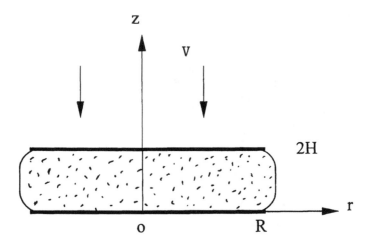

FIGURE 2.7 Schematic diagram of the squeezing flow viscometer.

plate plastimeter [19], for example, has been used as an empirical method of getting the material properties of highly viscous polymers such as unvulcanized rubber. The popularity of these devices stems from the simplicity of operation and the fact that particle-and-fiber-filled, solidlike materials that might otherwise slip or fracture in conventional viscometers can be tested with ease. Squeezing flow between parallel plates also simulates the processes of compression molding and stamping, in which we use pressure to fill a mold with a very viscous polymer charge containing fibers. The resulting flow field, however, is neither viscometric nor purely extensional. Also, the deformation is neither steady in time nor homogeneous in space. Furthermore, while the squeezing force can be predicted readily if the flow curve of the polymer is available [20], obtaining the flow curve from the measured force is possible only if we assume a form of the stress-versus-shear-rate relationship. These and other aspects of the theory have been reviewed in the literature [21]. In the absence of fluid inertia, the force-versus-plate-separation relation for power-law liquids is given by the Scott equation [21,22]:

$$
F = \left(-\frac{dH}{dt} \right)^n \left[\frac{2n + 1}{2n} \right]^n \frac{\pi K}{H^{2n+1}} \frac{R^{n+3}}{n + 3}
\tag{2.23}
$$

in which R is the plate radius, K is the fluid consistency index, and n is the power-law index. For Newtonian liquids, n equals unity, K equals the Newtonian viscosity, and Equation (2.23) is known as the Stefan equation.

The Scott equation can usually represent the response of unfilled polymeric fluids to compressive flow provided that squeezing is not rapid; for rapid squeezing (high Deborah number flows), the influence of fluid elasticity has to be taken into account as well, but this is not an easy exercise [21]. For particulate-and-fiber-filled materials, one often observes a yield stress, and the analysis leading to Eq. (2.23) can be repeated using either the Bingham plastic or Herschel–Bulkley fluid models [19,23]. The resulting expressions are much more involved when compared to Eq. (2.23), but one can use the experimental data to extract the model parameters in a straightforward manner. Note that Lipscomb and Denn [24] have objected to this analysis on the grounds that Bingham-type fluids cannot contain yield surfaces in complex geometries. As a practical matter, though, these theoretical objections are not very serious, and the procedure of Covey and Stanmore [23] works quite well.

Although squeezing flow viscometers are available commercially [2], we can easily attach the parallel plates to the crosshead of a tensile testing machine and build our own instrument. We can also conduct experiments at high temperatures with the use of an isothermal oven. Since the total strain that can be imposed is small, this viscometer is best suited for making measurements at low deformation rates; under these conditions, elastic effects also are not important. In closing this section, we mention that the use of a low-viscosity lubricating liquid layer on the surface of the plates changes the mode of deformation within the more viscous material to equal biaxial extension. This lubricated squeezing flow can be utilized for measuring the biaxial extensional viscosity [25].

REFERENCES

1. J. R. Van Wazer, J. W. Lyons, K. Y. Kim, and R. E. Colwell. Viscosity and Flow Measurement. Wiley Interscience, New York, 1963.
2. J. M. Dealy. Rheometers for Molten Plastics. Van Nostrand Reinhold, New York, 1982.
3. A. A. Collyer and D. W. Clegg (eds.). Rheological Measurement. Elsevier Applied Science, London, 1988.
4. K. Walters. Rheometry. Chapman and Hall, London, 1975.
5. B. Rabinowitsch. Ober die viscositat und elastizitat von solen. Z. Physik. Chem. (Leipzig) 145A:1-26 (1929).
6. E. B. Bagley. End corrections in the capillary flow of polyethylene. J. Appl. Phys. 28:624–627 (1957).
7. E. B. Bagley. The separation of elastic and viscous effects in polymer flow. Trans. Soc. Rheol. 5:355–368 (1961).
8. H. C. Brinkman. Heat effects in capillary flow I. Appl. Sci. Res. A2:120–124 (1951).
9. J. Kestin, M. Sokolov, and W. Wakeham. Theory of capillary viscometers. Appl. Sci. Res. 27:241–264 (1973).
10. G. D. Galvin, J. F. Hutton, and B. Jones. Development of a high-pressure, high-shear-rate capillary viscometer. J. Non-Newtonian Fluid Mech. 8:11-28 (1981).
11. R. B. Bird, W. E. Stewart, and E. N. Lightfoot. Transport Phenomena. Wiley, New York, 1960.
12. D. M. Binding and K. Walters. Elastico-viscous squeeze films, Part 3. The torsional-balance rheometer. J. Non-Newtonian Fluid Mech. 1:277-286 (1976).
13. B. Maxwell and R. P. Chartoff. A polymer melt in an orthogonal rheometer. Trans. Soc. Rheol. 9:41-52 (1965).
14. H. A. Barnes. A review of the slip (wall depletion) of polymer solutions, emulsions and particle suspensions in viscometers: its cause, character, and cure. J. Non-Newtonian Fluid Mech. 56:221-251 (1995).
15. A. J. Giacomin and J.M. Dealy. A new rheometer for molten plastics. SPE ANTEC Tech. Papers 32:711-714 (1986).
16. A. J. Giacomin, T. Samurkas, and J. M. Dealy. A novel sliding plate rheometer for molten plastics. Polym. Eng. Sci. 29:499-504 (1989).
17. L. L. Blyler, Jr. and J. H. Daane. An analysis of Brabender torque rheometer data. Polym. Eng. Sci. 7:178-181 (1967).
18. J. T. Chung. Torque rheometer technology and instrumentation. In: N. P. Cheremisinoff, ed. Encyclopedia of Fluid Mechanics. Vol. 7. Gulf, Houston, 1988, pp. 1081-1138.
19. J. R. Scott. Theory and application of the parallel-plate plastimeter. I.R.I. Trans. 7:169-186 (1931).

20. G. Brindley, J. M. Davies, and K. Walters. Elastico-viscous squeeze films. Part I. J. Non-Newtonian Fluid Mech. 1:19–37 (1976).
21. R. K. Gupta and A. B. Metzner. Squeezing flows. Adv. Transport Processes 6:27–77 (1989).
22. P. J. Leider and R.B. Bird. Squeezing flow between parallel disks. I. Theoretical analysis. Ind. Eng. Chem. Fundam. 13:336–341 (1974).
23. G. H. Covey and B.R. Stanmore. Use of the parallel-plate plastometer for the characterization of viscous fluids with a yield stress. J. Non-Newtonian Fluid Mech. 8: 249–260 (1981).
24. G. G. Lipscomb and M. M. Denn. Flow of Bingham fluids in complex geometries. J. Non-Newtonian Fluid Mech. 14:337–346 (1984).
25. S. Chatraei, C. W. Macosko, and H. H. Winter. Lubricated squeezing flow: a new biaxial extensional rheometer. J. Rheol. 25:433–443 (1981).

3

Shear Viscosity of Melts of Flexible Chain Polymers

I. INTRODUCTION

At a sufficiently low temperature, every polymer acts like a hard glassy material. To process such a polymer into a fiber or a film or a molded article, it is necessary to heat it to a temperature at which it either softens (if amorphous) or melts (if crystalline) and at which it can undergo a shaping operation. Because the shear viscosity of flexible chain polymers decreases as the temperature increases, polymer processing is done at high temperatures. The particular temperature picked is based on a number of considerations. The temperature has to be high enough that the polymer flows easily, but not so high that an excessive amount of heat has to removed when the plastic part is cooled back to room temperature, since this represents a penalty both in terms of energy consumption and processing time. In addition, the material must not thermally degrade at the chosen temperature. Besides temperature, the operating pressure and the rate of shear are also important. In injection molding, for instance, very high pressures are employed, especially in the packing stage. This results in an increase in the viscosity. During mold filling, on the other hand, the shear rate in the melt delivery system is typically about 1000 sec^{-1}, while at the mold gate it may be as high as 10^5 sec^{-1}. Under these conditions, the polymer may exhibit marked shear thinning, and the viscosity at high rates of shear may be several orders of magnitude smaller than the viscosity at low rates of shear. It is, therefore, necessary to measure the shear viscosity, called the *apparent viscosity*, as a function of temperature, pressure, and the shear rate.

Typical curves for the steady-shear viscosity of a molten polymer as a function of temperature and shear rate are shown in Figure 3.1 on logarithmic coordinates. Notice that at high temperatures, the melt has a large but constant viscosity at low rates of shear; this region is known as the lower Newtonian region, and it is characterized by a temperature-dependent zero shear rate viscosity η_0. However, over much of the usual accessible shear rate range, particularly at the lower temperatures, the viscosity decreases nearly linearly with shear rate. In this linear range, the so-called power law equation holds, in which the shear stress and the apparent viscosity are respectively given by:

$$\tau = K\dot{\gamma}^n \tag{3.1}$$

$$\eta_a = K\dot{\gamma}^{(n-1)} \tag{3.2}$$

where the constant K is called the consistency index and the constant n is known as the power-law index. For Newtonian liquids, n equals unity and K is the same as the viscosity; polymer melts, however, are pseudoplastic, and the value of n is less than 1. The power law generally does not hold accurately for more than

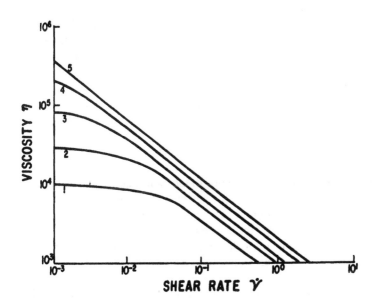

FIGURE 3.1. Typical log viscosity–log shear rate curves at five different temperatures. Curve 1 is for the highest temperature; curve 5 is for the lowest temperature. For a typical polymer, the temperature difference between each curve is approximately 10°C.

a couple of decades' change in the shear rate, but in the range where the equation is valid, the slope of the straight lines shown in Figure 3.1 is $(n - 1)$. Equation 3.2 obviously cannot represent the viscosity over the entire range of shear rates, because it predicts a meaningless increase in the viscosity as the shear rate is decreased. A popular alternative model that does not have this shortcoming is the Carreau equation [1]:

$$\frac{\eta_a}{\eta_0} = [1 + (\lambda_c \dot{\gamma})^2]^{(n-1)/2} \tag{3.3}$$

which has three constants: η_0, λ_c, and n. Where η_0 accommodates a limiting shear viscosity at low shear rates, the parameter λ_c determines the point of onset of shear thinning; as shown in Figure 3.1, the onset of shear thinning moves to larger values of the shear rate as the temperature of measurement is increased.

Polymer melts are obviously non-Newtonian, in the sense that their viscosity depends on the rate of shear. They are also non-Newtonian in the sense that they are viscoelastic; i.e., they have some properties of elastic solids. After inception of shear flow at a constant shear rate, for example, the shear stress takes some time to reach a steady state and can even overshoot the steady-state value. Similarly, upon cessation of steady shearing, the shear stress takes a finite amount of time, known as the *relaxation time*, to go to zero. As a consequence, the shear viscosity can depend on time even when the deformation rate is kept constant. We defer an examination of viscoelastic effects to later chapters; here we focus on the steady-shear viscosity alone.

For a given polymer, as explained earlier, the viscosity depends on processing variables such as the temperature, pressure, and the shear rate. However, the viscosity value changes as the chemical nature of the polymer changes. Thus, the shear viscosity of polyethylene differs from that of nylon, even when the values of the three processing variables are kept unchanged. Additionally, for a given polymer type, the viscosity is a function of the molecular weight, the molecular weight distribution and the presence of any chain branching. In the sections that follow, we systematically examine the influence of each of these variables. We present typical data, show how these data can be combined to give viscosity master curves, propose equations for data representation, and provide physical explanations for the recorded observations. Both model and commercial polymers are considered. As might be expected, data are most abundant for high-volume polymers such as polyethylene, polypropylene, and polystyrene. Note that while the polyolefins are crystalline, polystyrene is amorphous, and the molecular weight of all three polymers, when manufactured commercially, is high enough that their rheological behavior is quite non-Newtonian. Also note that these three polymers are normally synthesized using chain growth (addition) polymerization, and they are reasonably stable at processing temperatures. As opposed to this, other important polymers like polyesters and nylons are relatively Newtonian due to their lower molecular weight. Also, they are made by the process of step growth (condensation) polymerization; at processing temperatures, they are subject to hydrolysis or postcondensation, depending on the extent to which they are dried before being processed. Since most polymers tend to oxidize when kept at elevated temperatures for extended periods of time, it is normal to use an inert atmosphere of nitrogen or argon gas when making rheological measurements.

II. ZERO-SHEAR VISCOSITY

A major difference between the rheological behavior of polymer melts and low-molecular-weight materials is that the shear viscosity of polymeric fluids depends on the rate of shear. This was seen in Figure 3.1, and we will pursue this aspect in detail in the next section. Here we examine the zero shear viscosity or the limiting behavior at low shear rates. Note that it is sometimes necessary to extrapolate data to the zero-deformation rate, because the use of a viscometer at hand may not always allow for the unequivocal determination of the zero-shear viscosity. Alternately, dynamic mechanical data may be employed to compute the zero-shear viscosity, since [2]

$$\eta_0 = \lim_{\omega \to 0} \frac{G''(\omega)}{\omega} \tag{3.4}$$

where G'' is the loss modulus defined by Eq. (1.8). The variables that influence the zero-shear viscosity are polymer molecular weight, molecular weight distribution, chain branching, temperature, and pressure.

A. Influence of Molecular Weight and Molecular Weight Distribution

It is possible to study the influence of molecular weight on the melt viscosity of linear polymers, since the technique of anionic polymerization, which yields materials of narrow molecular weight distribution, allows for the synthesis of essentially monodisperse polymers such as polystyrene and polybutadiene. Of course, a polydisperse sample can be fractionated, but this is a tedious procedure that results in only small quantities of material. When isothermal, zero-shear-viscosity data are measured as a function of polymer molecular weight, it is found that below a critical molecular weight M_c that is independent of temperature, the zero-shear viscosity is roughly proportional to the molecular weight M_w. That is,

$$\eta_0 = K_1 M_w \qquad M_w < M_c \tag{3.5}$$

At molecular weights above M_c, the zero-shear viscosity depends upon M_w to a power equal to about 3.4 or 3.5:

$$\eta_0 = K_2 M_w^{3.4} \qquad M_w > M_c \tag{3.6}$$

This behavior is illustrated in Figure 3.2 for polystyrene melts at 217°C [3] and is representative of polystyrene in general [4]. It is also found to hold for other polymers [5]. To emphasize the generality of Eqs. (3.5) and (3.6), Graessley and coworkers reanalyzed available data on polyethylene and showed that the relationship of melt viscosity versus molecular weight was not influenced by the ability of the polymer to crystallize [6]. The truth of this conclusion was further demonstrated by presenting data on (amorphous) polybutadiene [7] and (crystalline) hydrogenated polybutadiene [8].

The transition between Eqs. (3.5) and (3.6) is usually not sharp. However, data can generally be represented by two straight lines that intersect at $M = M_c$; in Fig. 3.2, M_c equals approximately 38,000. For most polymers, M_c lies between about 10,000 and 40,000 and increases with increasing chain stiffness. Some typical values are given in Table 3.1. The change in slope of the plot of viscosity versus molecular weight shown in Fig. 3.2 is explained by saying that when M_w exceeds a critical molecular weight M_c, polymer chains begin to get entangled with each other and the entanglement density increases with increasing molecular weight [9,10]. As sketched in Figure 3.3, entanglements may be visualized as temporary crosslinks whose effect is to prevent chain crossability and to increase fluid friction vastly. This effect of entanglements is analogous to the fact that it requires more force to pull a strand from a ball of yarn when the strands are entangled and knotted than when they are not entangled. Thus, flow becomes much more difficult because forces applied to one polymer chain become transmitted to and distributed among many other chains.

TABLE 3.1   Entanglement Molecular Weight M_c

Polymer	M_c
Linear polyethylene (PE)	4,000
Polyisobutylene	17,000
Polyvinyl acetate	29,200
Polystyrene	38,000
Polydimethyl siloxane	35,200
Polymethyl methacrylate	10,400
Caprolactam polymer (linear)	19,200
Caprolactam polymer (tetra-branched)	22,000
Caprolactam polymer (octa-branched)	31,100

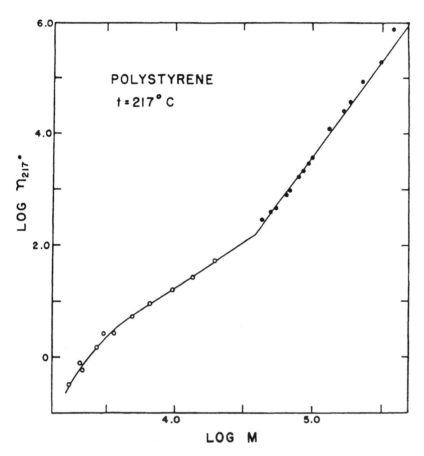

FIGURE 3.2. Double logarithmic plot of zero-shear viscosity as a function of the polymer molecular weight for polystyrene fractions. (From Ref. 3. Copyright © 1954 by John Wiley & Sons, Inc. Reprinted by permission of John Wiley & Sons, Inc.)

With the foregoing picture of chain entanglements in mind, it is logical to inquire whether the critical chain length or the number of monomer units needed for entanglement formation is the same for most polymers. It is found that although there is some variability in the value of the critical chain length Z_c amongst polymers, the variation in Z_c is much smaller than the variation in M_c [9]; Z_c corresponds to about 300–700 main chain atoms [11]. Furthermore, since the molecular weight is linearly related to the polymer chain length, M_w in Eqs. (3.5) and (3.6) can be replaced by the chain length. Indeed, when data on a number of polymers are plotted as the logarithm of the zero-shear viscosity versus the logarithm of the chain length, the change in slope occurs at essentially the same

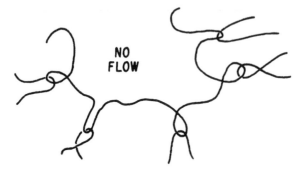

FIGURE 3.3 Schematic diagram of entanglement formation in quiescent polymer melts.

value of Z [12]. The significance of this observation is that the zero shear viscosity of any polymer can be predicted as a function of molecular weight if one is able to compute the temperature-dependent critical viscosity corresponding to M_c or Z_c in Figure 3.2. One way of doing this is through the group contribution method of Van Krevelen and Hoftyzer [13].

In order to explain the observed relationship between the zero-shear viscosity and the molecular weight on a fundamental level, one has to appeal to molecular models. In one of the earliest theories, due to Debye and Bueche [14–16], the shear viscosity was calculated by determining the amount of energy dissipated due to fluid friction in a steady laminar shearing flow at a constant shear rate. The analysis is very straightforward and leads to the expected proportionality between the zero-shear viscosity and the molecular weight; the constant of proportionality involves the monomer molecular weight and unknown quantities, such as polymer chain dimensions and the friction coefficient per monomer unit. In the presence of entanglements, it is assumed that the motion of one molecule is resisted by the neighboring molecules that are dragged along with it but move at a lower velocity. The neighboring molecules, in turn, snag other molecules, which also move but at a still slower velocity. When all the resistances are added, the result is Eq. (3.6). Even though this is exactly the anticipated result, the notion of one chain dragging several other chains leads to an incorrect dependence of the mean relaxation time on the molecular weight [17]. Correcting this internal inconsistency requires the use of a more sophisticated model. At the present time, the accepted picture of the motion of a polymer molecule in an entangled melt is one in which the surrounding molecules constrain the molecule of interest to lie within a tube so that it can move readily only along its own length. This slithering motion is known as *reptation* [17,18], and it is the central feature of a molecular theory developed by Doi and Edwards [19]. The theory is fairly involved, and working through the details requires a certain facility with mathematics. While an explicit expression can be derived for the zero-shear viscosity, the dependence on the molecular weight turns out to be to the third power rather than to the expected 3.4 power [20]. Nonetheless, the reptation model provides a remarkably consistent interrelationship between various viscoelastic functions, which has been confirmed with observations on linear polymers [21].

For polydisperse, linear polymers, we expect the zero-shear viscosity to depend on the molecular weight distribution or, at the very least, on one or more averages of the molecular weight distribution. Measurements on binary blends of narrow-molecular-weight-distribution polystyrene [22,23] and on samples of wide-molecular-weight-distribution polyethylene [24] at a variety of temperatures have demonstrated that the weight average molecular weight is the important molecular weight. It is found that Eqs. (3.5) and (3.6) remain valid when the weight average molecular weight is used in place of M_w. This unique dependence of the zero shear viscosity on $\overline{M}_w$ has also been observed for silicone polymers [25], polypropylene [26], and nylon 6 [27], and it leads to experimentally verifiable blending rules for the viscosity [25]. However, there are some published data that suggest that when the molecular weight distribution is very wide, as in commercial polymers, the melt viscosity may not be strictly dependent upon $\overline{M}_w$ [28,29]. Rather, the logarithm of the zero-shear viscosity may be proportional to an average molecular weight in between $\overline{M}_w$ and the next higher moment of the distribution curve beyond $\overline{M}_w$. This implies that the high-molecular-weight "tails" on the distribution curve can be especially important in affecting the zero-shear viscosity and other rheological properties.

B. Influence of Chain Branching

Although the discussion so far has centered on linear polymers, in commercial practice these are the exception rather than the rule. Indeed, even the stereoregular, high-density polyethylene formed using Ziegler–Natta catalysts has a small number of short branches. Only recently has the development of (soluble) metallocene catalysts that have an active center opened up the possibility of synthesizing long chain molecules with virtually no branching. Present-day polymers have a great variety of branched structures, which influences their rheological and mechanical properties. The branches can be long or short, and these, in turn, can have secondary branching. The branches can be randomly spaced along the backbone chain, or several branches can originate from a single point to give a star-shaped molecule. Some of these possibilities are sketched in Figure 3.4. While the comb structure could be considered to be representative of linear low-density polyethylene made by copolymerizing ethylene with a monomer such as butene-1, random branching is more characteristic of traditional low-density polyethylene. For controlled rheological studies, however, the specially synthesized star-branched molecule has proven popular [30]. An interesting variation

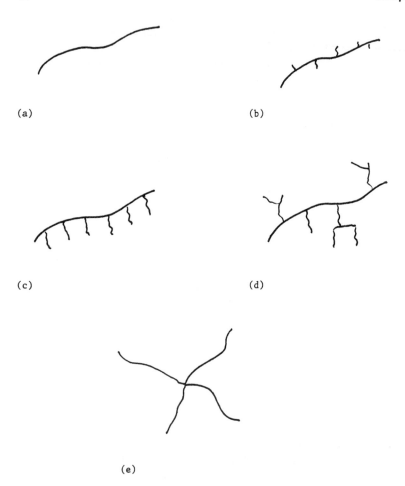

FIGURE 3.4 Chain branching in polymers: (a) linear molecule; (b) short branches; (c) comb structure; (d) long (random) branches; (e) star-branched molecule.

is a dendritic macromolecule: a treelike structure with branches emanating from a multifunctional core and from each unit of monomer [30a].

Short branches generally do not affect the viscosity of a molten polymer very much. On the other hand, long branches can have a very large effect. Branches that are long but that are still shorter than those required for entanglements decrease the zero-shear viscosity when compared to a linear polymer of the same molecular weight [31–33]. Such branching reduces the viscosity because branched molecules are more compact than linear molecules. It is found that the zero-shear viscosity can be considered the product of two terms, one the monomeric friction coefficient and the other a structure factor that is proportional to the number of interaction points [11,30]. Since individual molecules in a molten polymer assume the shape of a random coil, one has the situation depicted in Figure 3.5 [34], and it is clear that there is a lesser degree of molecular overlap, resulting in a lower viscosity due to fewer intermolecular interactions between branched macromolecules. However, if the branches are so long that they can participate in entanglements, the branched polymer may have a viscosity at low rates of shear that is significantly greater than that of a linear polymer of the same molecular weight [35,36]. This conclusion appears to be true even when the number of branches is below the detection limit of techniques, such as ^{13}C-NMR [37].

Thus, the molecular weight of branches is more important than their number. It is these branching differences that cause the zero-shear viscosity of high-density polyethylene to be higher than that of linear low-density polyethylene but lower than that of low-density polyethylene [38]. A logical consequence of these arguments is that the zero-shear viscosity of a low-molecular-weight branched polymer must be lower than that of a linear polymer of comparable molecular weight, but it should increase much more rapidly with molecular weight as the length of the branches is increased. This is seen to be true in Figure 3.6 for four-branched star polystyrenes; the slope of the plot for the branched polymer is 4.5 rather than the 3.4 for the linear polystyrene.

C. Influence of Temperature

As mentioned earlier, a polymer, whether amorphous or crystalline, is a hard, brittle, glassy solid at low temperatures. At sufficiently high temperatures, on the other hand, an uncrosslinked polymer turns into a viscous liquid. The transition between these two states is not sharp; there is an intermediate rubbery stage in which the material behaves like an elastomer or a soft elastic solid. The tempera-

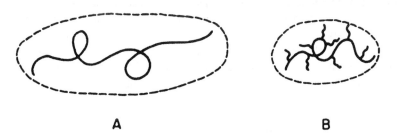

A   **B**

FIGURE 3.5 Oversimplified view of how the hydrodynamic volume varies for (A) a linear polymer and (B) a branched polymer of the same molecular weight. (Reproduced with permission from Ref. 34; copyright © 1981, Division of Chemical Education, Inc.)

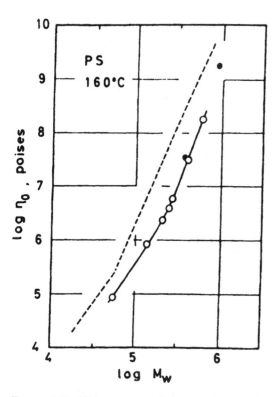

FIGURE 3.6 Molecular weight dependence of zero shear viscosity for star-branched polystyrenes. (Reprinted with permission from Ref. 39. Copyright 1971 American Chemical Society.)

ture at which a hard, glassy polymer becomes a rubbery material is called the glass transition temperature, T_g. In physical terms, the glass transition temperature is interpreted as the temperature below which only the vibrational motion of atoms or the short cooperative motion of a small collection of atoms can take place. Immediately above T_g, however, the polymer backbone can flex and large recoverable deformations are possible, because the molecule can uncoil but not flow. The rubbery region extends over a significant temperature range because flow is hampered by the presence of polymer chain entanglements that act like temporary crosslinks. In view of all this, it is not surprising that the viscosity of most polymers changes greatly with temperature.

For ease of flow, a polymer molecule must have enough thermal energy to make it mobile or able to get away from its neighbors, and, in addition, there must be enough space around it to allow it to move past other polymer molecules [16]. Below T_g, the latter condition is not met, and the polymer is a solid. Above

T_g, the magnitude of the shear viscosity is dependent on the availability of *free volume*, which, on a microscopic level, is that part of the sample volume not actually occupied by the polymer molecules. The free volume is zero at 0 K, and it increases with increasing temperature. If one assumes that the ratio of the volume of a polymer chain segment to the free volume associated with that segment is the same for all amorphous polymers at the glass transition temperature, it is possible to show that [13,16]:

$$\log \frac{\eta_0(T)}{\eta_0(T_g)} = \frac{-17.44(T - T_g)}{51.6 + (T - T_g)} \tag{3.7}$$

Equation (3.7) is known as the WLF equation, after Williams, Landel, and Ferry, who first proposed it [40,41]. The viscosity at T_g is often about 10^{12} Pa-sec. Equation (3.7) is found to be approximately applicable to all polymers in a temperature range between T_g and $T_g + 100$ K.

At temperatures far above the glass transition temperature or the melting point for crystalline polymers, there is ample free volume available, and the temperature dependence of the zero-shear viscosity is determined by energy barriers to motion. Under these circumstances, the viscosity follows the Andrade or Arrhenius equation to a good approximation:

$$\eta_0 = Ke^{E/RT} \tag{3.8}$$

In this equation, K is a constant characteristic of the polymer and its molecular weight, E is the activation energy for the flow process, R is the universal gas constant, and T is the temperature in degrees Kelvin. The activation energy generally is in the range between 2.09×10^7 to 2.09×10^8 J/kg-mole. Table 3.2 lists some typical values of the energy of activation for several polymers. The value of 16.7 kJ/g-mol for dimethyl silicone polymers is the smallest known; this low value is the result of the great flexibility of the silicone polymer chain. The energy of activation for flow increases as the size of the side groups increases and as the chain becomes more rigid [42,43].

To illustrate the enormous temperature dependence of the viscosity, let us use the Andrade equation with $E = 20,000$ cal/mole. On raising the temperature from 300 K to 310 K, the viscosity decreases by a factor of about 2.96! Figure 3.7 shows how the viscosity changes with temperature according to the WLF equation. The Andrade equation would give a nearly straight line on the same type of plot. If the viscosity and its temperature coefficient are the same for the WLF and Andrade equations at 100 K above T_g, the viscosity according to the Andrade equation would be given approximately by the dashed line in Figure 3.7. As already mentioned, the Andrade equation is satisfactory if the temperature is greater than $(T_g + 100)$; at lower temperatures, the WLF equation should be used with an adjustable empirical value for $\eta_0(T_g)$.

TABLE 3.2   Energy of Activation for Flow of Polymers

Polymer	Energy of activation E	
	kcal/g-mol	kJ/g-mol
Demethyl silicone	4	16.7
Polyethylene (high density)	6.3–7.0	26.3–29.2
Polyethylene (low density)	11.7	48.8
Polypropylene	9.0–10.0	37.5–41.7
Polybutadiene (cis)	4.7–8	19.6–33.3
Polyisobutylene	12.0–15.0	50–62.5
Polyethylene terephthalate	19	79.2
Polystyrene	25	104.2
Poly (α-methyl styrene)	32	133.3
Polycarbonate	26–30	108.3–125
Poly (1-butene)	11.9	49.6
Polyvinyl butyral	26	108.3
SAN (styrene acrylonitrile copolymer)	25–30	104.2–125
ABS (20% rubber) (acrylonitrile-butadiene-styrene copolymer)	26	108.3
ABS (30% rubber)	24	100
ABS (40% rubber)	21	87.5

For liquids obeying the Andrade equation, the energy of activation E is nearly independent of the temperature. However, for the WLF equation, the apparent energy of activation or the temperature coefficient of viscosity not only depends upon temperature but also upon the glass transition temperature:

$$E = \frac{Rd(\log \eta_0)}{d(1/T)} = \frac{4.12 \times 10^3 T^2}{(51.6 + T - T_g)^2} \tag{3.9}$$

Table 3.3 gives E for various values of T_g and $(T - T_g)$. The energy of activation becomes very large as the temperature approaches T_g, especially if T_g is large.

If the energy of activation of a particular polymer is not known, it can be estimated by assigning values to each of the groups making up the monomeric unit and adding up the values for all the groups [13,44]. The energy of activation E is given by

$$E = \left[\frac{\Sigma X_i}{M_0} \right]^3 \tag{3.10}$$

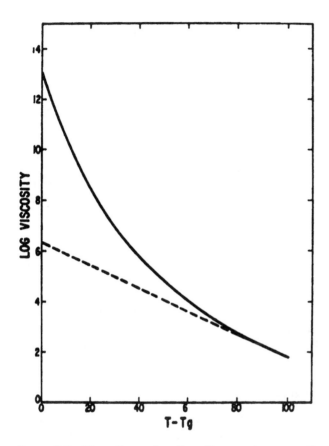

FIGURE 3.7   Viscosity as a function of temperature according to the WLF equation. The dotted line gives the approximate viscosity dependence according to the Arrhenius equation when matched to the WLF value at $(T_g + 100\ \text{K})$.

The molecular weight of the monomeric unit is M_0, and X_i is an empirical value assigned to group i. Values of X_i have been tabulated by van Krevelen and Hoftyzer [13,44].

Thus far, we have considered the influence of temperature on the steady-shear viscosity when this quantity is measured at constant temperature. Commercial polymer processing operations, however, can involve extremely large rates of change of temperature following a fluid element. If, under rapidly varying temperature conditions, the instantaneous value of material functions such as the shear viscosity is still given by Eq. (3.7) evaluated at the instantaneous value of the temperature, we say that the polymer melt is *thermorheologically simple* [45].

TABLE 3.3 WLF Energy of Activation for Viscous Flow

$T - T_g$	$T_g = 200$ K	$T_g = 250$ K	$T_g = 300$ K	$T_g = 350$ K
0	61.9	96.7	139.3	189.6
2	58.5	91.1	130.8	177.7
5	54.1	83.6	119.6	162.1
10	47.9	73.4	104.3	140.7
20	38.9	58.6	82.3	110.0
30	32.7	48.5	67.4	89.4
50	25.0	35.9	48.9	63.9
80	18.7	25.9	34.4	44.0
100	16.1	22.0	28.7	36.3

More generally, though, even if the deformation rate is kept constant while the temperature is raised or lowered, stress transients of fairly large magnitude can arise [46,47]. A partial explanation of this phenomenon is provided by the observation that the value of the glass transition temperature that appears in Eq. (3.7) is itself influenced by the rapid variation in temperature; for polystyrene melts, a T_g value that is 12–16°C higher than the equilibrium value has been measured for cooling rates in the range of 0.5–4.5°C/sec [46].

D. Influence of Pressure

Even though liquids are normally considered incompressible, this is not strictly true. During polymer processing operations such as extrusion and injection molding, the applied pressure can reach a thousand atmospheres [48]; such a large hydrostatic pressure results in a decrease in the free volume in the melt, and, according to several theories [49–51], should lead to a significant increase in the viscosity, especially near the polymer glass transition temperature. In this sense, an increase in pressure can be considered equivalent to a decrease in the temperature [52]. Indeed, the combination of high pressure and low temperature tends to induce crystallization in polymers such as polyethylene. It has been suggested that Eq. (3.8) for the temperature dependence of viscosity should be modified to [50]:

$$\eta_0 = K \exp\left(\frac{E}{RT} + \frac{CV_0}{V_f}\right) \tag{3.11}$$

in which V_0 is approximately the occupied volume at the glass transition temperature and V_f is the increase in volume due to thermal expansion above T_g. Thus,

$$V_f = \alpha V_0(T - T_g) \tag{3.12}$$

where α is the coefficient of thermal expansion. The value of the constant C in Eq. (3.11) is generally between 0.5 and 1.0.

There are not many data about the effect of pressure on the viscosity of polymer melts, and generally there is poor agreement between data from different sources. Viscosity increases either linearly with pressure or at a rate somewhat faster than the pressure. Westover has described a double-piston apparatus for measuring viscosity as a function of pressure up to 25,000 psi, and this is shown in schematic form in Figure 3.8 [53]. The hydrostatic pressure exerted on the polymer sample is determined by the pressure of the oil at the extreme ends of the hydraulic cylinders. The application of a known differential pressure across the die causes flow to take place through the die. The resulting volumetric flow rate is measured by knowing the volume swept out by the ram at the entrance side of the die. Data are analyzed according to the procedure described in Sec. 2.2. Results obtained on a low-density polyethylene sample at three different temperatures and two die sizes are displayed in Figure 3.9 [53]. It is seen that increasing pressure has a dramatic effect on the measured viscosity, particularly as the temperature is lowered. In other studies of the influence of pressure on viscosity, McGowan and Cogswell [54] employed a pressurized concentric cylinder viscometer, while Kamal and Nyun [55] proposed a method of analyzing capillary viscometer data to extract the pressure dependence of the viscosity. As mentioned in Chapter 2, a capillary instrument to make viscosity measurements at high pressures is manufactured commercially by Polymer Laboratories. On

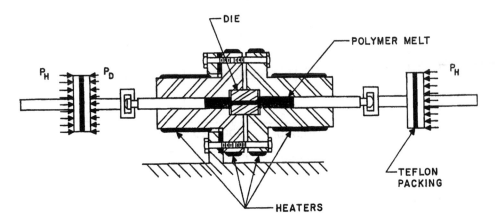

FIGURE 3.8 Schematic diagram of the hydrostatic pressure viscometer. (From Ref. 53.)

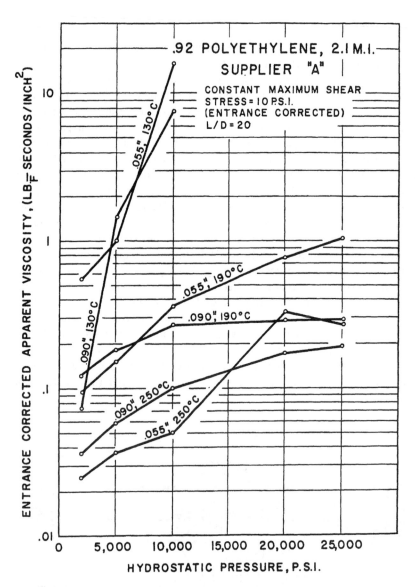

FIGURE 3.9 Apparent viscosity versus hydrostatic pressure at three temperatures and two die sizes. (From Ref. 53.)

the theoretical side, Utracki [56] has shown that the zero-shear viscosity data of McGowan and Cogswell [54] on polyethylene, polystyrene, and PMMA can be satisfactorily explained within the framework of the free volume theory. Recent developments have been described by Chakravorty et al. [56a].

III. SHEAR RATE–DEPENDENT VISCOSITY

An outstanding characteristic of polymer melts is their shear-thinning behavior whereby the apparent viscosity decreases as the rate of shear increases. This non-Newtonian behavior is of tremendous practical importance in the processing and fabrication of plastics and elastomers. First, the decreased viscosity makes the molten polymer easier to process or squirt through small channels, such as in the filling of a mold. In other words, the energy required to operate a large injection molding machine or extruder is reduced due to this phenomenon. Second, the decrease in viscosity is associated with the development of elasticity in the melt. This elasticity produces such phenomena as die swell, or the "puff-up" of extruded strands. Molecular orientation in molded objects also is closely related to polymer elasticity.

Typical curves for the apparent viscosity as a function of shear rate for a number of different temperatures are shown in Figure 3.1. Actual data, again as a function of shear rate but at a constant temperature, are shown in Figure 3.10 for monodisperse polystyrene melts of different molecular weights [57]. Figure 3.10 reveals that the zero-shear viscosity increases with molecular weight, as expected. Since all the molecular weights are larger than the critical value for entanglement formation, shear thinning is observed in each case. Because low-molecular-weight polymers have fewer entanglements than high-molecular-weight ones, it is not surprising that the point of onset of shear thinning moves to smaller values of the shear rate with increasing molecular weight. Furthermore, it is found that all the curves merge into a single line at high shear rates, implying a constant value of the power-law index.

The data displayed in Figures 3.1 and 3.10 can almost always be reduced to a single master curve if the ratio of the apparent viscosity to the zero-shear viscosity (i.e. the relative viscosity) is plotted in terms of an appropriate dependent variable. Such a treatment is also suggested by Eq. (3.3); it merely remains to define λ_r. In the simplest case, a unique curve is obtained if the logarithm of the relative viscosity is plotted against the logarithm of the product of the shear rate with the zero-shear viscosity, as shown in Figure 3.11; the effect of temperature largely is automatically compensated for by using the relative viscosity instead of the viscosity itself and by replacing the shear rate with the shear stress in the Newtonian region. Kraus and Gruver [58] found this to be true for random copolymers of butadiene and styrene. Since the ordinate as well as the abscissa are scaled with the zero-shear viscosity, whose behavior with both molecular

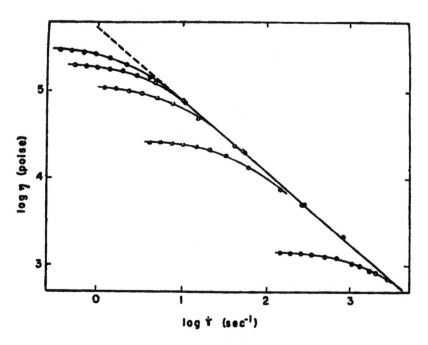

FIGURE 3.10 Viscosity as a function of shear rate for polystyrene melts of different molecular weights at 183°C From left to right, the curves are for molecular weights of 242,000, 217,000, 179,000, 117,000, and 48,500. (From Ref. 57.)

weight and temperature is well understood, obtaining the shear rate–dependent viscosity with the help of such a master curve is a straightforward matter. In other cases, better fit of the data is obtained if the abscissa scale is log ($\dot{\gamma}\eta_0 M_w^{\alpha}/\rho T$); the constant alpha is generally near unity [59], but it may vary somewhat. The temperature T is in degrees Kelvin, and ρ is the density of the polymer at each temperature [4,57]. Such a master curve is shown in Figure 3.12 employing the data of Figure 3.10 [57].

There are several theories and resulting equations that often are used to describe the shear rate dependence of the viscosity. These include the theories of Bueche [16] and of Graessley [10,60,61], with the latter theory being the more successful one. In the Graessley theory, the viscosity is calculated by dividing by the square of the shear rate the energy dissipated per unit time per unit volume due to steady shearing at a constant shear rate. The energy dissipation is itself computed by determining the fluid drag at each entanglement point and adding together the contributions from all the polymer–polymer entanglements in unit volume of fluid. A lowering in viscosity with increasing shear rate is attributed

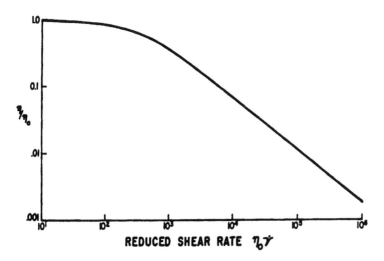

FIGURE 3.11 The master curve produced from the data of Figure 3.1 by vertical and horizontal shifting of the curves.

to shear-induced reduction in the entanglement density. The final result for the relative viscosity is

$$\frac{\eta}{\eta_0} = \frac{2}{\pi}\left[\cot^{-1}\alpha + \frac{\alpha(1 - \alpha^2)}{(1 + \alpha^2)^2}\right]$$

(3.13)

in which $\alpha = (\eta/\eta_0)(1/2\,\dot{\gamma}\theta_0)$, where θ_0 is a time constant for the formation of entanglements at low shear rates. As with Eq. (3.3) earlier, Eq. (3.13) suggests that a unique curve should be obtained if the relative viscosity is plotted against $\dot{\gamma}\theta_0$. This is generally found to be the case.

A. Influence of Polydispersity and Molecular Architecture

A distribution of molecular weights affects the value of the shear rate at which non-Newtonian behavior becomes apparent. The polymer with a broad distribution exhibits non-Newtonian flow at a lower rate of shear than a polymer that has the same zero rate of shear viscosity but a narrow distribution of molecular weights [59,62]. However, as illustrated in Figure 3.13 [34], shear thinning is generally more pronounced in the polymer possessing the narrower molecular weight distribution. The width of the distribution curve can be expressed by the polydispersity index, which increases as the distribution broadens. The shift in non-Newtonian behavior to lower shear rates as the distribution broadens has

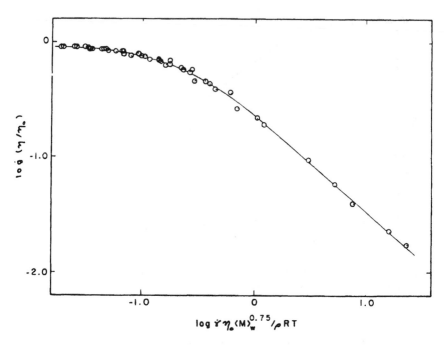

FIGURE 3.12 Logarithmic plot of reduced apparent viscosity against reduced shear rate, using the data of Figure 3.10. (From Ref. 57.)

important practical implications. Fractions or polymers with a very narrow distribution appear to have a higher viscosity than the same polymer with a normal or broad distribution under conventional molding conditions, where the rate of shear is high. Thus, polymers with a normal or broad distribution of molecular weights are generally easier to extrude or mold than a polymer with a narrow distribution.

The effect of chain branching on the shear rate–dependent viscosity is similar to the effect of chain branching on the zero-shear-rate viscosity. Thus, at high rates of shear, branched polymers in nearly all cases have lower viscosity than linear ones of the same molecular weight. Although, increasing the number of branches tends to decrease the viscosity, the molecular weight of the branches seems to be more important than their number [63].

In view of the foregoing, the manufacturer of polymers has the means of controlling the zero-shear-rate viscosity and the shear rate dependence of viscosity by variations in molecular weight, molecular weight distribution, and length of branches. These structural factors control the rheology, which, in turn, affects

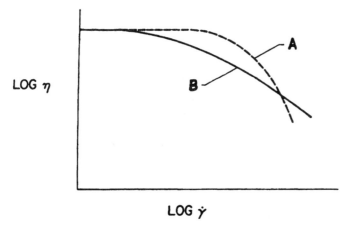

FIGURE 3.13   Plot of generalized viscosity versus shear rate for (A) a narrow-mo-lecular-weight-distribution polymer and (B) a branch-molecular-weight-distribution polymer. The average molecular weight in the two cases is assumed to be the same. (Reproduced with permission from Ref. 34; copyright © 1981, Division of Chemical Education, Inc.)

the processing or fabrication and the mechanical characteristics of finished objects [33].

B.   Master Curves for Temperature Dependence

Several empirical methods have been developed for predicting the temperature dependence and shear rate dependence of the shear viscosity of a polymer melt from a limited amount of experimental data. These have been motivated by the observation that isothermal curves for the variation of shear stress as a function of the shear rate at different temperatures can be made to superimpose by means of a horizontal shift, provided logarithmic coordinates are used. Such data on low-density polyethylene are shown in Figure 3.14 [64]; note that the shear rates have been calculated with the assumption of Newtonian flow behavior, although this does not influence the superposition procedure. The curve at one convenient temperature T_R can be chosen as the reference curve. In Figure 3.14 the curve at 200°C was selected as the reference curve. All the other curves can then be superimposed on the reference curve by shifting them horizontally along the shear rate axis. Curves for temperatures above the reference temperature are moved to the left; curves for lower temperatures are shifted to the right. The resulting master curve is shown in Figure 3.15. It is important to know how much each curve had to be shifted to superimpose upon the master curve and how this shifting

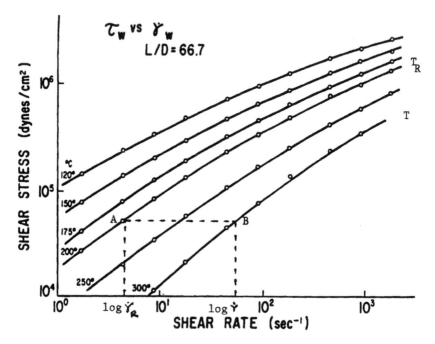

FIGURE 3.14 Flow curves at various temperatures for a low-density polyethylene. (From Ref. 64)

depended upon the temperature. With this information, the procedure can be reversed by using the master curve to calculate the flow behavior at any arbitrary temperature. Note that the temperature dependence of the extent of shifting can actually be related to the temperature dependence of the zero-shear viscosity with the help of the following argument.

Let us pick two points A and B having the same ordinate value, as shown in Figure 3.14. Let the abscissa of point A be $\log \dot{\gamma}_R$ and that of point B be $\log \dot{\gamma}$, and let the distance between them, measured from the reference curve to the curve of interest, be $\log a_T$. Clearly then

$$\log \dot{\gamma}_R - \log \dot{\gamma} = \log a_T \qquad (3.14)$$

and $a_T < 0$ when $T > T_R$. If both the points are in the linear (zero-shear) region,

$$\log \left[\frac{\eta_0(T)}{\eta_0(T_R)} \right] = \log \left[\frac{\tau(T,\dot{\gamma})/\dot{\gamma}}{\tau(T_R,\dot{\gamma}_R)/\dot{\gamma}_R} \right] = \log \left(\frac{\dot{\gamma}_R}{\dot{\gamma}} \right) = \log a_T \qquad (3.15)$$

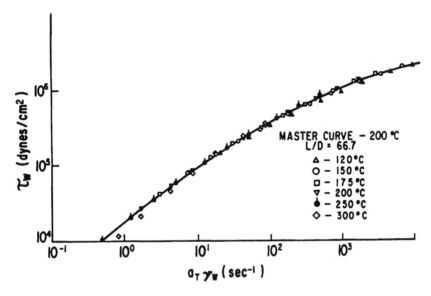

FIGURE 3.15 Master curve produced from the data of Figure 3.14. (Modified from Ref. 64.)

If we now evaluate Eq. (3.7) at $T = T_R$ and subtract the result from Eq. (3.7) as originally written, then

$$\log\left[\frac{\eta_0(T)}{\eta_0(T_R)}\right] = \log a_T = \frac{-C_1(T - T_R)}{C_2 + (T - T_R)} \qquad (3.16)$$

in which $C_2 = 51.6 + T_R - T_g$ and $C_1 = 17.44 \times 51.6/C_2$. Since the shift factor a_T is the same in the linear as well as the nonlinear region, the second equality in Eq. (3.16) is true in general. Even though the use of universal constants appearing in Eq. (3.7) gives reasonable results, better agreement with experimental data is achieved if C_1 and C_2 are specially chosen for each polymer [41]. At high temperatures, of course, Eq. (3.8) needs to be used in place of Eq. (3.7) in the derivation leading to Eq. (3.16).

Temperature–shear rate superposition is a very powerful technique, since for many polymers the shift factors are independent of polymer molecular weight [64]. Furthermore the same shift factors work for other viscoelastic functions as well.

C. Influence of Plasticizers

There are many instances where polymers contain plasticizing liquids at a mass concentration of only a percent or so. Plasticizers are typically added to improve the processibility of the polymer. An example is the addition of dioctyl phthalate to poly(vinyl chloride) to convert the polymer from a rigid material to a soft, flexible one. The major effect of the plasticizer is the lowering of the polymer glass transition temperature [65]. Liquids also have glass transitions, generally well below 0°C and in some cases far below −100°C. Often the glass transition temperature of a liquid is about two-thirds of its melting point in degrees Kelvin. The mixture glass transition temperature lies between the T_g values of the components and can be estimated by the following mixture rule [65]:

$$T_g = T_{gp}\phi_p + T_{gL}\phi_L \qquad (3.17)$$

where T_{gp} is the glass transition temperature of the polymer and T_{gL} is the corresponding value for the plasticizer. ϕ_p and ϕ_L are the volume fractions of the polymer and plasticizer respectively. Knowing the change in the glass transition temperature, we can calculate the change in viscosity with the help of Eq. (3.7).

IV. CONCLUDING REMARKS

This chapter has examined the influence of processing variables such as temperture, pressure, and shear rate and the influence of structural variables such as molecular weight, molecular weight distribution, and chain branching on the steady shear viscosity of melts of flexible chain polymers. Typical data were presented and explained, and it was shown how viscosity master curves could be generated. Many polymer processors, however, do not have access to sophisticated viscometers. They generally characterize polymer melts through a parameter called the *melt flow index* (MFI). This is the mass of polymer in grams extruded in 10 minutes through a capillary of specified length and diameter by the application of pressure using a dead weight as specified by ASTM D 1238. The MFI, therefore, is a single-point viscosity measurement at relatively low shear rate and temperature. Shenoy et al. [66] have shown that temperature-independent viscosity master curves can be generated for a given polymer type by plotting the product of the shear viscosity and MFI against the ratio of the shear rate to MFI. They have also provided master curves for several commodity thermoplastics [66], engineering plastics [67], and specialty polymers [68]. A knowledge of the MFI alone, in conjunction with these master curves and the WLF equation, provides the entire flow curve. (See also Ref. 69.)

We emphasize that we have not considered transient effects in this chapter. These are important in polymer processing operations where the time to reach steady state may be comparable to or even longer than the time scale of the processing operation being analyzed. These transient effects are examined in the chapter on liquid crystal polymer rheology, where the behavior of flexible chain polymers is compared and contrasted with the flow behavior of rigid chain polymers. Finally we point out that in the processing of semicrystalline polymers, the rheological behavior, in a temperature range between the melting point and the glass transition temperature, is influenced by the presence of crystallinity in the melt. There is a great paucity of data under these conditions, and it is common to account for the presence of crystals in an empirical manner [70].

REFERENCES

1. R. B. Bird, R. C. Armstrong, and O. Hassager. Dynamics of Polymeric Liquids. Vol. 1. 2nd ed. Wiley, New York, 1987, p. 171.
2. K. Walters. Rheometry. Chapman and Hall, London, 1975, p. 30.
3. T. G. Fox and P. J. Flory. The glass temperature and related properties of polystyrene. Influence of molecular weight. J. Polym. Sci. 14:315–319 (1954).
4. A. Casale, R.S. Porter, and J.F. Johnson. Dependence of flow properties of polystyrene on molecular weight, temperature, and shear. J. Macromol. Sci.-Rev. Macromol. Chem. C5:387–408 (1971).
5. T. G. Fox and V. R. Allen. Dependence of the zero shear melt viscosity and the related friction coefficient and critical chain length on measurable characteristics of chain polymers. J. Chem. Phys. 41:344–352 (1964).
6. V. R. Raju, G. G. Smith, G. Marin, J. R. Knox, and W. W. Graessley. Properties of amorphous and crystallizable hydrocarbon polymers. I. Melt rheology of fractions of linear polyethylene. J. Polym. Sci.: Polym. Phys. Ed. 17:1183–1195 (1979).
7. W. E. Rochefort, G. G. Smith, H. Rachapudy, V. R. Raju, and W. W. Graessley. Properties of amorphous and crystallizable hydrocarbon polymers. II. Rheology of linear and star-branched polybutadiene. J. Polym. Sci.: Polym. Phys. Ed. 17:1197–1210 (1979).
8. H. Rachapudy, G. G. Smith, V. R. Raju, and W. W. Graessley. Properties of amorphous and crystallizable hydrocarbon polymers. III. Studies of the hydrogenation of polybutadiene. J. Polym. Sci.: Polym. Phys. Ed. 17:1211–1222 (1979).
9. R. S. Porter and J. F. Johnson. The entanglement concept in polymer systems. Chem. Rev. 66:1–27 (1966).
10. W. W. Graessley. The entanglement concept in polymer rheology. Adv. Polym. Sci. 16: 1–179 (1974).
11. G. C. Berry and T. G. Fox. The viscosity of polymers and their concentrated solutions. Adv. Polym. Sci. 5: 261–357 (1968).
12. T. G. Fox, S. Gratch, and S. Loshaek. Viscosity relationships for polymers in bulk and in concentrated solution. In: F. R. Eirich, ed., Rheology. Vol. 1. Academic Press, New York, 1956, pp. 431–493.
13. D. W. van Krevelen and P. J. Hoftyzer. Properties of Polymers, 2nd ed. Elsevier Scientific, Amsterdam, 1976, pp. 331–369.
14. P. Debye. The intrinsic viscosity of polymer solutions. J. Chem. Phys. 14:636–639 (1946).
15. F. Bueche. Viscosity, self-diffusion and allied effects in solid polymers. J. Chem. Phys. 20:1959–1964 (1952).
16. F. Bueche. Physical properties of polymers. Wiley, New York, 1962.

17. P. G. de Gennes. Scaling Concepts in Polymer Physics, Cornell University Press, Ithaca, NY, 1979.
18. P. G. de Gennes. Reptation of a polymer chain in the presence of fixed obstacles. J. Chem. Phys. 55:572–579 (1971).
19. M. Doi and S. F. Edwards. The Theory of Polymer Dynamics, Clarendon Press, Oxford, 1986.
20. W. W. Graessley. Some phenomenological consequences of the Doi–Edwards theory of viscoelasticity. J. Polym. Sci.: Polym. Phys. Ed. 18:27–34 (1980).
21. W. W. Graessley. Entangled, linear, branched and network polymer systems—molecular theories. Adv. Polym. Sci. 47:67–117 (1982).
22. J. F. Rudd. The effect of molecular weight distribution on the rheological properties of polystyrene. J. Polym. Sci. 44:459–474 (1960).
23. V. R. Allen and T. G. Fox. Viscosity–molecular weight dependence for short chain polystyrene. J. Chem. Phys. 41:337–343 (1964).
24. R. A. Mendelson, W. A. Bowles, and F. L. Finger. Effect of molecular structure on polyethylene melt rheology. II. Shear-dependent viscosity. J. Polym. Sci.: Part A-2 8:127–141 (1970).
25. E. M. Friedman and R. S. Porter. Polymer viscosity–molecular weight distribution correlations via blending: for high molecular weight poly(dimethyl siloxanes) and for polystyrenes. Trans. Soc. Rheol. 19:493–508 (1975).
26. W. Minoshima, J. L. White, and J. E. Spruiell. Experimental investigation of the influence of molecular weight distribution on the rheological properties of polypropylene melts. Polym. Eng. Sci. 20:1166–1176 (1980).
27. H. M. Laun. Prediction of elastic strains of polymer melts in shear and elongation. J. Rheol. 30:459–501 (1986).
28. S. Saeda, J. Yotsuyanagi, and K. Yamaguchi. The relation between melt flow properties and molecular weight of polyethylene. J. Appl. Polym. Sci. 15:277–292 (1971).
29. A. Rudin and K. K. Chee. Zero shear viscosities of narrow and broad distribution polystyrene melts. Macromolecules 6:613–624 (1973).
30. W. W. Graessley. Effect of long branches on the flow properties of polymers. Accounts Chem. Res. 10:332–339 (1977).
30a. C. J. Hawker, P. J. Farrington, M. E. Mackay, K. L. Wooley, and J. M. J. Frechet. Molecular ball bearings? The melt viscosity of dendritic macromolecules. J. Am. Chem. Soc. 117:4409–4410 (1995).
31. R. A. Mendelson. Flow properties of polyethylene melts. Polym. Eng. Sci. 9:350–355 (1969).
32. R. A. Mendelson, W. A. Bowles, and F. L. Finger. Effect of molecular structure on polyethylene melt rheology. I. Low-shear behavior. J. Polym. Sci.: Part A-2 8:105–126 (1970).
33. J. Miltz and A. Ram. Flow behavior of well-characterized polyethylene melts. Polym. Eng. Sci. 13:273–279 (1973).
34. G. L. Wilkes. An overview of the basic rheological behavior of polymer fluids with an emphasis on polymer melts. J. Chem. Education 58:880–892 (1981).
35. G. Kraus and J. T. Gruver. Rheological properties of multichain polybutadienes. J. Polym. Sci.: Part A 3:105–122 (1965).
36. V. R. Raju, H. Rachapudy, and W. W. Graessley. Properties of amorphous and crys-

tallizable hydrocarbon polymers. IV. Melt rheology of linear and star-branched hydrogenated polybutadiene. J. Polym. Sci.: Polym. Phys. Ed. 17: 1223–1235 (1979).

37. J. M. Carella. Comments on the paper "Comparison of the Rheological Properties of Metallocene-Catalyzed and Conventional High-Density Polyethylenes." Macromolecules 29:8280–8281 (1996).

38. D. R. Saini and A. V. Shenoy. Viscoelastic properties of linear low density polyethylene melts. Eur. Polym. J. 19:811–816 (1983).

39. T. Masuda, Y. Ohta, and S. Onogi. Rheological properties of anionic polystyrenes. III. Characterization and rheological properties of four-branch polystyrenes. Macromolecules 4:763–768 (1971).

40. M. L. Williams, R. F. Landel, and J. D. Ferry. The temperature dependence of relaxation mechanisms in amorphous polymers and other glass-forming liquids. J. Am. Chem. Soc. 77: 3701–3706 (1955).

41. J. D. Ferry, Viscoelastic Properties of Polymers. 3rd ed. Wiley, New York, 1980.

42. H. Schott. Dependence of activation energy for viscous flow of polyhydrocarbons on bulk of substituents. J. Appl. Polym. Sci. 6: S29–S30 (1962).

43. R. S. Porter and J. F. Johnson. Temperature dependence of polymer viscosity. The influence of polymer composition. J. Polym. Sci.: Part C 15:373–380 (1966).

44. D. W. van Krevelen and P. J. Hoftyzer. Newtonian shear viscosity of polymeric melts. Angew. Makromol. Chem. 52:101–109 (1976).

45. M. J. Crochet and P. M. Naghdi. On thermomechanics of polymers in the transition and rubber regions. J. Rheol. 22:73–89 (1978).

46. D. A. Carey, C. J. Wust, Jr., and D. C. Bogue. Studies in nonisothermal rheology: Behavior near the glass transition temperature and in the oriented glassy state. J. Appl. Polym. Sci. 25:575–588 (1980).

47. R. K. Gupta and A. B. Metzner. Modeling of nonisothermal polymer processes. J. Rheol. 26:181–198 (1982).

48. F. N. Cogswell. Polymer Melt Rheology. George Godwin, London, 1981.

49. M. H. Cohen and D. Turnbull. Molecular transport in liquids and glasses. J. Chem. Phys. 31:1164–1169 (1959).

50. P. B. Macedo and T. A. Litovitz. On the relative roles of free volume and activation energy in the viscosity of liquids. J. Chem. Phys. 42:245–256 (1965).

51. I. C. Sanchez. Towards a theory of viscosity for glass-forming liquids. J. Appl. Phys. 45:4204–4215 (1974).

52. F. N. Cogswell. The influence of pressure on the viscosity of polymer melts. Plastics and Polymers 41(Feb.):30–43 (1973).

53. R. F. Westover. Effect of hydrostatic pressure on polyethylene melt rheology. SPE Trans. 1:14–20 (1961).

54. J. C. McGowan and F. N. Cogswell. The effect of pressure and temperature upon the viscosity of liquids with special reference to polymeric liquids. Br. Polym. J. 4: 183–198 (1972).

55. M. R. Kamal and H. Nyun. The effect of pressure on the shear viscosity of polymer melts. Trans. Soc. Rheol. 17:271–285 (1973).

56. L. A. Utracki. A method of computation of the pressure effect on melt viscosity. Polym. Eng. Sci. 25:655–668 (1985).

56a. S. Chakravorty, M. Rides, C. R. G. Allen, and C. S. Brown. Polymer melt viscosity increases under pressure: Simple new measurement method. Plast.. Rubber Comp. Proc. Applic. 25:260–261 (1996).

57. R. A. Stratton. The dependence of non-Newtonian viscosity on molecular weight for "monodisperse" polystyrene. J. Colloid Interf. Sci. 22:517–530 (1966).

58. G. Kraus and J. T. Gruver. Melt viscosity of random copolymers of butadiene and styrene prepared by anionic polymerization. Trans. Soc. Rheol. 13:315–322 (1969).

59. R. L. Ballman and R. H. M. Simon. The influence of molecular weight distribution on some properties of polystyrene melt. J. Polym. Sci.: Part A 2:3557–3575 (1964).

60. W. W. Graessley. Molecular entanglement theory of flow behavior in amorphous polymers. J. Chem. Phys. 43:2696–2703 (1965).

61. W. W. Graessley. Viscosity of entangling polydisperse polymers. J. Chem. Phys. 47:1942–1953 (1967).

62. C. K. Shih. The effect of molecular weight and molecular weight distribution on the non-Newtonian behavior of ethylene-propylene-diene polymers. Trans. Soc. Rheol. 14:83–114 (1970).

63. V. C. Long. G. C. Berry. and L. M. Hobbs. Solution and bulk properties of branched polyvinyl acetates IV—melt viscosity. Polymer 5:517–524 (1964).

64. R. A. Mendelson. Prediction of melt viscosity flow curves at various temperatures for some olefin polymers and copolymers. Polym. Eng. Sci. 8:235–240 (1968).

65. L.E. Nielsen and R.F. Landel. Mechanical Properties of Polymers and Composites. 2nd ed. Marcel Dekker. New York. 1994. p. 21.

66. A. V. Shenoy. S. Chattopadhyay. and V. M. Nadkarni. From melt flow index to rheogram. Rheol. Acta 22:90–101 (1983).

67. A. V. Shenoy. D. R. Saini. and V. M. Nadkarni. Rheograms for Engineering thermoplastics from melt flow index. Rheol. Acta 22:209–222 (1983).

68. D. R. Saini and A. V. Shenoy. Melt rheology of some specialty polymers. J. Elastomers Plast. 17:189–217 (1985).

69. A. V. Shenoy and D. R. Saini. Thermoplastic Melt Rheology and Processing. Marcel Dekker. New York. 1996.

70. K. F. Zieminski and J. E. Spruiell. Computer simulation of melt spinning. J. Appl. Polym. Sci. 35:2223–2245 (1988).

4

Shear Viscosity of Polymer Solutions

I. INTRODUCTION

The shear flow of polymer solutions, both dilute and concentrated, and employing either water or an organic liquid as the solvent, is encountered in a wide variety of technologically important situations. Common examples include the addition of 2–3% of a polymer such as polymethyl methacrylate to a mineral oil in order to formulate multigrade motor oils; the polymer serves to reduce the dependence of oil viscosity on temperature and also to lower the pour point. More concentrated solutions, containing up to 30% or 40% polymer, are used to manufacture textile fibers by the process of (wet or dry) solution spinning. This is because, though polymers such as cellulose acetate cannot be melted, they can be dissolved in a solvent such as acetic acid or acetone. In the petroleum industry, aqueous solutions of xanthan gum polysaccharides and partially hydrolyzed polyacrylamides in a concentration range of a few hundreds of ppm are utilized for enhanced oil recovery by water flooding; the thickening effect of the added polymer reduces the mobility difference between the injected water and the oil in place, thereby improving both areal and vertical sweep efficiencies. The thickening effect of polymers is also put to good use in the food industry, where xanthan gum is a common additive in salad dressings and other products. Paints and coatings are other applications of polymer solutions. Note that high-molecular-weight polymers can be fairly shear sensitive, and chain scission takes place quite easily. It is for this reason that, to achieve a desired enhancement in viscosity, one often adds a large amount of low-molecular-weight polymer instead of a small amount of high-molecular-weight polymer to a solvent if large deformation rates are likely to be encountered.

We obviously expect the viscosity of a polymer solution to depend on concentration. In addition, as in the case of a polymer melt, we expect that the viscosity of a polymer dissolved in an organic solvent will be a function of temperature, pressure, shear rate, molecular weight, and molecular architecture. Typical viscosity–shear rate curves for polystyrene solutions of different concentrations are shown in Figure 4.1 [1]; the concentrations are such that polymer–polymer entanglements are present in each instance. These curves show that dilute solutions remain Newtonian in behavior to higher rates of shear than do more concentrated solutions. This is because the presence of the solvent reduces the number of chain entanglements per unit volume. The absolute lower limit of the solution viscosity is of course the solvent viscosity; for the solvent used in Figure 4.1, this is 1.3 centipoise. Since the various curves in Figure 4.1 are qualitatively similar to the corresponding curves in Figure 3.1, they can again be represented using the power-law model or the Carreau model, Eqs. (3.2) and (3.3) respectively. If a more viscous solvent is used, the solution viscosity does tend to approach the

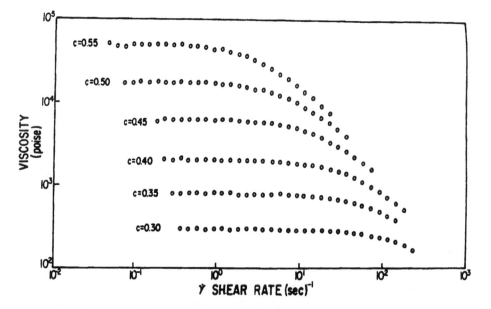

FIGURE 4.1   Shear rate dependence of polystyrene solutions in *n*-butyl benzene at different concentrations (g/cc) at 30°C. Molecular weight = 411,000. (From Ref. 1.)

solvent viscosity at high enough shear rates, giving an upper Newtonian region. Now the Carreau equation is written as

$$\frac{\eta_a - \eta_\infty}{\eta_0 - \eta_\infty} = [1 + (\lambda_t \dot{\gamma})^2]^{(n-1)/2} \tag{4.1}$$

in which η_∞ is the limiting viscosity at high shear rates, usually taken to be the solvent viscosity. If the solvent is extremely viscous, the solution and solvent viscosities may be almost the same, independent of the shear rate. This situation results in a constant viscosity, elastic liquid [2,3].

The treatment in this chapter parallels that of the previous chapter in that the influence of each process and structural variable on the viscosity of polymer solutions is examined in a systematic manner. While polymer concentration is a new variable, additional variables arise if the polymer contains ionizable groups. In a high dielectric medium such as water, a polymer such as partially hydrolyzed polyacrylamide can ionize to give a polyion that assumes an extended conformation due to electrostatic repulsion between charged groups. The presence of salt, however, tends to screen these charges resulting in a more coiled conformation. Changes in conformation lead to changes in viscosity. Water-soluble polymers can also be modified by the incorporation of hydrophobic moieties. These can associate with each other, resulting in the formation of a network that again alters the solution viscosity. Note that water-soluble polymers are also commonly employed as flocculants in water treatment, and block copolymers are used to stabilize a solid-in-liquid suspension sterically; these applications are considered in Chapter 10 on the rheology of suspensions.

II. INTRINSIC VISCOSITY—THE LIMIT OF INFINITE DILUTION

If we think of an infinitely dilute polymer solution to be a suspension of isolated, noninteracting spheres in a Newtonian liquid, we can immediately relate the relative viscosity η_R to the polymer volume fraction ϕ through the Einstein result [4]:

$$\eta_R = \frac{\eta}{\eta_s} = 1 + 2.5\phi \tag{4.2}$$

in which η and η_s are the solution and solvent viscosities, respectively. The polymer volume fraction, in turn, can be related to the mass concentration c (see Ref. 5, for example) to yield

$$\phi = \frac{vcN_A}{M} \tag{4.3}$$

where v is the volume occupied by a single polymer molecule, N_A is Avogadro's number, and M is the polymer molecular weight. In writing Eq. (4.3), it is assumed that the polymer sample is monodisperse and that each molecule assumes a spherical shape in solution. Introducing Eq. (4.3) into Eq. (4.2), rearranging the result, and extrapolating to zero concentration gives the intrinsic viscosity or the limiting viscosity number $[\eta]$ as

$$[\eta] = \lim_{c \to 0} \frac{\eta_R - 1}{c} = 2.5 \frac{vN_A}{M} \qquad (4.4)$$

Experimental data can generally be represented in terms of the Huggins equation:

$$\frac{\eta_R - 1}{c} = [\eta] + k[\eta]^2 c \qquad (4.5)$$

where k is known as the Huggins constant. Equation (4.4) shows that the intrinsic viscosity is directly related to the size of a polymer molecule in solution. If the coil size changes due, say, to changes in temperature, pH, or ionic strength, the intrinsic viscosity must also change. Furthermore, if the coil size or sphere volume can be related to the polymer molecular weight, one can employ Eq. (4.4) for the measurement of the molecular weight. This connection between the sphere radius r and the polymer chain length is usually made with the help of a molecular model. If, for example, one takes a linear, flexible polymer to be freely jointed chain of n links, each of length l, one can show that [5]

$$r \propto \sqrt{n}\, l \qquad (4.6)$$

provided there are no excluded volume effects that are long-range interactions due to attraction and repulsion forces between widely separated chain segments or between polymer segments and solvent molecules. This happens at what is known as the theta condition [6].

Because the polymer molecular weight is proportional to n and the sphere volume is proportional to the third power of r, a combination of Eqs. (4.4) and (4.6) reveals that

$$[\eta] \propto M^{1/2} \qquad (4.7)$$

which is valid under theta conditions alone; these generally arise when a poor solvent is used. In a good solvent, polymer–solvent interactions result in coil expansion [7], and we find that

$$[\eta] = KM^a \qquad (4.8)$$

which is known as the Mark–Houwink–Sakurada equation. The values of the constants K and a depend on the polymer, the solvent, and the temperature; these are determined experimentally using monodisperse polymer fractions, and extensive tabulations are available [7,8]. The value of the exponent a typically varies from 0.5 at the theta temperature to 0.8 for a very good solvent. According to Eqs. (4.5) and (4.8), the zero-shear viscosity of a dilute polymer solution should increase with increasing molecular weight at a fixed mass concentration.

In terms of actually measuring the relative viscosity as a function of concentration for the purpose of computing the intrinsic viscosity, any of the viscometers described in Chapter 2 can be used, provided that the shear rate is low enough that we are in the lower Newtonian region. This requirement is generally satisfied if the solutions are so dilute that the relative viscosity does not appreciably exceed unity. In such a case, it is customary to employ (inexpensive) glass capillary viscometers of the type shown in Figure 4.2. This particular instrument is known as a suspended level Ubbelhode viscometer. In use, one fills bulb A with the polymer solution, allows the liquid to drain through the capillary under the influence of gravity, and notes the time taken for the liquid level to move between the two marks made on bulb A. The process is repeated for the pure solvent and also for the polymer solution at different concentrations. Note that proper clean-

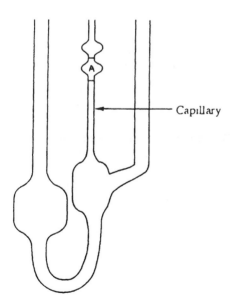

FIGURE 4.2 An Ubbelhode viscometer.

ing of the capillary and good temperature control are required to obtain consistent data.

If we recall that the flow of a Newtonian liquid through a capillary is described by the Hagen–Poiseuille equation [Eq. (2.5)], then it is evident that the relative viscosity is given by the ratio of the flow rate of the pure solvent to that of the solution. The volume of bulb A, however, is fixed, and the flow rate is inversely proportional to the efflux or drainage time t, with the result that

$$\frac{\eta}{\eta_s} = \frac{t\rho}{t_s\rho_s} \tag{4.9}$$

and we may calculate the relative viscosity with ease. If it is found that the shear rate is not low enough, extrapolation of the data to lower shear rates may be necessary. Also, if the polymer sample is not monodisperse, the molecular weight calculated with the help of Eq. (4.8) will be an average value intermediate between the number-average molecular weight and the weight-average molecular weight. Such data can still be useful for quality control purposes.

In closing this section, we note that Saini and Shenoy [9] have proposed a simple method for the generation of temperature-independent master curves for polymer solution viscosity as a function of shear rate using data obtained with the help of glass capillary viscometers of the type considered here.

III. ZERO-SHEAR VISCOSITY AS A FUNCTION OF MOLECULAR WEIGHT AND CONCENTRATION

The zero-shear viscosity of a polymer-solvent system can change quite considerably in going from pure solvent to undiluted polymer. In the extreme case of a low-viscosity liquid added to a polymer near its glass transition temperature, the viscosity may change by a factor of 10^{15} in covering this concentration range. Even at temperatures well above T_g, the viscosity of a polymer solution has been found to vary from the first power to greater than the fourteenth power of the concentration of the polymer, depending upon the concentration range, the polymer–solvent system, temperature, and other experimental conditions [10]. Typical experimental data are shown in Figure 4.3 [11].

In terms of obtaining a rough estimate of the viscosity of a polymer solution at a temperature well above T_g, the data of Figure 4.3 suggest the following mixture rule:

$$\log \eta_0 = \phi_p \log \eta_p + \phi_L \log \eta_L \tag{4.10}$$

The viscosities of the pure polymer and the liquid are η_P and η_L, respectively; ϕ_P and ϕ_L are the corresponding volume fractions. The difficulty in finding a

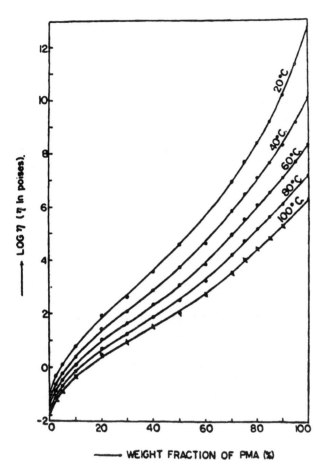

FIGURE 4.3 Viscosity of polymethyl acrylate dissolved in diethyl phthalate. (Reprinted with permission from Ref. 11. Copyright 1962 American Chemical Society.)

single (empirical) equation that can accurately represent data over the entire concentration and temperature range is related to the fact that the physics of the situation changes as the polymer concentration changes. In the limit of infinite dilution, long-chain polymer molecules behave in the fashion of impenetrable spheres, while as bulk polymer we have an entangled mass of fluid. At an intermediate concentration, the spheres bump up against each other, and there is coil overlap but no entanglement. These three regions are roughly differentiated by the value of the product $[\eta]c$. According to Eq. (4.4), $[\eta]M/N_A$ is the volume of a polymer molecule multiplied by 2.5, whereas cN_A/M is obviously the number of polymer molecules per unit volume. As a consequence, $[\eta]c$, which is the product of these two quantities, represents the volume fraction of polymer multiplied by 2.5. If this number is small compared to unity, the polymer solution is considered dilute; if in the range of 1–10, the solution is low to moderately concentrated; if above 10, the solution is entangled.

Based on the foregoing picture, it is reasonable to expect that different theories and correspondingly different equations will portray polymer–solution viscosity data in the different concentration regimes. We have already seen that the Huggins equation [Eq. (4.5)], fits data in the dilute regime. It is found that data in the coil-overlap region can often be represented with the help of the Martin equation [12]:

$$\frac{\eta_{sp}}{[\eta]c} = \exp\left(k[\eta]c\right) \tag{4.11}$$

in which η_{sp} is the specific viscosity, defined as $(\eta_R - 1)$. A feature of both the Huggins equation and the Martin equation is that the relative viscosity is postulated to be a unique function of $[\eta]c$. This observation was exploited by Lyons and Tobolsky [13], who proposed an empirical equation for the viscosity of polymer solutions over the entire concentration range, provided that the molecular weight of the polymer was below the value needed for entanglement formation. In the Lyons–Tobolsky equation, the relative viscosity is defined with respect to the solvent viscosity, and, for this reason, the accuracy of the theory is expected to be poor for extremely concentrated solutions [14]. However, the equation has been found to be satisfactory in a number of cases [15]. The Lyons–Tobolsky equation is:

$$\frac{\eta_{sp}}{[\eta]c} = \exp\frac{k[\eta]c}{1 - bc} \tag{4.12}$$

A new constant, b, appears in Eq. (4.12), and it can be considered an empirical constant, or it can be calculated from melt data on the pure polymer.

Besides empirical equations, there are several fundamental theories that seek to relate the zero-shear viscosity of a polymer solution to the mass concentration and the polymer molecular weight. If the polymer concentration is such that there are no chain entanglements present, an explicit expression can be derived for the zero-shear viscosity by examining the behavior of a single, isolated polymer molecule.

In the Bueche theory [16], the energy dissipated per unit volume due to fluid friction in a shear field is calculated by multiplying the energy dissipated by the motion of a single polymer molecule times the number of polymer molecules per unit volume. This quantity then leads to the following expression for the polymer contribution to the zero-shear viscosity:

$$\eta_0 = \frac{cN_A \zeta nl^2}{36M_0} \tag{4.13}$$

in which c is the mass concentration, N_A is Avogadro's number, M_0 is the molecular weight of a monomer unit, ζ is the friction coefficient per monomer unit (similar in concept to Stokes' law for drag on spheres), and n and l are the same quantities as in Eq. (4.6). Since n is proportional to molecular weight M, the zero-shear viscosity is predicted to be proportional to the product cM.

In the bead-spring theories, each polymer molecule in solution is idealized as $(N + 1)$ spheres, each of mass m, connected by N massless springs [17]. The polymer solution then is a noninteracting suspension of these stringy entities in a Newtonian liquid. Under the influence of flow, the springs get stretched and oriented, and the tension in the spring makes a contribution to the stress in the liquid. By adding together the contributions from all the (identical) polymer molecules, we get the complete stress, from which the viscosity can be determined. In the Rouse model, the polymer molecules are taken to be free draining (not impenetrable), and the viscosity contributed by the polymer molecules under theta conditions is found to be [18]

$$\eta_0 = \frac{cMN_A l^2 \zeta}{36M_0^2} \tag{4.14}$$

Again the zero-shear viscosity is predicted to increase linearly with cM.

When experimental data for the zero-shear viscosity of polymer solutions are examined, it is found that the dependence of the viscosity on cM is stronger than the linear relation predicted by Eqs. (4.13) and (4.14). It turns out that this is due to the variation of the friction coefficient ζ with polymer concentration. Note that changes in polymer concentration lead to changes in the free volume. When data are corrected to give viscosities at a constant value of the monomeric friction coefficient (or equivalently the same free volume), linear behavior is established. This is shown in Figure 4.4 [19]. An examination of Figure 4.4 also reveals that linearity between the viscosity and cM persists only up to a critical value of cM. Beyond this critical value, the slope of the viscosity-versus-cM curve increases to about 3.4 on logarithmic coordinates. The behavior of polymer solutions, therefore, is entirely analogous to that of polymer melts, for which c equals the melt density. The change in slope in Figure 4.4 is again due to the formation of chain entanglements. The critical molecular weight for entanglement formation now depends on the polymer concentration and is given by

$$(Mc)_{sol} = \left[\frac{\rho}{c} \right] Mc \tag{4.15}$$

where Mc is the molecular weight needed for entanglement formation in the melt.

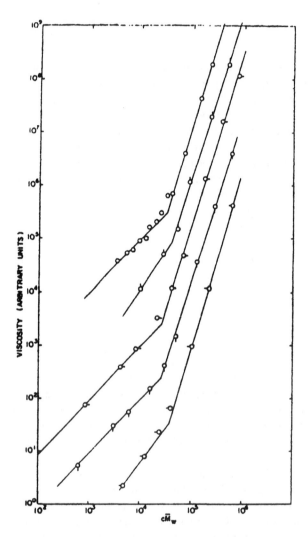

FIGURE 4.4 Viscosity versus the product *cMw* for polystyrene. The uppermost curve is for the molten polymer at 217°C. Progressively lower curves are for 0.55 g/ml polymer in *n*-butyl benzene, 0.415 g/ml and 0.31 g/ml in di-octyl phthalate, and 0.255 g/ml in *n*-butyl benzene. Data at the various concentrations have been shifted vertically to avoid overlap. (From Ref. 19.)

The idea that the viscosity of a polymer solution is determined by the free volume has been employed by Kelley and Bueche [20] to predict the viscosity of concentrated (entangled) polymer solutions using the viscosity of the pure polymer as the reference. The final equations are:

$$\frac{\eta}{\eta_p} = \phi_p^4 \exp\left(\frac{1}{\phi_p f_p + \phi_L f_L} - \frac{1}{f_p}\right) \tag{4.16}$$

$$f_p = 0.025 + 4.8 \times 10^{-4}(T - T_{gp}) \tag{4.17}$$

$$f_L = 0.025 + \alpha_L(T - T_{gL}) \tag{4.18}$$

The free volumes of the polymer and the liquid are f_p and f_L, respectively. The difference between the coefficients of expansion of the liquid above and below T_g is α_L. Generally, α_L is of the order of 10^{-3}. The glass transition temperatures of the polymer and solvent are T_{gp} and T_{gL}, respectively. Figure 4.5 shows a plot of Eq. (4.16) for several values of f_p and f_L. The figure shows the tremendous sensitivity of the relative viscosity to small changes in f_p and f_L. The concentration dependence of relative viscosity is especially great at temperatures close to the T_g of the pure polymer, as represented by curve D. Very small amounts of a liquid or plasticizer can reduce the viscosity of the polymer by one or more decades. Equations (4.17) and (4.18) give approximate theoretical values of f_p and f_L. However, because of the extreme sensitivity of viscosity to the free volumes, in practical situations it may be more convenient to consider f_p and f_L as empirical constants. Also, since Eq. (4.16) was derived for entangled macromolecules, it may not hold for dilute polymer solutions.

IV. INFLUENCE OF STRUCTURAL FACTORS, TEMPERATURE, AND PRESSURE ON THE ZERO-SHEAR VISCOSITY

The influence of polydispersity on the zero-shear viscosity of polymer solutions is the same as the influence of this variable on the zero-shear viscosity of polymer melts (seen earlier in Sec. II.A. of Chap. 3) In other words, viscosity data on polydisperse polymers superpose with the data shown in Figure 4.4 when the weight-average molecular weight is used in place of Mw. The effect of chain branching is similar (see Sect. II.B orf Chap. 3) In a given solvent, the branched polymer has a smaller radius of gyration and a smaller intrinsic viscosity than a linear polymer of the same molecular weight. At low and intermediate concentrations of polymer in the same solvent, there are no chain entanglements, and the viscosity of the branched polymer is lower than that of the linear polymer; it is found that [21]

$$\eta_B(c,M) = \eta_L(c,gM) \tag{4.19}$$

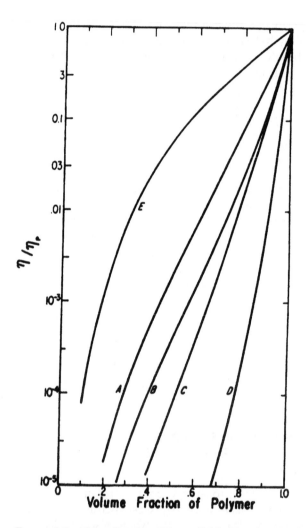

FIGURE 4.5   Viscosity of solutions divided by the viscosity of the polymer at the same temperature as a function of the concentration of the polymer according to the theory of Kelley and Bueche. Values of f_P and f_L, respectively, for the different curves are A: 0.1, 0.2; B: 0.1, 0.3; C: 0.05, 0.1; D: 0.05, 0.2; E: $f_p = f_L$.

where g is the square of the ratio of the branched to unbranched radii of gyration and the subscripts B and L denote branched and linear, respectively. At high concentrations, deviations from Eq. (4.19) begin to occur when the branches are long enough to form entanglements. Now, measured values of the viscosity of branched polymers can be as much as 100 times the viscosity of linear polymers at comparable cgM values [21].

The temperature dependence of the zero-shear viscosity of polymer solutions follows the WLF equation. [Eq. (3.7)] at low temperatures and the Arrhenius equation [Eq. (3.8)] at high temperatures. Recall that the WLF equation involves the polymer glass transition temperature as the reference temperature. For polymer solutions, this quantity can be estimated with the help of Eq. (3.17), or it may be measured using dilatometry, since the specific volume of polymers and their solution changes slope on going through the glass transition temperature [22]. Note that if a different temperature is used as the reference temperature, the appropriate form of the WLF equation is Eq. (3.16). Reference 23 may be consulted for additional details.

The pressure dependence of the viscosity of polymer solutions is important in a number of applications, including elastohydrodynamic lubrication, where the pressure often exceeds 1 GPa [24]. Bair and Winer [25] constructed a falling-body viscometer to measure the high-temperature, high-pressure viscosity of a number of liquid lubricants; they found that the isothermal viscosity could increase by several orders of magnitude upon increasing the pressure from atmospheric to 0.5 GPa, especially at lower temperatures. These data were explained by Yasutomi et al. [24], who modified the WLF equation to include the effect of pressure on the glass transition temperature and the free volume. The modified form of the WLF equation is

$$\log \frac{\eta_0(T,p)}{\eta_0(T_g)} = \frac{c_1 [T - T_g(p)] F}{c_2 + [T - T_g(p)] F} \tag{4.20}$$

where the viscosity at the glass transition temperature is taken to be independent of pressure. The glass transition temperature itself depends on pressure according to

$$T_g = T_{g0} + A_1 \ln(1 + A_2 p) \tag{4.21}$$

while the free-volume expansion coefficient F is given by

$$F = 1 - B_1 \ln(1 + B_2 p) \tag{4.22}$$

in which A_1, A_2, B_1, and B_2 are constants. For a given liquid, these constants have to be determined by a regression analysis of sufficient viscosity–pressure data. Wu et al. [26] further developed the theory and demonstrated good agreement with data on blends of polymers with mineral oils.

V. SHEAR RATE–DEPENDENT VISCOSITY

The flow curve of a polymer solution depends on polymer concentration, polymer molecular weight, solvent viscosity, temperature, and the applied shear rate. The influence of concentration and shear rate on the measured viscosity is shown in Figure 4.1. This figure is qualitatively similar to Figures 3.1 and 3.10, which show the influence of temperature and molecular weight, respectively, on the curve of melt viscosity versus shear rate. Consequently, it is not surprising that, for a given polymer, data superposition that is independent of concentration, molecular weight, solvent viscosity, and temperature can be obtained by plotting the ratio of the solution viscosity to the zero-shear viscosity in terms of an appropriate dependent variable. In the simplest case, we can simply plot the relative viscosity against $\dot{\gamma}/\dot{\gamma}_0$, where $\dot{\gamma}$ represents the shear rate at which the viscosity has fallen to 90% of its zero-shear value. Tam and Tiu [27] showed that this procedure worked satisfactorily for aqueous solutions of both flexible chain polymers such as polyacrylamide and polyethylene oxide, and semirigid polymers, such as xanthan gum and carboxy methyl cellulose. If the solvent viscosity η_s is significant in comparison to the solution viscosity, better data superposition may be obtained by plotting $(\eta - \eta_s)/(\eta_0 - \eta_s)$. For predictive purposes, one needs to know $\dot{\gamma}_0$; this may be taken to be the reciprocal of a characteristic time λ_E defined as [28]:

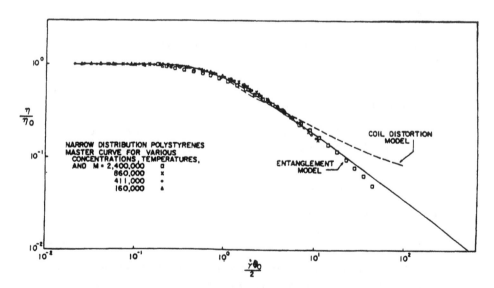

FIGURE 4.6   Master curve for polystyrene solutions in *n*-butyl benzene. The solid line represents Eq. (3.13). (From Ref. 1.)

$$\lambda_t = \frac{\eta_0 - \eta_*}{c} \frac{M}{RT} \tag{4.23}$$

which appears to work well for polyethylene oxide dissolved in water or in a mixture of water and glycerol.

As in the case of polymer melts, the decrease in solution viscosity with increasing shear rate is considered to be the result of shear-induced reduction in the number of entanglements. The variation of the relative viscosity of a polymer solution with shear rate should, therefore, again be described by Eq. (3.13). This is indeed found to be true, and Figure 4.6 shows that a master curve is obtained when viscosity data on monodisperse polystyrenes for a variety of concentrations, molecular weights, and temperatures are plotted as suggested by this equation [1]; also included in Figure 4.6 are the data shown earlier in Figure 4.1. Since data on undiluted polymers too lie on the the same master curve, the influence of polydispersity and chain branching on the non-Newtonian viscosity of polymer solutions is identical to that discussed for polymer melts in Section III.A of Chap. 3.

VI. VISCOSITY OF POLYELECTROLYTES

Proteins and many other biological macromolecules are multiply charged compounds, as are some commonly encountered synthetic polymers, such as partially hydrolyzed polyacrylamide used in enhanced oil-recovery operations. Furthermore, by adding charged groups to a synthetic polymer such as polystyrene we can obtain a water-soluble polymer, polystyrene sulfonate in this particular case. The rheological behavior of these polyelectrolytes in solution can be vastly different from that of the uncharged polymers considered thus far [29,30]. This is because of interactions between different segments of a given polymer molecule, between different molecules, and between polymer and solvent. These interactions, especially at lower concentrations, alter the apparent size of the polymer in solution as determined by intrinsic viscosity measurements. A polyion can dissociate in a solvent such as water, and this results in like charges being distributed along the length of the polymer molecule. Electrostatic repulsion among the like charges leads to coil expansion and to an increase in the solution viscosity relative to the unexpanded coil. As the polymer concentration is increased, the solution viscosity tends to increase, because the number of coils increases. However, due to the presence of like charges, the coils repel each other, and their size decreases even as their number increases. Consequently, one finds that as the polymer concentration increases, the overall solution viscosity increases at a slower rate compared to uncharged polymers. Fuoss [31] found that

$$\eta - \eta_* \sim c^{1/2} \tag{4.24}$$

as opposed to the linear increase predicted by Eq. (4.4) This effect can be reversed by the addition of salt or another electrolyte that shields the charges. If a sufficient amount of salt is added, the charged macromolecule can even assume a random coil conformation and behave similar to a nonionic polymer in a theta solvent. This behavior carries over to the entangled regime as well, where the zero-shear viscosity increases with the 3/2 power of the concentration instead of the 3.4 power. This is illustrated in Figure 4.7 by the data of Fernandez Prini and Lagos on aqueous solutions of polystyrene sulfonate with sodium counterions [32]. These observations, and also the change in the zero-shear viscosity with polymer concentration in the presence of salt, have been explained by Rubinstein et al. [33] and Dobrynin et al. [34] with the help of dynamic scaling laws.

An unusual observation with polyelctrolytes is that, contrary to the behavior of nonionic polymers, the point of onset of shear thinning moves to higher shear rates as the polymer concentration in solution is increased. This is shown in Figure 4.8 for aqueous solutions of the sodium salt of sulfonated polystyrene [35]. The addition of salt makes the polymer size smaller and lowers the viscosity level, but the macromolecules become less sensitive to orientation and deformation, leading to a still-lower level of shear thinning [36]. These twin facts have been used to advantage by Tam and Tiu [37] to develop low- (but constant-) viscosity, highly elastic liquids by dissolving a few hundred ppm of polyacrylamide of 8–10 million molecular weight in saline solutions. The relaxation times

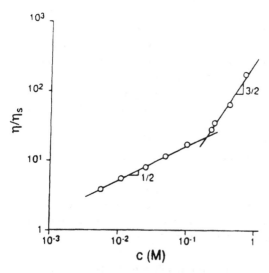

Figure 4.7 Concentration dependence of the relative viscosity for sodium polystyrene sulfonate in the absence of added salt. (Reprinted with permission from Ref. 34. Copyright 1995 American Chemical Society.)

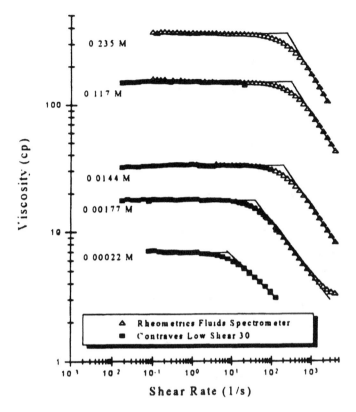

FIGURE 4.8 Shear rate dependence of the viscosity of aqueous solutions of the sodium salt of sulfonated polystyrene. Concentrations indicated are in moles of monomer per liter. (From Ref. 35.)

of these ideal, elastic liquids are similar to those of polymer melts at processing temperatures, and this makes these constant-viscosity liquids valuable for experimentally simulating polymer melt behavior at room temperature.

The size of a charged polymer in solution can also be changed by changing the pH; this again influences both the intrinsic viscosity and the zero shear viscosity. The latter quantity is shown in Figure 4.9 as a function of pH for aqueous solutions of a partially hydrolyzed polyacrylamide [38]; the hydrolysis of the nonionic polymer is generally effected with the help of an alkali such as NaOH or KOH. In solution, the pH level is changed by adding either 1 M HCl or 1 M NaOH. The maximum in the viscosity comes about because the polymer coil is most expanded at a slightly alkaline pH. At low pH values, the added hydrogen ions favor the formation of neutralized carboxyl groups on the polymer chain,

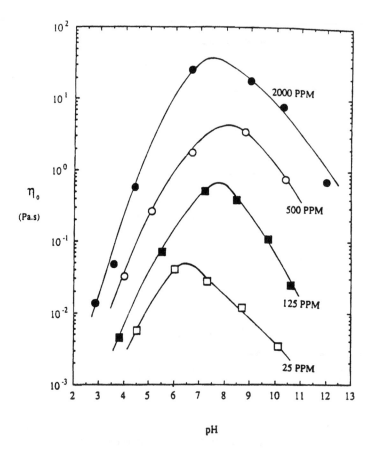

FIGURE 4.9 Effect of pH on the zero-shear viscosity of aqueous polyacrylamide solutions. (From Ref. 38.)

making the polymer resemble a nonionic polymer that has a small coil size. At high pH values, the excess sodium ions are attracted to the carboxyl groups along the polymer chain, which again causes the polymer to coil up due to a local reduction in the negative charge on the chain.

VII. SURFACTANT SOLUTIONS AND ASSOCIATIVE THICKENERS

Synthetic detergents are surface-active agents that are small molecules with typical molecular weights ranging from 200 to 400. In very dilute solution, these surfactants are well dispersed, their behavior is Newtonian, and their zero-shear viscosity is only marginally greater than the viscosity of the solvent. They would ordinarily be of little interest to the rheologist, except that under appropriate conditions they can behave like polymer solutions and exhibit viscoelasticity, a non-Newtonian viscosity, and even a yield stress [39]. The presence of a yield stress is particularly advantageous in endowing stability against creaming or sedimentation to an emulsion prepared using the surfactant as an emulsifier. These non-Newtonian properties arise due to the ability of surfactant molecules to form large aggregates of different shapes.

Thus, as the concentration of surfactant molecules in solution is increased, a critical micelle concentration (CMC) is reached at which surfactant globules begin to be formed. At this stage, the solutions are still isotropic and of low viscosity. Upon increasing the concentration beyond the CMC, the globules grow into rodlike aggregates. The rodlike micelles can increase in length with further increases in surfactant concentration and become large enough to form entanglements with each other. If salt is added to this solution, a three-dimensional network can form; this is shown schematically in Figure 4.10 [39].

A consequence of gelation is that the structural relaxation time can be as large as 10 sec, but it can be reduced significantly with the help of other additives without destroying the network structure. A mixture of cetylpyridinium chloride

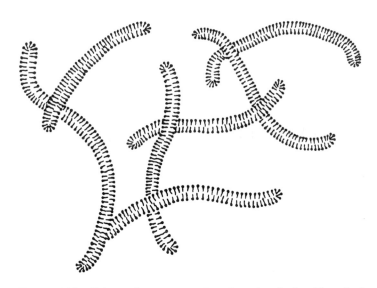

FIGURE 4.10   Schematic representation of a network of rodlike micelles in a surfactant solution. (From Ref. 39.)

(CPyCl) and sodium salicylate (NaSal) in aqueous solution is an example of this progressive evolution of structures, and Figure 4.11 displays the zero-shear-rate viscosity of this system as a function of salt concentration [40]. The increase in viscosity in region II is due to the formation of entanglements between rodlike micelles. The other maxima and minima result from changes in the network relaxation time with changes in salt concentration. Usually the shear viscosity can be written as a product of a shear modulus and a relaxation time. In the present instance, measurements show that it is the relaxation time that varies while the modulus remains unchanged as we go from region II to region V.

Associative thickeners are low-molecular-weight, water-soluble polymers that are similar to surfactants in that they increase the viscosity of water by associ-

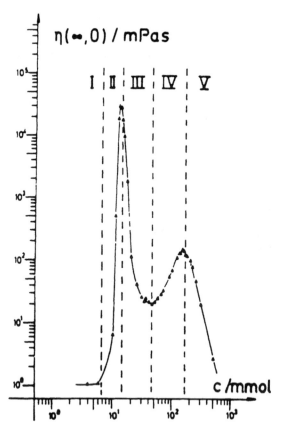

FIGURE 4.11   Zero-shear viscosity of a 15-mmol aqueous solution of CPyCl as a function of NaSal concentration. (Reprinted with permission from Ref. 40. Copyright 1988 American Chemical Society.)

ating with themselves and with any suspended solids or drops. Two types of associative thickeners have achieved commercial acceptance, especially in the paints and coatings industry. These are the hydrophobically modified ethylene urethane oxide rheology modifiers (HEUR) and the hydrophobically modified alkali swellable emulsions (HASE). The former type is the condensation product of the reaction between polyethylene glycol and a diisocyanate, and end-capped by hydrophobic groups such as long-chain alkanols. The typical molecular weight is about 40,000–50,000, and HEUR can be thought of as a double-ended surfactant molecule [41]. The chemical structure of a model HEUR molecule is shown in Figure 4.12 [42], and it has a hydrophilic middle but hydrophobic ends. While HEUR thickeners are nonionic, the HASE variety is anionic. Here, thickening takes place above a pH of 7 due to repulsion among carboxylate anions that are distributed along the polymer backbone. A major advantage of these thickeners is that the increase in viscosity is not accompanied by an increase in elasticity. Thus, paints that are thickened in this manner do not spatter as much as they would if cellulosic polymers had been employed.

In solution, HEUR molecules form a network by associating as micelles and also by adsorbing onto the surface of any suspended solids or drops. The resulting increase in viscosity can be used to retard the settling of suspended material. The network, however, is easily disrupted, and the thickening action at low-shear rates depends on the surface area available for adsorption, while the viscosity at high shear rates is solely a function of polymer concentration. In the absence of suspended matter, intra- and intermolecular interactions make the zero-shear viscosity of HEUR thickeners increase rapidly with increasing concentration, more so than with unmodified polymers of similar molecular weight; the rate of viscosity increase becomes greater as the end-cap chain length is increased [43]. The viscosity generally remains Newtonian for an extended range of shear rates and then increases slightly and finally decreases. Sometimes thixotropy is observed. These and other observations have been explained by Annable et al. [43] with the help of transient-network models.

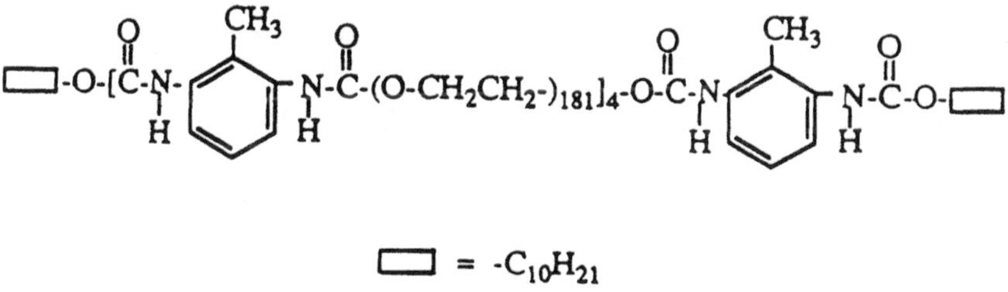

$$\square\text{-O-[C-N-}\bigcirc\text{-N-C-(O-CH}_2\text{CH}_2\text{-)}_{181}]_4\text{-O-C-N-}\bigcirc\text{-N-C-O-}\square$$

$$\square = \text{-C}_{10}\text{H}_{21}$$

FIGURE 4.12 Structure of a model HEUR thickener. (From Ref. 42.)

HASE-associative thickeners are modifications of traditional alkali-soluble thickeners that are carboxyl functional copolymers produced by free-radical polymerization of monomers such as methacrylic acid and ethyl acrylate [44]. The modification consists of the attachment of hydrophobic groups along the polymer backbone. These thickeners are essentially insoluble in water at low pH but dissolve at pH values above 7. Solution viscosity increases because of chain expansion due to repulsion between the carboxylate anions along the polymer backbone and also due to the formation of an intermolecular network resulting from hydrophobic interactions. The variation in zero-shear viscosity with pH and polymer concentration is shown in Figure 4.13 for aqueous solutions of an HASE-model polymer prepared by the Union Carbide Corporation [44]. The solution viscosity is independent of polymer concentration and comparable to the solvent viscosity at low pH. However, as the pH increases beyond 5, the viscosity increases suddenly, by as much as six orders of magnitude for the most concentrated solution. Thereafter, the change in viscosity with pH is quite gradual. A possible explanation for this behavior is provided is Figure 4.14. At low pH, the polymer exists as a dispersion rather than as a solution. When an alkali such as ammonium hydroxide is added to raise the pH, acid groups on the polymer chain are neutralized and the polymer dissolves. With increasing pH, the viscosity increases because (1) the polymer chain expands due to charge repulsion, and (2) intermolecu-

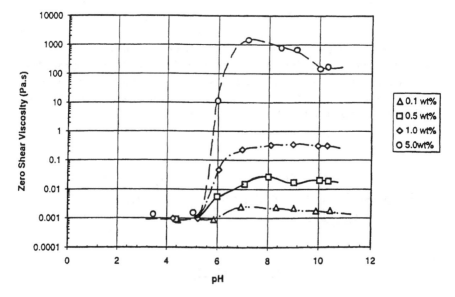

Figure 4.13 Zero-shear viscosity of HASE 5142 at different pHs. (From Ref. 44.)

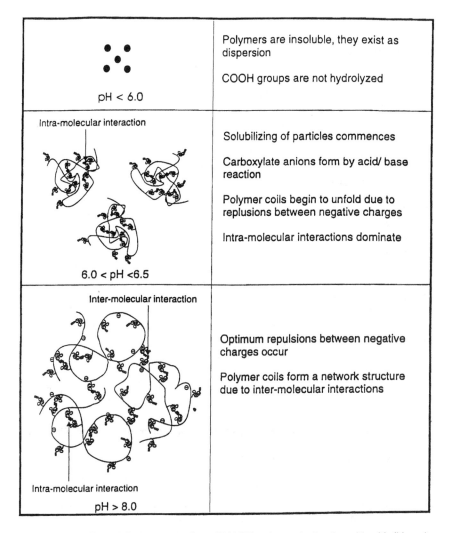

pH < 6.0	Polymers are insoluble, they exist as dispersion COOH groups are not hydrolyzed
Intra-molecular interaction 6.0 < pH <6.5	Solubilizing of particles commences Carboxylate anions form by acid/ base reaction Polymer coils begin to unfold due to replusions between negative charges Intra-molecular interactions dominate
Inter-molecular interaction Intra-molecular interaction pH > 8.0	Optimum repulsions between negative charges occur Polymer coils form a network structure due to inter-molecular interactions

FIGURE 4.14   Pictorial representation of HASE polymer behavior with pH. (Unpublished results of Dr. K. C. Tam.)

lar interactions lead to network formation. At high shear rates, these intermolecular contacts are disrupted, and substantial shear thinning is observed. The state of the art in the area of associative thickeners has been documented in a series of symposia edited by Glass [45–48].

REFERENCES

1. W. W. Graessley, R. L. Hazleton, and L. R. Lindeman. The shear-rate dependence of viscosity in concentrated solutions of narrow-distribution polystyrene. Trans. Soc. Rheol. 11:267–285 (1967).
2. D. V. Boger. A highly elastic constant-viscosity fluid. J. Non-Newt. Fluid Mech. 3: 87–91 (1977/78).
3. G. Prilutski, R. K. Gupta, T. Sridhar, and M. E. Ryan. Model viscoelastic liquids. J. Non-Newt. Fluid Mech. 12:233–241 (1983).
4. A. Einstein. Eine Neuve Bestimmung der Molekuldimension. Ann. Physik 19:289–306 (1906).
5. A. Kumar and R. K. Gupta. Fundamentals of Polymers. McGraw-Hill, New York, 1998.
6. P. J. Flory. Principles of Polymer Chemistry. Cornell University Press, Ithaca, NY, 1953.
7. M. Kurata and W. H. Stockmayer. Intrinsic viscosities and unperturbed dimensions of long chain molecules. Fortschr. Hochpolym.-Forsch. 3:196–312 (1963).
8. M. Kurata and Y. Tsunashima. viscosity–molecular weight relationships and unperturbed dimensions of linear chain molecules. In: J. Brandrup and E. H. Immergut, eds. Polymer Handbook. 3rd ed. Wiley, New York, 1989. p. VII/1.
9. D. R. Saini and A. V. Shenoy. Quick estimation of dilute polymer solution rheology and activation energy. J. Appl. Polym. Sci. 33:41–48 (1987).
10. T. E. Newlin, S. E. Lovell, P. R. Saunders, and J. D. Ferry. Long-range intermolecular coupling in concentrated poly-n-butyl methacrylate solutions and its dependence on temperature and concentration. J. Colloid Sci. 17:10–25 (1962).
11. H. Fujita and E. Maekawa. Viscosity behavior of the system polymethyl acrylate and diethyl phthalate over the complete range of composition. J. Phys. Chem. 66: 1053–1058 (1962).
12. C. W. Macosko. Rheology. VCH, New York, 1994, p. 481.
13. P. F. Lyons and A. V. Tobolsky. Viscosity of polypropylene oxide solutions over the entire concentration range. Polym. Eng. Sci. 10:1–3 (1970).
14. K. S. Gandhi and M. C. Williams. Solvent effects on the viscosity of moderately concentrated polymer solutions. J. Polym. Sci.: Part C 35:211–234 (1971).
15. F. Rodriguez. Suitability of equations for viscosity correlation over the entire concentration range. J. Polym. Sci.: Part B 10:455–459 (1972).
16. F. Bueche. Viscosity, self-diffusion and allied effects in solid polymers. J. Chem. Phys. 20:1959–1964 (1952).
17. R. B. Bird, C. F. Curtiss, R. C. Armstrong, and O. Hassager. Dynamics of Polymeric Liquids. 2nd ed. Vol. 2. Wiley, New York, 1987.
18. P. E. Rouse, Jr. A theory of the linear viscoelastic properties of dilute solutions of coiling polymers. J. Chem. Phys. 21:1272–1280 (1953).

19. W. W. Graessley. The entanglement concept in polymer rheology. Adv. Polym. Sci. 16:1–179 (1974).
20. F. N. Kelley and F. Bueche. Viscosity and glass temperature relations for polymer–diluent systems. J. Polym. Sci. 50:549–556 (1961).
21. W. W. Graessley. Effect of long branches on the flow properties of polymers. Accounts Chem. Res. 10:332–339 (1977).
22. J. D. Ferry. Viscoelastic Properties of Polymers. 3rd ed. Wiley, New York, 1980.
23. A. A. Tager and V. E. Dreval. Newtonian viscosity of concentrated solutions of polymers. Russian Chem. Rev. 36:361–373 (1967).
24. S. Yasutomi, S. Bair, and W. O. Winer. An application of a free volume model to lubricant rheology I—dependence of viscosity on temperature and pressure. J. Tribology 106:291–303 (1984).
25. S. Bair and W. O. Winer. Some observations in high pressure rheology of lubricants. J. Lubrication Technol. 104:357–364 (1982).
26. C. S. Wu, E. E. Klaus, and J. L. Duda. Development of a method for the prediction of pressure-viscosity coefficients of lubricating oils based on free-volume theory. J. Tribology 111:121–128 (1989).
27. K. C. Tam and C. Tiu. Steady and dynamic shear properties of aqueous polymer solutions. J. Rheol. 33:257–280 (1989).
28. M. Oritz, D. DeKee, and P. J. Carreau. Rheology of concentrated poly(ethylene oxide) solutions. J. Rheol. 38:519–539 (1994).
29. H. Morawetz. Macromolecules in Solution. 2nd ed. Wiley, New York, 1975.
30. M. Hara, ed. Polyelectrolytes. Marcel Dekker, New York, 1993.
31. R. M. Fuoss. Polyelectrolytes. Discuss. Faraday Soc. 11:125–134 (1951).
32. R. F. Fernandez Prini and A. E. Lagos. Tracer diffusion, electrical conductivity, and viscosity of aqueous solutions of polystyrenesulfonates. J. Polym. Sci., Part A 2: 2917–2928 (1964).
33. M. Rubinstein, R. H. Colby, and A. V. Dobrynin. Dynamics of semidilute polyelectrolyte solutions. Phys. Rev. Lett. 73:2776–2779 (1994).
34. A. V. Dobrynin, R. H. Colby, and M. Rubinstein. Scaling theory of polyelectrolyte solutions. Macromolecules 28:1859–1871 (1995).
35. D. C. Boris and R. H. Colby. Shear viscosity of polyelectrolyte solutions. Proc. XIIth Int. Congr. Rheol., Quebec City, Canada, 1996, pp. 195–196.
36. A. Ait-Kadi, P.J. Carreau, and G. Chauveteau. Rheological properties of partially hydrolyzed polyacrylamide solutions. J. Rheol. 31:537–561 (1987).
37. K. C. Tam and C. Tiu. A low viscosity, highly elastic ideal Fluid. J. Non-Newt. Fluid Mech. 31:163–177 (1989).
38. K. C. Tam and C. Tiu. Effect of pH on the excluded volume of high molecular weight anionic polyacrylamide. Polym. Commun. 31:102–104 (1990).
39. H. Hoffmann and G. Ebert. Surfactants, micelles and fascinating phenomena. Angew. Chem. Int. Ed. Engl. 27:902–912 (1988).
40. H. Rehage and H. Hoffmann. Rheological properties of viscoelastic surfactant systems. J. Phys. Chem. 92:4712–4719 (1988).
41. J. Prideaux. Rheology modifiers and thickeners in aqueous paints. Surface coatings International. J. Oil Color Chemists' Assoc. 76:177–182 (1993).
42. D. L. Lundberg, J. E. Glass, and R. R. Eley. Viscoelastic behavior among HEUR thickeners. J. Rheol. 35:1255–1274 (1991).

43. T. Annable, R. Buscall, R. Ettelaie, and D. Whittlestone. The rheology of solutions of associating polymers: comparison of experimental behavior with transient network theory. J. Rheol. 37:695–726 (1993).
44. K. C. Tam, M. L. Farmer, R. D. Jenkins, and D. R. Bassett. Rheological properties of hydrophobically modified alkali-soluble polymers: effects of ethylene-oxide chain length. J. Polym. Sci. B: Polym. Phys. 36:2275–2290 (1998).
45. J. E. Glass, ed. Water-Soluble Polymers: Beauty with Performance. Adv. Chem. Series 213. American Chemical Society, Washington, DC, 1986.
46. J. E. Glass, ed. Polymers in Aqueous Media: Performance through Association. Adv. Chem. Series 223. American Chemical Society, Washington, DC, 1989.
47. J. E. Glass and D. N. Schulz, eds. Polymers as Rheology Modifiers, ACS Symposium Series 462. American Chemical Society, Washington, DC, 1991.
48. J. E. Glass, ed. Hydrophilic Polymers: Performance with Environmental Acceptability. Adv. Chem. Series 248. American Chemical Society, Washington, DC, 1996.

5

Normal Stress Differences in Polymers During Shear Flow

I. INTRODUCTION

Normal stresses were defined in Chap. 1. Unequal normal stresses in shear flow are not found with Newtonian liquids but are characteristic of non-Newtonian polymer melts and solutions. Normal stress differences are primarily manifestations of the elasticity of polymeric materials, and this elasticity results from the orientation and extension of polymer chain segments. Chain extension and orientation is aided by the presence of entanglements and results in the storage of energy within the polymeric fluid.

If we consider steady shear flow in a Cartesian coordinate system, it is reasonable to expect that the typical polymer molecule will stretch due to the presence of a velocity gradient and will also tend to align itself with the flow field; this will result in the presence of a force f along the polymer chain axis, as shown in Figure 5.1. If the polymer molecule lies in the x_1-x_2 plane, the angle that the force vector makes with the x_1 axis will be less than 45° and will decrease progressively with increasing shear rate. Thus, for flexible macromolecules, the component of the external force f along x_1 will always be greater than the corresponding component along x_2. As a result, the polymer contribution to the normal stress T_{11} will be larger than the contribution to T_{22}, and this will make the first normal stress difference, N_1, a positive quantity. The second normal stress difference, N_2, was thought to be zero for many years, but it is now accepted to be nonzero and negative but significantly smaller than N_1 in magnitude. At very low shear rates, both of the normal stress differences increase with the square of the rate of shear, so the corresponding normal stress coefficients become constant in

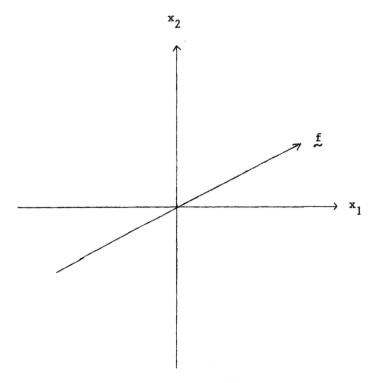

FIGURE 5.1 Force acting on a typical polymer molecule during shear flow.

the limit of vanishingly low deformation rates. The low shear rate limiting value of the first normal stress coefficient can be obtained from dynamic mechanical experiments as [1]:

$$\psi_{10} = \lim_{\gamma \to 0} \psi_1 = 2 \lim_{\omega \to 0} \frac{G'}{\omega^2} \tag{5.1}$$

Equation (5.1) is a useful result, since it is often easier to measure the storage modulus as compared to the first normal stress difference. In general, the normal stress coefficients decrease with increasing shear rate; however, the high shear rate limit is uncertain, owing to the difficulty of making accurate normal stress measurements at high shear rates.

Figure 5.2 shows representative first normal stress difference data for polymer melts as a function of shear rate at three constant temperatures [2]. These data are for a commercial sample of polystyrene, and were obtained on a Weissenberg rheogoniometer employing cone-and-plate fixtures. Clearly, stress levels decrease

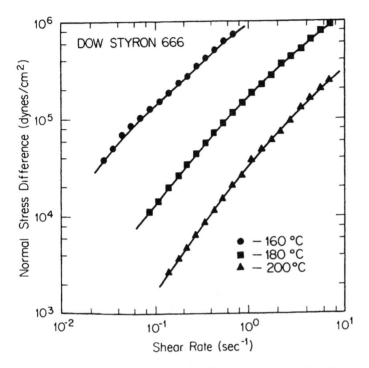

FIGURE 5.2 Principal normal stress difference as a function of shear rate and temperature for molten polystyrene. (From Ref. 2.)

on increasing the temperature of measurement. Also, on logarithmic coordinates, the slope of each of the three curves increases as the shear rate decreases. However, data do not extend to low-enough shear rates for us to observe the expected quadratic dependence of N_1 on the shear rate. Most of the early work on polymer solutions was carried out on concentrated solutions of polyisobutylene, and Figure 5.3 shows the steady-state shear stress and both normal stress differences for this polymer dissolved in decalin [3]. Measurements were made at room temperature and utilized cone-and-plate as well as parallel-plate fixtures. The variation of the first normal stress difference with shear rate is qualitatively similar to the behavior of molten polymers. Note that the first normal stress difference curve lies below the shear stress curve at low shear rates, but this trend is reversed upon increasing the shear rate. This happens because of the stronger dependence of N_1 on the shear rate compared with the dependence of τ on $\dot{\gamma}$. The ratio of N_1 to the shear stress is often used as a measure of fluid elasticity; this ratio tends to zero as the shear rate tends to zero. Values of the second normal stress differ-

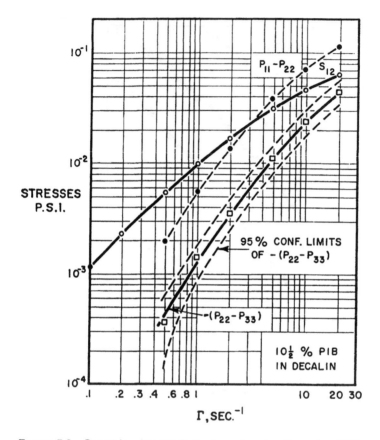

FIGURE 5.3 Stress levels, steady laminar shearing flows, for 10.5% polyisobutylene in decalin. (From Ref. 3.)

ence are not as accurate as those of the first normal stress difference, but these are clearly of opposite sign and are almost an order of magnitude smaller. Stress levels as well as the precision of measurement decrease as the polymer concentration in solution is decreased [3]. Since both of the normal stress coefficients are even functions of the shear rate and since they become constant at small shear rates, the variation of either of them with shear rate can be represented mathematically as:

$$\psi_i = \frac{a_i}{1 + b_i \dot{\gamma}^2} \qquad i = 1,2 \qquad (5.2)$$

where a_i and b_i are constants. This form also emerges from most rheological models, whether fundamental or empirical and whether based on network theory or dilute solution theory [4,5].

Upon inception of flow at a constant shear rate, steady-state stress levels are neither attained instantly nor reached monotonically. Unless the shear rate is extremely low, the shear stress as well as the normal stresses exhibit overshoots relative to their steady-state values, and there may be more than one oscillation [6,7]. The percentage by which the various viscometric functions overshoot the steady-state value increases with increasing shear rate, and several minutes may be needed for the transients to die out. Peak N_1 values observed by Meissner for a polyethylene melt were as much as four times the steady-state value [6]. The shear stress transients are similar to the normal stress transients, but the maximum in the shear stress is relatively smaller than the maximum in the first normal stress difference and occurs before it [6]. For molten polyethylene, the shear stress maximum is observed at a total strain of 7; the normal stress maximum occurs at a total strain of 14 [8]. For polybutadiene solutions, Menezes and Graessley [9] found that while the ratio of the strain values at which N_1 and the shear

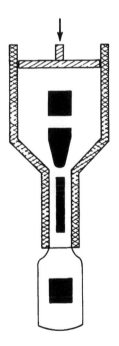

FIGURE 5.4 Schematic diagram illustrating the phenomenon of die swell.

stress were a maximum was approximately 2.5, the strain at the overshoot peak in N_1 was 5 at low shear rates and increased progressively with increasing shear rate. Transient effects arise on cessation of shearing as well. Stresses decay monotonically to zero, with the shear stress relaxing faster than the normal stress differences; the rate of decay depends on the shear rate prior to cessation of shearing [7].

Normal stresses produce a number of phenomena not found with Newtonian liquids. For example, when a polymer is extruded from an orifice, a capillary, or a slit, the diameter or thickness of the resulting strand is considerably greater than the diameter of the hole from which it came. This phenomenon, shown in Figure 5.4, is called *die swell*. That die swell is caused by the presence of N_1 resulting from fluid being sheared in the capillary is easy to explain. As liquid emerges into the atmosphere, the total stress in the flow direction is the negative of the atmospheric pressure. A positive N_1 implies that the total stress in the radial direction is negative (compressive) and, in magnitude, is greater than the pressure of the atmosphere. In other words, the tube wall pushes down on the liquid being sheared, and the liquid pushes up on the tube with an equal stress

NORMAL STRESS

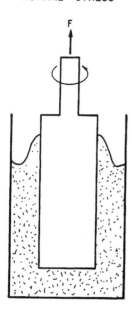

FIGURE 5.5   Creep of a polymeric liquid up the rotating inner cylinder of a coaxial cylinder rheometer.

that exceeds the atmospheric pressure in magnitude. When the fluid emerges into the atmosphere, there is no constraining tube: the fluid snaps back like a stretched rubber band, and we observe die swell.

Another unexpected phenomenon due to normal stresses is the Weissenberg effect, or the ascending of polymer liquids up rotating shafts, as illustrated in Figure 5.5. In contrast, rotating shafts in Newtonian liquids cause a depression of the liquid surface because of centrifugal forces. In cone-and-plate rheometers, and in other rotating systems of similar geometry, the normal stresses tend to force the cone and plate apart; a measurement of this force is the most reliable method of obtaining the first normal stress difference, and this is the principle by which N_1 is determined in the commercial instruments described in Chapter 2. If holes are drilled in the plates parallel to the rotating axis, liquid will be forced up through the holes. This is the basis of one early technique for calculating normal stresses by measuring the height to which a liquid will ascend up a tube for a given speed of rotation [10,11]. The first normal stress difference is proportional to the height of the liquid in the tube, and the liquid level is a maximum for a tube at the center of rotation and decreases toward the edge of the rotating disk. Instead of measuring liquid levels, we may use pressure transducers

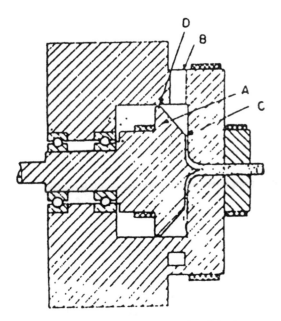

FIGURE 5.6 Normal stress extruder. A is the rotor, polymer is fed at point B, and the gap between the plates is given by the spacing at point C. (From Ref. 14. Reprinted with permission from MODERN PLASTICS, a publication of Chemical Week Associates, New York.)

located along the radius of the plate [12]. As shown in the next section, the normal stresses are proportional to the square of the shearing stress, so a doubling of the speed of rotation will increase the normal stress by a factor of nearly 4.

Normal stresses can be important in a number of polymer processing situations. In wire coating, normal stresses help produce a smooth coating of uniform thickness up to the point where melt fracture instability takes over, and at the same time they keep the wire properly centered if the second normal stress difference is negative [13]. If a polymer melt is sheared between a rotating plate and a stationary plate, the tendency of the fluid to push up against the plates under the influence of a positive first normal stress difference can be utilized to extrude the melt continuously. This can be done by punching a hole at the center of the stationary plate, as shown in Figure 5.6. This screwless, normal stress extruder was originally proposed by Maxwell and Scalora [14], and it is now available commercially from Custom Scientific Instruments in Cedar Knolls, NJ. A major advantage of the Maxwell extruder is that only gram quantities of polymer are required, and this is helpful when working with experimental materials. Even though the residence time is short, polymer blending can be carried out and shear-sensitive as well as hygroscopic polymers accommodated.

II. LOW-SHEAR-RATE BEHAVIOR OF THE FIRST NORMAL STRESS DIFFERENCE

The commercial availability of very sophisticated rotational viscometers in the last two decades has made the determination of the first normal stress difference a straightforward procedure for both polymer melts and polymer solutions. In this respect, cone-and-plate geometry is the preferred one, but, as mentioned in Chapter 2, secondary flows and instabilities limit the shear rate range of measurement to a maximum of a few hundred reciprocal seconds. While these deformation rates are significantly smaller than those often encountered in industrial operations, enough data are available to allow us to make definite statements concerning the influence of variables such as temperature, shear rate, concentration, and molecular structure on the measured values of N_1.

As is evident from Figures 5.2 and 5.3, the first normal stress difference of a polymer melt or a polymer solution at a fixed concentration depends on both temperature and shear rate. It is a remarkable fact, however, that all the data for a particular fluid collapse onto a single straight line when the logarithm of N_1 is plotted against the logarithm of the shear stress. This is shown in Figure 5.7 for the melt data of Figure 5.2 and in Figure 5.8 for data on liquid M1 [15], a 0.244% solution of high-molecular-weight polyisobutylene in a mixed solvent of kerosene and low molecular weight polybutene [16]. The slope of the line in Figure 5.7 is 1.66, and that of the line in Figure 5.8 is 2. In general, we can say that

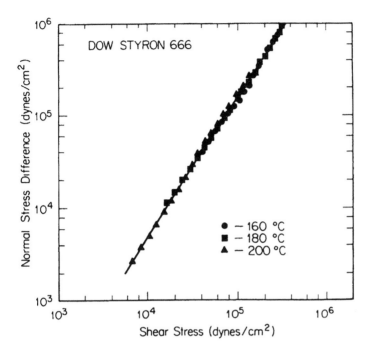

FIGURE 5.7 Principal normal stress difference as a function of shear stress for the data of Fig. 2. (From Ref. 2.)

$$N_1 = A\tau^a \tag{5.3}$$

in which A and a are constants and τ is the shear stress. The maximum observed value of the slope a, and this corresponds to behavior in the linear viscoelastic range, is 2. Another remarkable feature of first normal stress difference data when plotted versus shear stress rather than against shear rate is the narrowing of differences among different polymers [17] or among different molecular weight fractions of the same polymer [18]. The utility of Eq. (5.3), though, goes beyond data correlation alone. Since it is relatively easy to compute or to measure the shear stress, one often uses Eq. (5.3) to determine the first normal stress difference, even in polymer processing situations [19]. Furthermore, the ratio of the first normal stress difference in shear to twice the corresponding shear stress is a dimensionless group known as the *recoverable shear*; a critical value of this group is frequently employed to predict the onset of elastic flow instabilities, such as melt fracture [20]. Note that the ratio of the recoverable shear to the shear rate is considered to be a characteristic fluid time, because it has the dimensions of time and a value that tends to a constant as the shear rate tends to zero. This

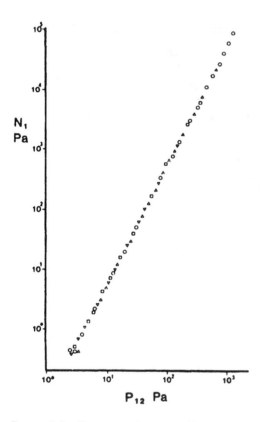

FIGURE 5.8 First normal stress difference as a function of shear stress for fluid M1. The different symbols are for different temperatures that vary from 20°C to 50°C in intervals of 10°C. (From Ref. 15, with permission from Elsevier Science.)

characteristic time, also called the *fluid relaxation time*, is a measure of fluid elasticity, and it represents the time duration over which stresses persist after cessation of shearing; this interpretation follows from the predictions of some of the most common rheological models. On dividing the relaxation time by the duration time of a process of interest, one gets a dimensionless number called the *Deborah number*. Fluid elasticity is important only for those processes for which the Deborah number is large [21].

   If the dependence of shear stress on variables such as temperature and shear rate is known, the dependence of the first normal stress difference in shear on the same variables follows directly from Eq. (5.3). In particular, N_1 decreases with increasing temperature, and the polymer becomes less elastic as well, because the relaxation time goes down owing to the fact that the exponent a in Eq. (5.3) exceeds unity. Additionally, Eq. (5.3) predicts that time–temperature superposition must hold for the first normal stress difference when plotted as a function of the shear rate, just as it does for the shear stress (see Figs. 3.14 and 3.15); the temperature shift factors in the two cases are identical. Since the melt viscosity of monodisperse, linear polymers varies with the 3.4 or 3.5 power of the molecular weight at low shear rates (see Sect. II.A in Chap. 3), we can expect the zero-shear-rate first normal stress difference or the first normal stress coefficient to increase with molecular weight to about the power 7; nylon 6 melt data of Laun [22], shown in Fig. 5.9, fulfill this expectation. Regarding the influence of molecular weight distribution, it appears that broadening the distribution of a linear polymer, while keeping the average molecular weight and the shear stress unchanged, results in an increase in the normal stress difference but a decrease in the exponent a in Eq. (5.3) [2, 23]. According to Minoshima et al. [18], data on molten polystyrenes suggest that the constant A in Eq. (5.3) may be written as

$$A = A' \left(\frac{M_z}{M_w} \right)^{3.5} \tag{5.4}$$

where A' is another constant and M_z is the z-average molecular [24]. The effect of the introduction of long-chain branching is unclear. While the polyethylene data of Mendelson et al. [25] show a decrease in the first normal stress difference, the reverse is found to be true for branched polyethylene terephthalate [26].

The preceding remarks have, to a large extent, been made with reference to the behavior of polymer melts. However, the shear rate, temperature, molecular weight, and molecular weight distribution dependence of the first normal stress difference of polymer solutions is similar to that of melts [27]. Now, though, polymer concentration and solvent viscosity are additional variables; increasing the value of either quantity results in an increase in N_1. Low-shear-rate data on polyisobutylene solutions [28] suggest that the group $N_1 c^2 / \tau^2$ is essentially constant. At higher shear rates, data can be represented as dimensionless plots on logarithmic coordinates. The result is a master curve that is independent of polymer concentration, polymer molecular weight, and solvent viscosity. Figure 5.10 shows such a plot of the dimensionless first normal stress difference of poly (ethylene oxide) solutions as a function of the dimensionless shear rate taken from the work of Oritz et al. [29]. N_1 is made dimensionless by multiplying it with $\lambda_f / (\eta - \eta_s)$ while the shear rate is made dimensionless by multiplying it with λ_f. Here η is the solution viscosity, η_s is the solvent viscosity, and λ_f is the same time constant defined earlier by Eq. (4.23). Note that the normal stress data on polymer melts can also be plotted in the form of temperature-and-molecular-weight-independent master curves; such unified curves are useful for estimating the normal stress response of chemically similar polymers [30].

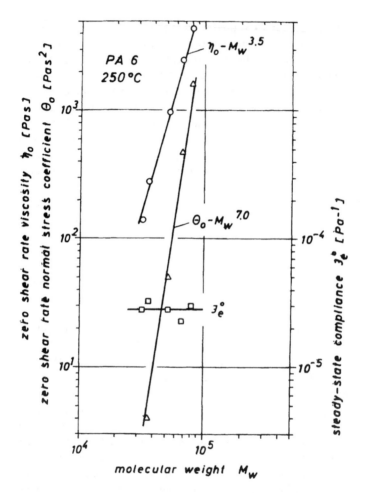

FIGURE 5.9 Zero-shear viscosity and the first normal stress coefficient of a nylon 6 melt as a function of molecular weight. (From Ref. 22.)

In order to explain the observed relationship between the first normal stress difference function of polymeric fluids and the various material and processing variables, we can take one of two approaches. In the phenomenological approach, we can relate the three-dimensional stress to the history of deformation in a general way and then examine limiting cases. For simple fluids with fading memory, the first order approximation for slow flows gives the Newtonian fluid [31]. The next-higher approximation, known as the second-order fluid, possesses viscoelas-

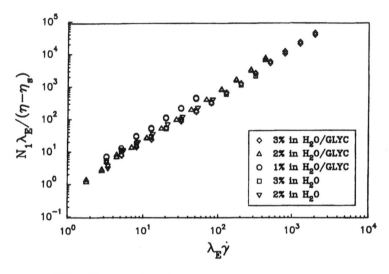

FIGURE 5.10 Dimensionless first normal stress difference versus dimensionless shear rate for PEO solutions in two different solvents. The average molecular weight is 1.8 million. (From Ref. 29.)

ticity. For this fluid, the normal stress differences are nonzero; in particular, the first normal stress difference is predicted to increase quadratically with shear rate. As we have seen, this behavior is observed at low shear rates, but transient effects, such as stress relaxation, cannot be represented using the second-order fluid model. However, stress relaxation, stress overshoots, and a less-than-quadratic increase in the first normal stress difference at high shear rates can be explained with the help of more general constitutive equations derived using a continuum mechanics approach [7,31]. These equations, though, suffer from the handicap that model parameters cannot be related to polymer structural variables; to accomplish this, we have to employ a molecular approach. Three different molecular models are popular with rheologists: bead-spring models for dilute solutions, network models for melts, and reptation models for concentrated solutions and melts [32].

The essential elements of bead-spring theories were mentioned in Sec. III of Chap. 4. In the simplest situation, each polymer molecule is modeled as a dumbbell that consists of two equal masses connected by an infinitely extensible, linear, elastic spring [33,34]. The use of this model results in a closed-form, two-constant, constitutive equation for the polymer contribution to the stress in the solution. This equation is called the *upper convected Maxwell equation*. It captures all the main qualitative features of polymer viscoelasticity, and for this reason it has been used extensively to model polymer processing operations [35]. The Maxwell equation predicts that the first normal stress difference is given by:

$$N_1 = \frac{2\theta\tau^2}{\eta} \qquad (5.5)$$

in which θ and η are, respectively, the constant-relaxation time and shear viscosity. Thus Eq. (5.5) is the same as Eq. (5.3) if a is taken to be 2. However, the constants in Eq. (5.5) have a physical meaning. In particular, the relaxation time can be related to molecular variables as [36]:

$$\theta = \frac{K\eta_s M^{3/2}\alpha^3}{T} \qquad (5.6)$$

where K is a constant and α is a coil expansion factor whose value depends on solvent quality and temperature. Clearly, the combination of Eqs. (5.5) and (5.6) reveals the influence of shear rate, temperature, molecular weight, solvent viscosity, and solvent quality on the first normal stress difference in shear. These equations have been verified for dilute polymer solutions up to moderate shear rates, and Tam et al. [36, 37] have used this theoretical framework to formulate a large number of constant-viscosity, ideal elastic liquids; typical data are displayed in Figure 5.8. These liquids, which can have relaxation times comparable to those of polymer melts at processing temperatures, are known as *Boger fluids* [38], after D. V. Boger, who first formulated them; their constitutive behavior was first explained by Prilutski et al. [39]. At high shear rates, the first normal stress difference of dilute polymer solutions deviates from the predictions of Eq. (5.5). As a consequence, this equation has been modified by changing some of the model assumptions, such as by making the dumbells finitely extensible and the spring law nonlinear [40].

Network models employed for representing the flow behavior of polymer melts owe their origin to the theory of rubber elasticity. Unlike vulcanized rubber, the network junctions are temporary rather than permanent crosslinks. It is remarkable, however, that the simplest constitutive equation that emerges from the use of network theory is again the upper convected Maxwell equation (also known as the *Lodge rubberlike liquid* in the case of polymer melts). To accommodate values of the exponent a in Eq. (5.3) that become less than 2 with increasing shear rate, different assumptions can be made concerning the rates of creation and destruction of network points. This is still an area of active research, and a very large number of network-type models have been proposed in the literature [32]. The introduction of reptation models was an attempt to describe better the dynamics of concentrated polymer solutions and polymer melts. A discussion of reptation models is beyond the scope of this book, and the reader is referred to the technical literature for details [41,42].

III. MEASUREMENT OF N_1 AT HIGH SHEAR RATES

Due to problems of edge fracture and sample loss, the upper limit on shear rate for a cone-and-plate viscometer is about 10 sec^{-1} for polymer melts and a few hundred reciprocal seconds for polymer solutions. To obtain data at higher deformation rates, one uses capillary and slit devices to infer the first normal stress difference from a measurement of quantities such as the exit pressure, the jet thrust, die swell, and the hole pressure. Although these techniques are firmly established in the rheological literature and commercial instruments are available, the theory relating N_1 to the measured quantity involves assumptions that may not be strictly valid in all cases. One, therefore, has to exercise some caution in data interpretation.

A. Exit Pressure Method

Consider the flow of a viscous polymer through a long slit die, as shown in Fig. 5.11. As the polymer emerges into the atmosphere, it swells due to the presence

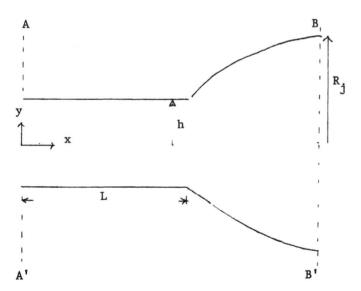

FIGURE 5.11 A polymeric fluid emerges into the atmosphere from a long slit die. A momentum balance is done between sections AA′ and BB′.

of velocity rearrangements that are complete at section BB'. If we assume that fluid inertia can be neglected and that the velocity profile in the die is fully developed all the way to the exit, located at $x = L$, a momentum balance between sections AA' and BB' reveals that [43]:

$$N_1(\dot{\gamma}_w) = P_e\left(1 + \frac{d \ln P_e}{d \ln \tau_w}\right) \tag{5.7}$$

In Eq. (5.7), $\dot{\gamma}_w$ is the shear rate at the die wall, given by an equation similar to Eq. (2.2), τ_w is the shear stress at the die wall, and P_e is the exit pressure. The exit pressure is a nonzero value obtained by extrapolating to $x = L$ actual measurements of the wall normal stress (or pressure) made along the die wall for different x values; flush-mounted pressure transducers are used, and the pressure profile must be linear in x if the flow is truly viscometric all the way to the die exit. This method of measuring N_1 has been extensively explored by Han [44 and references therein], who has helped commercialize the Seiscor/Han viscometer based on this principle. Typical data on a polystyrene melt are displayed in Fig. 5.12, which also contains cone-and-plate data at low shear rates. It is seen that the exit pressure data points lie on a straight line obtained by extrapolating the rheometer results to higher shear rates. Furthermore, this straight-line plot is independent of the temperature of measurement and suggests that Eq. (5.3) remains valid to very high shear rates; similar conclusions are reached by examining data on other polymers [44].

In closing this subsection, we mention that Han has noted that Eq. (5.7) does not yield correct results if measurements are made at wall shear stresses below a critical value of about 25 kPa. This technique has been critically analyzed by Boger and Denn [43], who have questioned the validity of the fully developed velocity profile assumption. This, together with the need to differentiate experimental data, has led them to conclude that results obtained using Eq. (5.7) may only represent an upper bound on the value of N_1.

B. Die Swell

This phenomenon, illustrated in Figure 5.4, is important in fabrication processes such as controlling the thickness of extruded sheets and in the making of bottles by blow molding. Shown in Figure 5.13 is the general behavior of the ratio of the jet diameter to the tube diameter as a function of the shear rate for polymer melts. Newtonian liquids and polymers at very low shear rates might be expected to have a die swell of 1.0, but actually they have a die swell of 1.12; this number can be computed by numerically solving the Navier–Stokes equation. Additional die swell first becomes noticeable at rates of shear at which the viscosity of the fluid starts to become non-Newtonian. The die swell increases with the rate of

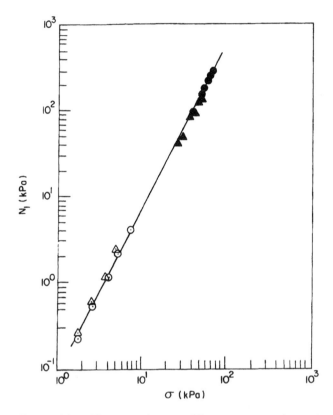

FIGURE 5.12   First normal stress difference versus shear stress for molten polysty-rene at 200°C and 220°C. Filled symbols are exit pressure data. (From Ref. 44.)

shear in a manner similar to the first normal stress difference, and this suggests that we might be able to determine N_1 from a measurement of die swell. However, die swell is not found to be a unique function of shear rate. For a given rate of shear, the die swell decreases as the length-to-diameter ratio of the capillary increases and ultimately attains an equilibrium value. This is due to the fact that extensional flow at the capillary entrance imparts greater molecular orientation compared to shear flow within the capillary. Thus, in a long capillary, some of the molecular orientation imparted to the polymer in the entrance region can relax out in the capillary itself. Also, at a constant shear rate, die swell (measured under isothermal conditions) tends to decrease somewhat as the temperature is raised because of the faster relaxation of molecular orientation at higher temperatures. But in analogy with the behavior of N_1, equilibrium die swell for a given polymer

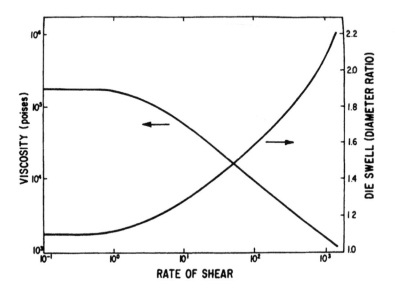

FIGURE 5.13 General behavior of die swell and viscosity as a function of the rate of shear in a capillary or similar duct.

sample appears to be a unique function of the shear stress that is independent of the temperature of measurement [45]. Additionally, the variation of die swell with polymer structural variables such as molecular weight, molecular weight distribution, and long-chain branching seems to follow the same trends as in the case of the first normal stress difference [23,24].

Various expressions have been proposed for relating die swell to the first normal stress difference, and several of these have been evaluated [24]. The most successful one appears to be based on a theory due to Tanner [46]. By assuming that the polymer leaving the tube is an elastic material that obeys Hooke's law, Tanner related die swell to the stresses acting on the polymer within the capillary. The final result is:

$$\frac{D_j}{D} = 0.12 + \left(1 + \frac{S_R^2}{2}\right)^{1/6} \tag{5.8}$$

in which D_j is the jet diameter, D is the capillary diameter, S_R is recoverable shear, and the constant 0.12 is introduced to obtain agreement with Newtonian fluid behavior at low shear rates. If die swell is measured as a function of the wall shear stress, the first normal stress difference can be estimated with the help of Eq. (5.8). However, as noted by Middleman [45], due to the form of Eq. (5.8), small errors in measuring die swell can lead to large errors in the calculated value of the recoverable shear. In this context, it must be remembered that the die swell values required in Eq. (5.8) are isothermal values; if extrusion is done in room-temperature air, it is necessary to anneal the extrudates to allow a complete recovery before measuring the jet diameter.

While many data sets agree with the predictions of Eq. (5.8), it has to be realized that this equation is a rather approximate one. Better predictions of die swell can be made using more realistic constitutive equations and more sophisticated mathematics, but the results show a great sensitivity to the nature of the viscoelastic constitutive equation employed [47]. Thus, one needs an accurate constitutive equation to relate die swell to N_1. However, if we had such an equation, we could use it directly to determine N_1; it would not be necessary to measure die swell at all!

C. Torsional Balance

In the parallel-plate viscometer described in Sec. V of Chap. 2, we use two parallel coaxial disks, separated by a fixed distance h, to shear the fluid of interest, and we determine a combination of N_1 and N_2 by measuring the force tending to pull the disks apart. In the torsional balance rheometer [48], we apply a known downward force to the upper disk and measure the resulting value of h as a function of the speed of rotation of the lower disk. Thus, Eqs. (2.17)–(2.19) still apply, and data analysis is done in the usual manner. The major advantage here is that smaller gap values can be attained, down to about 0.0003 cm, giving shear rates of the order of 10^5 sec^{-1}. In the conventional setup, such small gaps cannot be set with sufficient precision to obtain the same level of accuracy. Small gaps also reduce potential viscous heating problems.

This instrument has been used to measure the normal stresses in dilute polymer solutions, such as lubricants, but the two normal stress differences cannot be determined separately using this technique.

D. Hole Pressure Measurements

Unless flush-mounted pressure transducers are employed for flow through a slit die, the fluid pressure measured by drilling a hole through the channel wall will be lower than the true value. As shown in Fig. 5.14, this hole pressure error arises due to the bending of streamlines toward the bottom of the hole. The reduction in pressure, $p_1 - p_2$, can be determined with the help of a flush-mounted transducer located at the same axial position but on the opposite wall of the slit. This pressure difference is found to be a monotonically increasing function of the first normal stress difference within the flowing liquid [49]. Indeed, it can be theoretically calculated if assumptions are made about the constitutive behavior of the fluid. In particular, for the steady, isothermal, creeping flow of a second-

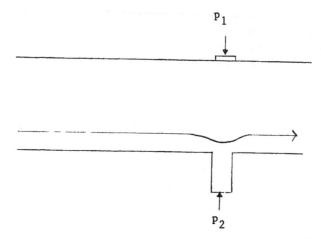

FIGURE 5.14 Influence of a pressure tap on streamline flow through a channel of rectangular cross section.

order fluid flowing over a transverse slot, the hole pressure error equals one-fourth of N_1 [50]. To analyze the situation for a more general viscoelastic fluid, several simplifying assumptions have to be made. According to Higashitani and Pritchard, the final results are [49,51]:

$$N_1 = 2n(p_1 - p_2) \tag{5.9}$$

where

$$n = \frac{d \log(p_1 - p_2)}{d \log \tau_w} \tag{5.10}$$

and, as expected, $n = 2$ for a second-order fluid.

The validity of the various assumptions in the Higashitani–Pritchard analysis and the modifications of this analysis have been discussed by Lodge [49]. A commercial instrument, employing hole pressure measurements and known as the Lodge Stressmeter, is now available from the Bannatek Company in Madison, WI. It has been used to make measurements on multigrade motor oils at engine operating conditions—up to 10^6 sec^{-1} in shear rate and 150°C in temperature.

E. Jet Thrust Technique

In the momentum balance for the exit pressure method, Figure 5.11, fluid inertia could be neglected due to the very viscous nature of the liquids involved. For low-viscosity polymer solutions, especially at high Reynolds numbers, we can actually neglect the exit pressure but not the inertia of the fluid. If we also assume, as before, that the flow is fully developed all the way to the tube exit, the momentum balance now leads to (for a circular-cross-section tube of radius R) [43]:

$$N_1(\dot{\gamma}_w) + 0.5 N_2(\dot{\gamma}_\omega) = \frac{1}{\pi R^2}\left(T_L + \frac{\tau_w}{2}\frac{\partial T_L}{\partial \tau_w}\right) \qquad (5.11)$$

in which T_L is the reduced thrust, defined as

$$T_L = 2\pi \int_0^R \rho v^2(r,0)r\, dr - T \qquad (5.12)$$

and T is the jet thrust on the tube. As long as all of the fluid enters the tube perpendicular to the jet, the magnitude of the thrust is $\rho\pi R_j^2\, v_j^2$; here, R_j is the jet radius and v_j its velocity. In practical terms, the jet thrust can be measured from the deflection of cantilever springs that carry the capillary tube. Thus, if jet thrust data are measured as a function of the wall shear stress, a combination of the two normal stress differences can be determined.

The history of the jet thrust method goes back to the very early 1960s. It has been a very popular technique due to its potential for providing high shear rate normal stress data on polymer solutions. Unfortunately, however, it now appears that the assumption of fully developed flow made in the derivation of Eq. (5.11) is in error. This makes the first normal stress difference calculated from jet thrust measurements higher than the true value by a substantial amount [52]. Since there is no theoretical or experimental way to estimate the extent of velocity rearrangement at the tube exit [43], the future utility of this method is in doubt.

IV. MEASUREMENT OF THE SECOND NORMAL STRESS DIFFERENCE

Compared to the volume of data available on the first normal stress difference, the amount of quantitative information about the second normal stress difference is quite meager. This is due primarily to two reasons. First, at a given shear rate, the size of N_2 is about an order of magnitude less than the size of N_1. Consequently, when a cone-and-plate viscometer is employed, either by itself or in combination with a parallel-plate instrument, the use of Eqs. (2.14)–(2.19) necessitates the taking of small differences between two large numbers, leading to very significant errors; one is often forced to conclude that N_2 is zero to within experimental error [53]. Second, it was thought that N_2 did not play much of a role in commercial polymer processing, and this fact resulted in minimal industrial interest in this material function [54]. Recently, however, it has been recognized that the accurate determination of the second normal stress difference may be important for both theoretical and practical reasons. On the theoretical side, such a measurement can help us in testing the quantitative validity of constitutive equations proposed for nonlinear viscoelasticity, while, on the practical side, N_2 may actually control important rheological phenomena. It has been suggested by Tanner and Keentok [55], and verified by Lee et al. [56], that edge fracture in a cone-and-plate viscometer occurs when N_2 exceeds a critical value. Also, Shaqfeh and coworkers have demonstrated that the second normal stress difference can strongly stabilize Taylor–Couette flow [57]. More recently, Levitt et al. have concluded that N_2 plays a significant role in morphology development during the preparation of polymer blends [58].

A. Use of Standard Viscometers

According to Eq. (2.16), a graph of the total stress exerted by the fluid on the plate surface in a cone-and-plate viscometer when plotted against the logarithm of the radial position should be a straight line of slope $(N_1 + 2N_2)$. This slope, together with an independent measurement of N_1, can, in principle, yield N_2. As mentioned earlier, this is not an easy measurement, but it can be carried out. Christiansen and coworkers used the procedure outlined here, but employed a set of four extremely sensitive, custom-built, flush-mounted miniature pressure transducers in their cone-and-plate apparatus [27]. They also checked the internal consistency of all their data. Their second normal stress difference results on concentrated polystyrene solutions as a function of the shear stress are shown in Figure 5.15, and these resemble the functional behavior of N_1 versus shear stress seen earlier. In particular, the straight lines in Fig. 5.15a have a slope of 2, and the average value of the ratio of $-N_2$ to N_1 is 0.25 (see Fig. 5.15b). In addition, they found that for the monodisperse samples, N_2 increases with molecular weight to the 6.8–7.8 power; for polydisperse samples, N_2 increases with the width of the molecular weight distribution. It appears that long-chain branching also enhances the value of N_2 [59]. Lee and White [17] have published data on a number of molten polymers, and, while these data are relatively less accurate, they too show that N_2 is negative and considerably smaller than N_1. More recently, Meissner et al. [60] measured the ratio $-N_2/N_1$ for a low-density polyethylene at low shear rates using a cone-and-plate viscometer. They used a novel design in which the plate was partitioned into an inner disk connected to the normal force transducer while the outer ring was fixed to the frame of the instrument; they found that the ratio of the two normal stress differences was 0.24.

In closing this subsection, we mention that the principle of time–temperature superposition can be applied to second normal stress difference data to give a master curve using temperature shift factors obtained from the WLF equation. This allows us to compare data obtained under different conditions from two different instruments or to extend the shear rate range of measurement at a given temperature [61].

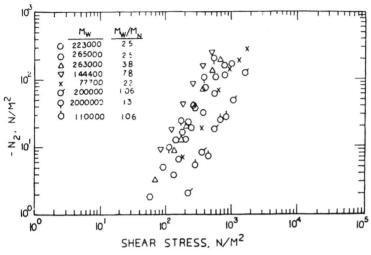

(a)

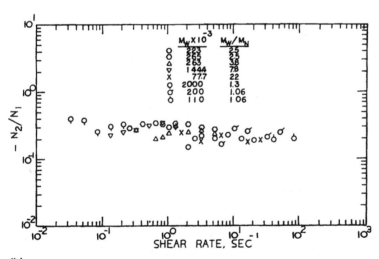

(b)

FIGURE 5.15 (a) The second normal stress difference versus shear stress for mono- and polydisperse polystyrene solutions in butylbenzene. (b) Normal stress difference ratio for these solutions. The concentration is 450 g/L except 250 g/L for the 2 million Mw polymer and 600 g/L for the 77,700 Mw polymer. (From Ref. 27.)

B. Rod Climbing

As shown in Figure 5.5, polymeric liquids climb up a rotating shaft, and this effect has been put to good use by Beavers and Joseph [62]. These authors analyzed the Weissenberg effect and solved the equations of motion using a perturbation method to relate the height of climb h at an angular velocity ω to a climbing constant β defined as $(0.5\psi_1^0 + 2\psi_2^0)$. The final result is:

$$h = h_\cdot + \frac{4r\beta\omega^2}{2\sqrt{\rho g}(4\sqrt{\gamma} + r\sqrt{\rho g})} \tag{5.13}$$

where $h_\cdot$ is the value of h in the absence of rotation, r is the rod radius, ρ is the liquid density, γ is the liquid surface tension, and g is the acceleration due to gravity. Here ψ_1^0 and ψ_2^0 are, respectively, the zero-shear-rate values of the first and second normal stress coefficients. Magda et al. [63] have used this technique to determine the second normal stress difference of an organic Boger liquid [39] and found it to be negative and about 30 times smaller than the first normal stress difference. Although this is a simple method, it is limited to low shear rates, due to the assumptions made in the derivation of Eq. (5.13).

C. The Tilted Trough

If we consider gravity-driven flow down a semicircular inclined channel, we find that the fluid surface is flat if the fluid is Newtonian but is curved if the fluid is viscoelastic; a negative second normal stress difference results in a convex shape, while a positive N_2 implies a concave surface. Kuo and Tanner [64] have published an analysis that allows for the determination of N_2 as a function of the wall shear stress from a measurement of the surface profile. The theory, however, is somewhat involved, and the reader is referred to the original paper for details.

V. CONCLUDING REMARKS

In the past, accurate measurement of the normal stress differences of polymeric fluids in shear has been a fundamental problem in polymer rheology. As the techniques presented in this chapter reveal, much progress has been made in this regard, and a great deal is now known about the relationship between the normal stress differences and variables such as shear rate, temperature, molecular weight, molecular weight distribution, and long-chain branching. The experimental determination of N_1 is now considered a routine measurement useful for fluid characterization and the development of rheological models that might be utilized for process modeling and simulation. In this context, attempts have also been made to reverse the process and to use specific constitutive equations to predict the first normal stress difference function from the measured flow curve [65,66].

Although some success has been achieved, the procedure cannot work in all cases. The Boger fluid, for example, has a constant viscosity, and it can be tailored to give different levels of elasticity [37]. Recall also the celebrated case of three polyethylenes that had the same viscosities and molecular weight distributions but behaved differently when used to make blown film [67]; the rheological differences among the polymers showed up in the first normal stress differences and in the extensional viscosities.

In the future, we can expect more user-friendly equipment with better software, an increased range of measurement, and greater accuracy and sensitivity. In addition, improved transducers should make it possible to measure the second normal stress difference with greater ease and to make reliable transient normal stress measurements, whether during step strain experiments or during startup and shutdown of shearing at a constant shear rate. Progress has also been made in the use of optical techniques for stress measurements. Fluid flow orients polymer molecules, and this makes the optical properties anisotropic. By relating the stress tensor to the refractive index tensor through the stress optic law, we can measure all the material functions with this noninvasive technique [68,69]. Indeed, commercial viscometers are beginning to incorporate this feature, and researchers are increasingly using birefringence measurements along with mechanical stress measurements.

REFERENCES

1. K. Walters. Rheometry. Chapman and Hall, London, 1975.
2. K. Oda, J. L. White, and E. S. Clark. Correlation of normal stresses in polystyrene melts. Polym. Eng. Sci. 18:25–28 (1978).
3. R. F. Ginn and A. B. Metzner. Measurement of stresses developed in steady laminar shearing flows of viscoelastic media. Trans. Soc. Rheol. 13:429–453 (1969).
4. R. J. Gordon and A. E. Everage. Jr., Bead-spring model of dilute polymer solutions: continuum modifications and an explicit constitutive equation. J. Appl. Polym. Sci. 15:1903–1909 (1971).
5. N. Phan-Thien. A nonlinear network viscoelastic model. J. Rheol. 22:259–283 (1978).
6. J. Meissner. Modifications of the Weissenberg rheogoniometer for measurement of transient rheological properties of molten polyethylene under shear. Comparison with tensile data. J. Appl. Polym. Sci. 16:2877–2899 (1972).
7. R. B. Bird, R. C. Armstrong, and O. Hassager. Dynamics of Polymeric Liquids. 2nd ed., Vol. 1. Wiley, New York, 1987.
8. M. H. Wagner. Analysis of time-dependent non-linear stress-growth data for shear and elongational flow of a low-density branched polyethylene melt. Rheol. Acta 15: 136–142 (1976).
9. E. V. Menezes and W. W. Graessley. Nonlinear rheological behavior of polymer systems for several shear-flow histories. J. Polym. Sci.:Polym. Phys. Ed. 20:1817–1833 (1982).

10. A. S. Lodge, Elastic Liquids, Academic Press, New York, 1964.
11. B. D. Coleman, H. Markovitz, and W. Noll. Viscometric Flows of Non-Newtonian Fluids. Springer-Verlag, New York, 1966.
12. E. B. Christiansen and W. R. Leppard. Steady-state and oscillatory flow properties of polymer solutions. Trans. Soc. Rheol. 18:65–86 (1974).
13. Z. Tadmor and R. B. Bird. Rheological analysis of stabilizing forces in wire-coating dies. Polym. Eng. Sci. 14:124–136 (1974).
14. B. Maxwell and A. J. Scalora. The elastic melt extruder—works without screw. Modern Plastics 37:107, October 1959.
15. N. E. Hudson and J. Ferguson. The shear flow properties of M1. J. Non-Newt. Fluid Mech. 35:159–168 (1990).
16. D. A. Nguyen and T. Sridhar. Preparation and some properties of M1 and its constituents. J. Non-Newt. Fluid Mech. 35:93–104 (1990).
17. B. C. Lee and J. L. White. An experimental study of rheological properties of polymer melts in laminar shear flow and of interface deformation and its mechanisms in two-phase stratified flow. Trans. Soc. Rheol. 18:467–492 (1974).
18. W. Minoshima, J. L. White, and J. E. Spruiell. Experimental investigation of the influence of molecular weight distribution on the rheological properties of polypropylene melts. Polym. Eng. Sci. 20:1166–1176 (1980).
19. J. Greener and G. H. Pearson. Orientation residual stresses and birefringence in injection molding. J. Rheol. 27:115–134 (1983).
20. C. J. S. Petrie and M. M. Denn. Instabilities in polymer processing. AIChE J. 22:209–236 (1976).
21. A. B. Metzner, J. L. White, and M. M. Denn. Behavior of viscoelastic materials in short-time processes. Chem. Eng. Prog. 62:81–92 (Dec. 1966).
22. H. M. Laun. Prediction of elastic strains of polymer melts in shear and elongation. J. Rheol. 30:459–501 (1986).
23. R. Racin and D. C. Bogue. Molecular weight effects in die swell and in shear rheology. J. Rheol. 23:263–280 (1979).
24. C. D. Han. Rheology in Polymer Processing. Academic Press, New York, 1976.
25. R. A. Mendelson, W. A. Bowles, and F. L. Finger. Effect of molecular structure on polyethylene melt rheology. I. Low-shear behavior. J. Polym. Sci.: Part A-2 8:105–126 (1970).
26. J. L. White and H. Yamane. A collaborative study of the rheological properties and unstable melt spinning characteristics of linear and branched polyethylene terephthalates. Pure Appl. Chem. 57:1441–1452 (1985).
27. H. W. Gao, S. Ramachandran, and E. B. Christiansen. Dependence of the steady-state and transient viscosity and first and second normal stress difference functions on molecular weight for linear mono- and polydisperse polystyrene solutions. J. Rheol. 25:213–235 (1981).
28. R. I. Tanner. A correlation of normal stress data for polyisobutylene solutions. Trans. Soc. Rheol. 17:365–373 (1973).
29. M. Oritz, D. De Kee, and P. J. Carreau. Rheology of concentrated poly(ethylene oxide) solutions. J. Rheol. 38:519–539 (1994).
30. A. V. Shenoy and D. R. Saini. An approach to the estimation of polymer melt elasticity. Rheol. Acta 23:608–616 (1984).

31. G. Astarita and G. Marrucci. Principles of Non-Newtonian Fluid Mechanics. McGraw-Hill, London, 1974.
32. R. G. Larson. Constitutive Equations for Polymer Melts and Solutions. Butterworths, Boston, 1988.
33. R. B. Bird, H. R. Warner, Jr., and D. C. Evans. Kinetic theory and rheology of dumbbell suspensions with Brownian motion. Adv. Polym. Sci. 8:1–90 (1971).
34. A. Kumar and R. K. Gupta. Fundamentals of Polymers. McGraw-Hill, New York, 1998.
35. M. J. Crochet, A. R. Davies, and K. Walters. Numerical Simulation of Non-Newtonian Flow. Elsevier, Amsterdam, 1984.
36. K. C. Tam and C. Tiu. A low viscosity, highly elastic ideal fluid. J. Non-Newt. Fluid Mech. 31:163–177 (1989).
37. K. C. Tam, T. Moussa, and C. Tiu. Ideal elastic fluids of different viscosity and elasticity levels. Rheol. Acta 28:112–120 (1989).
38. D. V. Boger. A highly elastic constant-viscosity fluid. J. Non-Newt. Fluid Mech. 3:87–91 (1977/78).
39. G. Prilutski, R. K. Gupta, T. Sridhar, and M. E. Ryan. Model viscoelastic liquids. J. Non-Newt. Fluid Mech. 12:233–241 (1983).
40. R. B. Bird, C. F. Curtiss, R. C. Armstrong, and O. Hassager. Dynamics of Polymeric Liquids. 2nd ed., Vol. 2. Wiley, New York, 1987.
41. M. Doi and S. F. Edwards. The Theory of Polymer Dynamics. Oxford University Press, Oxford, 1986.
42. M. Doi. Introduction to Polymer Physics. Oxford University Press, Oxford, 1996.
43. D. V. Boger and M. M. Denn. Capillary and slit methods of normal stress measurements. J. Non-Newt. Fluid Mech. 6:163–185 (1980).
44. C. D. Han. Slit rheometry. In: A. A. Colyer and D. W. Clegg, eds. Rheological Measurement. Elsevier, London, 1988, pp. 25–48.
45. S. Middleman. Fundamentals of Polymer Processing. McGraw-Hill, New York, 1977.
46. R. I. Tanner. A theory of die-swell. J. Polym. Sci.: Part A-2 8:2067–2078 (1970).
47. R. I. Tanner. Recoverable elastic strain and swelling ratio. In: A. A. Collyer and D. W. Clegg, eds. Rheological Measurement. Elsevier, London, 1988, pp. 93–118.
48. D. M. Binding and K. Walters. Elastico-viscous squeeze films, part 3. The torsional balance rheometer. J. Non-Newt. Fluid Mech. 1:277–286 (1976).
49. A. S. Lodge. Normal stress differences from hole pressure measurements. In: A. A. Collyer and D. W. Clegg, eds. Rheological Measurement. Elsevier, London, 1988, pp. 345–382.
50. R. I. Tanner and A. C. Pipkin. Intrinsic errors in pressure-hole measurements. Trans. Soc. Rheol. 13:471–484 (1969).
51. K. Higashitani and W. G. Pritchard. A kinematic calculation of intrinsic errors in pressure measurement made with holes. Trans. Soc. Rheol. 16:687–696 (1972).
52. J. M. Davies, J. F. Hutton, and K. Walters. A critical re-appraisal of the jet-thrust technique for normal stresses, with particular reference to axial velocity and stress rearrangement at the exit plane. J. Non-Newt. Fluid Mech. 3:141–160 (1977/78).
53. K. Walters. The second-normal-stress difference project. IUPAC Macro-83. Bucha-

rest, Romania. Plenary and invited lectures, Part 2, IUPAC Macromolecular Division, 1983, pp. 227–237.

54. K. Walters. Fundamental concepts, In: K. Walters, ed. Rheometry: Industrial Applications. Research Studies Press, Chichester, Eng., 1980. pp. 1–29.

55. R. I. Tanner and M. Keentok. Shear fracture in cone-plate rheometry. J. Rheol. 27: 47–57 (1983).

56. C. S. Lee, B. C. Tripp, and J. J. Magda. Does N_1 or N_2 control the onset of edge fracture? Rheol. Acta 31:306–308 (1992).

57. E. S. G. Shaqfeh, S. J. Muller, and R. G. Larson. The effects of gap width and dilute solution properties on the viscoelastic Taylor–Couette instability. J. Fluid Mech. 235:285–317 (1992).

58. L. Levitt, C. Macosko, and S. D. Pearson. Influence of normal stress difference on polymer drop deformation. Polym. Eng. Sci. 36:1647–1655 (1996).

59. C.-S. Lee, J. J. Magda, and K. L. DeVries. Measurements of the second normal stress difference for star polymers with highly entangled branches. Macromolecules 25:4744–4750 (1992).

60. J. Meissner, R. W. Garbella, and J. Hostettler. Measuring normal stress differences in polymer melt shear flow. J. Rheol. 33:843–864 (1989).

61. G. A. Alvarez, A. S. Lodge, and H.-J. Cantow. Measurement of the first and second normal stress differences: correlation of four experiments on a polyisobutylene/decalin solution "D1". Rheol. Acta 24:368–376 (1985).

62. G. S. Beavers and D. D. Joseph. The rotating rod viscometer. J. Fluid Mech. 69: 475–511 (1975).

63. J. J. Magda, J. Lou, S. G. Baek, and K. L. DeVries. Second normal stress difference of a Boger fluid. Polymer 32:2000–2009 (1991).

64. Y. Kuo and R. I. Tanner. On the use of open-channel flows to measure the second normal stress difference. Rheol. Acta 13:443–456 (1974).

65. S. I. Abdel-Khalik, O. Hassager, and R. B. Bird. Prediction of melt elasticity from viscosity data. Polym. Eng. Sci. 14:859–867 (1974).

66. M. H. Wagner. Predictions of primary normal stress difference from shear viscosity data using a single integral constitutive equation. Rheol. Acta 16:43–50 (1977).

67. J. Meissner. Basic parameters, melt rheology, processing and end-use properties of three similar low density polyethylene samples. Pure Appl. Chem. 42:553–612 (1975).

68. J. L. S. Wales. The Application of Flow Birefringence to Rheological Studies of Polymer Melts. Delft University Press, Rotterdam, 1976.

69. H. Janeschitz-Kriegl. Polymer Melt Rheology and Flow Birefringence. Springer-Verlag, Berlin, 1983.

6

Dynamic Mechanical Properties

I. INTRODUCTION

A dynamic mechanical experiment is one in which a polymer sample is subjected to a sinusoidal strain γ of infinitesimal amplitude γ_0 and fixed angular frequency ω; $\omega = 2\pi f$, where f is the frequency in hertz. Therefore,

$$\gamma = \gamma_0 \sin \omega t \tag{6.1}$$

Here the stress response τ is found to be sinusoidal but, in general, is out of phase by an angle δ. Thus,

$$\tau = \tau_0 \sin(\omega t + \delta) \tag{6.2}$$

or

$$\tau = (\tau_0 \cos \delta) \sin \omega t + (\tau_0 \sin \delta) \cos \omega t \tag{6.3}$$

This is known as linear behavior, and it is observed regardless of the mode of deformation, whether shear, extension, or flexure; the maximum strain amplitude for which Eq. (6.2) holds cannot be predicted a priori but has to be determined by experiment. On dividing the stress by the strain amplitude, we get the modulus G:

$$G = G'(\omega) \sin \omega t + G''(\omega) \cos \omega t \tag{6.4}$$

where $G' = (\tau_0 \cos \delta)/\gamma_0$ and $G'' = (\tau_0 \sin \delta)/\gamma_0$. The term G', called the *storage modulus*, is the same quantity defined by Eq. (1.7), and it is the in-phase component of the modulus and represents the energy stored and recovered per cycle

117

[1]. Correspondingly, the term G'', called the *loss modulus*, was defined earlier by Eq. (1.8) and it is the out-of-phase component of the modulus and represents the energy dissipated as heat per cycle of deformation. The ratio of G'' to G' is tan δ and is an alternate measure of energy dissipation.

The basic essentials of most instruments used for measuring dynamic properties are schematically illustrated in Figure 6.1. The polymer sample is deformed in shear or in tension by some oscillating driver, which may be either mechanical or electromagnetic in nature. The amplitude of the sinusoidal deformation is measured by a strain transducer, which may be a linear variable-differential transformer (LVDT), a variable-resistance gage, or some type of optical transducer. The force deforming the sample is measured by the small deformation of a relatively rigid spring or torsion bar to which is attached a stress transducer; this may be another LVDT. A few of the commercial instruments and others that are widely used are described in the Refs. 1–5. A special type of *oscillating rheometer*, known as the orthogonal rheometer, consists of two parallel disks the space between which is filled with polymer and whose axes are not quite collinear (see Fig. 2.4b). As one disk rotates relative to the other, an eccentric oscillatory motion is set up in the polymer [6].

Dynamic mechanical experiments are conducted routinely on both polymeric solids and polymeric liquids, and these are useful for material characterization and quality control purposes. The experiments may be conducted as a function of temperature at a fixed frequency or as a function of frequency under isothermal conditions. The former mode of operation is particularly well suited for work on solids; and Figure 6.2 shows representative data on an (amorphous)

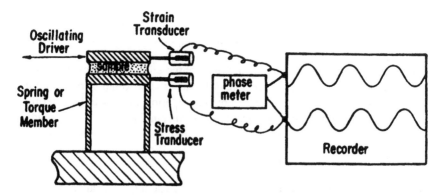

Figure 6.1 Schematic diagram of an oscillating rheometer for measuring the dynamic properties of fluids.

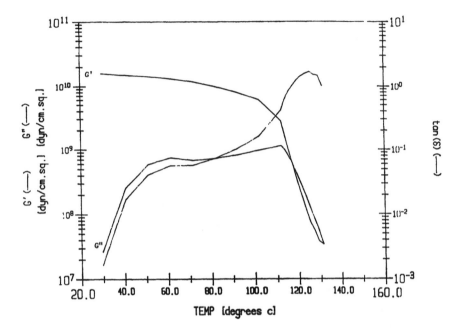

FIGURE 6.2 Storage modulus, loss modulus, and damping factor as a function of temperature for a polymethyl methacrylate sample subjected to torsional deformation at 1 Hz.

sample of polymethyl methacrylate subjected to torsional deformation at a frequency of 1 Hz. At low temperatures, the polymer is glassy, and it acts essentially as an elastic solid; this is confirmed by the fact that $G' \gg G''$. Under these conditions, the storage modulus equals the shear modulus, or stiffness; this value is typically on the order of 10^9 Pa. As the temperature approaches the glass transition temperature, approximately 125°C in the present case, the storage modulus falls rapidly by about three orders of magnitude while the loss modulus and tan δ increase rapidly and exhibit maxima. Thus, these data can be used to identify thermal transitions such as the glass transition temperature and secondary dispersions involving the motion of side chains of the polymer. Indeed, this technique has been used extensively to probe the morphology of semicrystalline polymers [2,7 and references therein]. Note, though, that T_g measured using dynamic mechanical analysis is usually slightly larger than that measured with the help of a differential scanning calorimeter; typically T_g increases with increasing frequency or a reduction in the time scale of measurement [2].

For molten polymers, it is more common to measure the storage and loss moduli as a function of frequency while keeping temperature unchanged. The characteristic response of a high-molecular-weight molten polymer to small-amplitude oscillatory shear deformation over 12 decades of frequency is sketched in Figure 6.3a [8]. It is seen that at the lowest frequencies (corresponding to small values of the Deborah number), the response is liquidlike, with the loss modulus exceeding the storage modulus. However, in this region, known as the

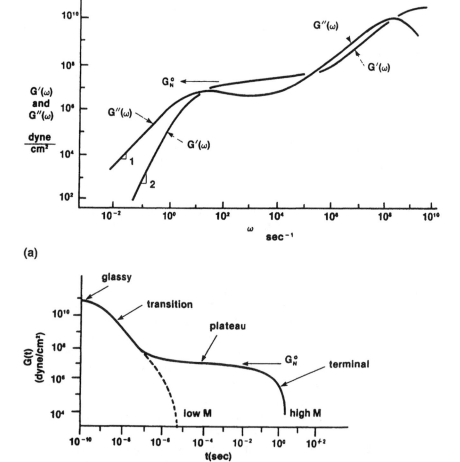

FIGURE 6.3 (a) Isothermal storage and loss moduli of a molten polymer as a function of frequency. (b) Stress relaxation behavior of the same molten polymer. (Reprinted with permission from Ref. 8. Copyright 1984 American Chemical Society.)

terminal region, the double logarithmic curve of G' has a slope of 2 while that of G'' has a slope of unity. Consequently, the two moduli soon become equal and display a plateau region; this is typical rubbery behavior and arises from the presence of chain entanglements. At still higher frequencies, there is a transition region that gives way to glassy behavior. Here the storage modulus becomes constant while the loss modulus goes through a maximum. The melt acts like a glassy solid at very high frequencies (corresponding to large values of the Deborah number) because it is only the small parts of the molecule that can respond over such short time scales. As we shall see later, dynamic mechanical data can be used to predict other small-strain properties of the polymer, and this is one major reason why we carry out oscillatory experiments. The truth of this assertion can actually be seen by comparing storage modulus versus frequency data with the stress relaxation modulus as a function of time. In a stress relaxation experiment, a constant strain is imposed on the polymer, and the time-dependent stress is monitored; Figure 6.3b displays the result [8]. Clearly, Figures 6.3a and 6.3b are similar, and one can be obtained from the other by equating time to the reciprocal of the frequency.

II. DEPENDENCE OF POLYMER MELT DYNAMIC MECHANICAL PROPERTIES ON TEMPERATURE, MOLECULAR WEIGHT, AND POLYMER STRUCTURE

Using commercial instruments, it is generally not possible to obtain dynamic data over the wide frequency range illustrated in Figure 6.3a. Instead, we make measurements at different constant temperatures and then employ the principle of time–temperature superposition to construct a master curve at a chosen reference temperature; this procedure has been described in detail in Section III.B of Chapter 3. Shown in Figure 6.4 are isothermal, storage modulus data as a function of angular frequency for a 167,000-molecular-weight, narrow-distribution polystyrene melt at a variety of temperatures [9]. These data can be made to collapse into a single curve by means of a horizontal shift, and the result is displayed in Figure 6.5 at a reference temperature of 160°C. Figure 6.5 also shows similar master curves for other molecular weight fractions, ranging from 8,900 (curve L9) to 581,000 (curve L18) [9]. Corresponding data for the loss modulus are given in Figure 6.6 [9]. The values of the shift factors needed to obtain these master curves depend on temperature alone, and these can be determined using the WLF equation [Eq. (3.16)] Clearly, changes in temperature are equivalent to changes in frequency or time.

Figures 6.5 and 6.6 reveal that both the elastic modulus and the loss modulus increase rapidly with frequency and with molecular weight up to a plateau

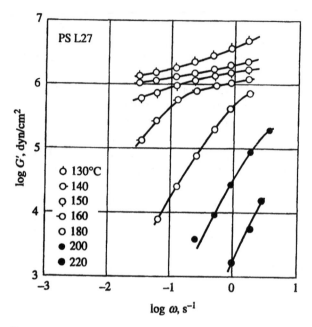

FIGURE 6.4 Frequency dependence of G' for narrow-distribution polystyrene L27, molecular weight 167,000, at various temperatures. (Reprinted with permission from Ref. 9. Copyright 1970 American Chemical Society.)

value of about 10^5 Pa. Increasing the polymer molecular weight shifts the flow or terminal region to progressively smaller frequencies. Low-molecular-weight polymers that have few if any entanglements have little or no plateau region in the log G' or log G'' versus log ωa_T plots. High-molecular-weight polymers have plateau regions that cover several decades of frequency. While the G' curve exhibits a sharp transition from the rubbery to the terminal zone, the G'' curve goes through a maximum. These plateaus exist in a frequency range above which the entanglements do not have time to slip and relax out the stress. These values of G' and G'' in the plateau regions are essentially independent of molecular weight, because the effective molecular weight between entanglements becomes constant. Entanglements that cannot relax out behave as crosslinks, and storage modulus data together with the kinetic theory of rubber elasticity can be used to estimate the molecular weight between entanglements [10].

The upturn in all the curves in Figures 6.5 and 6.6 at reduced frequencies greater than 10^3 radians/second is due to the beginning of the glass transition region. The various curves, except those corresponding to the lowest molecular weights, appear to come together here. The polymer melt no longer behaves as

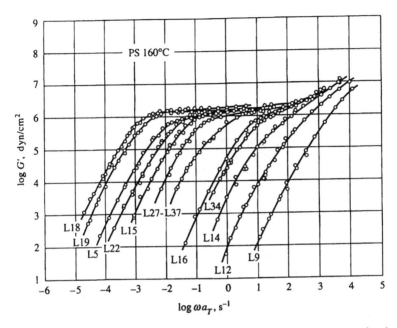

FIGURE 6.5 Master curves of G' for narrow-distribution polystyrenes having different molecular weights. The reference temperature is 160°C. (Reprinted with permission from Ref. 9. Copyright 1970 American Chemical Society.)

a fluid but begins to take on the characteristics of a glass at these high frequencies, and many kinds of molecular motions are no longer able to take place in time intervals less than 10^{-3} seconds.

On going from monodisperse to polydisperse polymers, we find that the shapes of the G' and G'' functions change considerably in the terminal and plateau regions but not in the transition region. In particular, as the width of the molecular weight distribution increases, the changeover from the rubbery to the terminal region in the storage modulus curve becomes much more gradual due to the presence of high-molecular-weight fractions, and even the plateau disappears [11]; this is seen in Figure 6.7 [12]. Simultaneously, G'' loses the peak that monodisperse polymers have in the frequency range in which G' tends to become constant [11]. The introduction of long-chain branching generally decreases the elasticity or G' of the melts [12,13]; typical data are again shown in Figure 6.7, which compares the behavior of narrow-distribution, broad-distribution, and four-branch star polystyrenes [12].

The fairly strong influence that polydispersity has on the dynamic mechanical properties of linear polymers has prompted efforts aimed at determining the

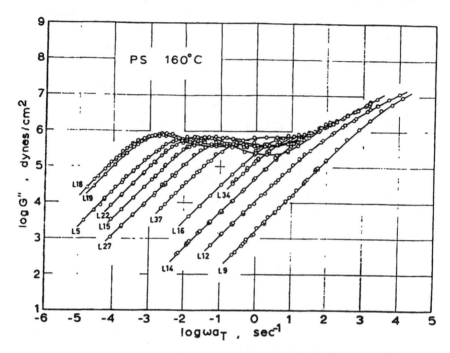

FIGURE 6.6 Master curves of G'' for narrow-distribution polystyrenes having different molecular weights. The reference temperature is 160°C. (Reprinted with permission from Ref. 9. Copyright 1970 American Chemical Society.)

molecular-weight distribution from linear viscoelasticity data [14]. It is well known that conventional methods of molecular-weight distribution measurement require that the polymer sample be soluble in some solvent at ambient temperatures [15]. However, many industrially important polymers, especially fluoropolymers, are fairly insoluble in common solvents; this has been one major motivation for the development of rheological methods for molecular weight distribution determination [15a]. To be successful in this endeavor, we need mixing rules that relate the composition and properties of the components in a polydisperse sample to the measured behavior of the mixture. Such mixing rules can be derived with the help of rheological constitutive equations, but the mathematics can be quite formidable. Consequently the reader is referred to the technical literature for details [14]. The method, though, appears to work, and commercial instruments now come equipped with appropriate software; an example is the Orchestrator software package from Rheometric Scientific.

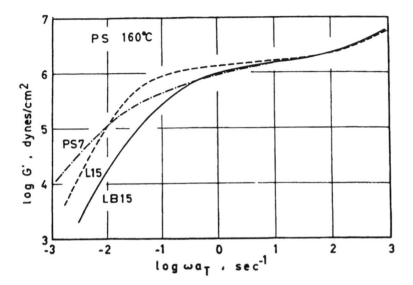

FIGURE 6.7 A comparison of G' curves for three different polystyrenes: L15 is monodisperse and linear, PS7 is broad distribution and linear, and LB15 is a four-branch star. The weight-average molecular weight in each case is similar. (Reprinted with permission from Ref. 12. Copyright 1971 American Chemical Society.)

III. BOLTZMANN SUPERPOSITION PRINCIPLE

The dynamic mechanical properties of a given polymer, whether represented in the form of storage modulus or loss modulus, can be utilized to predict the behavior of the same polymer in a different flow field. This is done by appealing to the Boltzmann superposition principle, whose essence can be understood with the help of the following argument [15].

If a viscoelastic material is subjected to a small, instantaneous strain γ at time zero, the stress τ relaxes over a period of time, and at time t we have

$$\tau(t) = G(t)\gamma \tag{6.5}$$

in which $G(t)$ is the stress relaxation modulus. Further, if a series of small strains $\gamma_1, \gamma_2, \ldots,$ is imposed on the material at times $t_1, t_2, \ldots,$ in the past, the stress at the present time t is a linear combination of the stresses resulting from each of these strains; i.e.,

$$\tau(t) = G(t - t_1)\gamma_1 + G(t - t_2)\gamma_2 + \cdots \tag{6.6}$$

where $(t - t_1)$ is the time elapsed since the imposition of strain γ_1. Converting the sum in Eq. (6.6) to an integral gives

$$\tau(t) = \int_{-\infty}^{t} G(t - s) \, d\gamma = \int_{-\infty}^{t} G(t - s)\dot{\gamma} \, ds \tag{6.7}$$

where s is a past time, $\dot{\gamma}$ is the rate of deformation, and $G(t - s)$ is the yet-to-be-determined stress relaxation modulus. Equation (6.7) can be generalized to three dimensions to give the Boltzmann superposition principle:

$$\tau_{ij}(t) = \int_{-\infty}^{t} G(t - s)\dot{\gamma}_{ij} \, ds \tag{6.8}$$

with $\dot{\gamma}_{ij}$ being the components of the rate of deformation tensor.

The Boltzmann superposition principle is the embodiment of the theory of linear viscoelasticity, and it is valid for both steady and transient deformations, provided the extent of deformation is small. To apply it to a small-amplitude sinusoidal deformation with $\gamma = \gamma_0 \sin \omega s$, we first change the independent variable in Eq. (6.8) from s to t', where $t' = t - s$. Thus,

$$\tau(t) = \int_{0}^{\infty} G(t')\dot{\gamma}(t - t') \, dt' \tag{6.9}$$

$$\dot{\gamma} = \omega\gamma_0 \cos[\omega(t - t')] \tag{6.10}$$

Inserting Eq. (6.10) into Eq. (6.9) and using a trigonometric identity, we get

$$\frac{\tau(t)}{\gamma_0} = \omega \cos(\omega t) \int_{0}^{\infty} G(t') \cos(\omega t') \, dt'$$
$$+ \omega \sin(\omega t) \int_{0}^{\infty} G(t') \sin(\omega t') \, dt' \tag{6.11}$$

which when compared to Eq. (6.4) yields the following expressions for the storage and loss moduli, respectively:

$$G'(\omega) = \omega \int_{0}^{\infty} G(t') \sin(\omega t') \, dt' \tag{6.12}$$

$$G''(\omega) = \omega \int_{0}^{\infty} G(t') \cos(\omega t') \, dt' \tag{6.13}$$

Equations (6.12) and (6.13) allow us to obtain the limiting behavior of the two moduli at very low and very high frequencies. The results are [15]:

$$\lim_{\omega \to 0} G'(\omega) = \left[\int_{0}^{\infty} t' G(t') \, dt' \right] \omega^2 \tag{6.14}$$

$$\lim_{\omega \to 0} G''(\omega) = \left[\int_0^\infty G(t') \, dt' \right] \omega \qquad (6.15)$$

$$\lim_{\omega \to \infty} G'(\omega) = G(0) \qquad (6.16)$$

$$\lim_{\omega \to \infty} G''(\omega) = 0 \qquad (6.17)$$

Equations (6.14) and (6.15) reveal that double logarithmic plots at low frequencies should have a slope of 2 for the storage modulus and a slope of unity for the loss modulus. This is shown in Fig. 6.3a. However, we may not always observe these limits, due to the inability to measure small stresses accurately at very low frequencies.

In a steady-shearing flow at a constant shear rate $\dot{\gamma}$, the Boltzmann superposition principle [Eq. (6.9)] predicts that

$$\tau = \dot{\gamma} \int_0^\infty G(t') \, dt' \qquad (6.18)$$

Since the ratio of the shear stress to the shear rate is the viscosity, a combination of Eqs. (6.15) and (6.18) says that

$$\eta_0 = \lim_{\omega \to 0} \frac{G''(\omega)}{\omega} \qquad (6.19)$$

and the viscosity appearing in Eq. (6.19) is the zero-shear-rate viscosity, because the theory is valid for small deformations only. This requires that the product of deformation rate with time be small, and since time can be large, the deformation rate must be small. Recall that Eq. (6.19) was seen earlier as Eq. (3.4). In Chap. 3, it was also mentioned that the zero-shear viscosity was independent of the polydispersity index and depended uniquely on the weight-average molecular weight. This is possible only if the low-frequency limit of G'' is a unique function of $\overline{M}_w$; such is, indeed, found to be the case [11].

In closing this section, we mention that linear viscoelasticity theory [Eq. (6.8)] always gives a zero first normal stress difference in shear flow. However, Eq. (5.1) relates the low-frequency limit of the storage modulus to the first normal stress difference in steady laminar shearing flow. This relationship is based on the predictions of the second-order approximation to a simple fluid with fading memory [16], and it has been confirmed by experimental results obtained by a very large number of authors [9].

IV. STRESS RELAXATION MODULUS

In order to predict the stress response of a linear viscoelastic material in a specific flow field through the use of Eq. (6.9), we need the stress relaxation modulus

$G(t')$. This may be obtained entirely by experiment or by a combination of experiment and theoretical models. In the latter case, mechanical analogs have proven to be particularly popular. Specifically, we consider the stress in a polymer to be composed of an elastic contribution and a viscous contribution, and, in the simplest situation, we represent the polymer as a spring and dashpot in series. This is shown schematically in Figure 6.8. The spring is Hookean, for which the stress and strain are related by

$$\tau = g\gamma \qquad (6.20)$$

whereas the dashpot is Newtonian, so

$$\tau = \eta \frac{d\gamma}{dt} \qquad (6.21)$$

Here g is the spring modulus and η is the dashpot viscosity; the linear combination of the spring and dashpot is called a *Maxwell element*.

It is clear that at any time, the total strain γ in the Maxwell element is a sum of the individual strains in the spring and dashpot; i.e.,

$$\gamma = \gamma_s + \gamma_d \qquad (6.22)$$

where the subscripts s and d denote the spring and the dashpot, respectively. Differentiating Eq. (6.22) with respect to time yeilds

$$\frac{d\gamma}{dt} = \frac{d\gamma_s}{dt} + \frac{\gamma_d}{dt} \qquad (6.23)$$

Because the applied stress τ equals the stress in both the spring and the dashpot, the terms on the right side of Eq. (6.23) can be obtained from Eqs. (6.20) and (6.21), with the following result

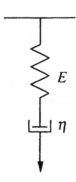

FIGURE 6.8 A Maxwell element.

$$\frac{d\gamma}{dt} = \frac{1}{g}\frac{d\tau}{dt} + \frac{\tau}{\eta} \tag{6.24}$$

If the polymer is initially stress free, i.e., $\tau = 0$ at $t = -\infty$, then the solution of Eq. (6.24) is

$$\tau(t) = g \int_{-\infty}^{t} e^{-(t-s)/\theta} \left(\frac{d\gamma}{ds}\right) ds \tag{6.25}$$

in which s is a dummy variable of integration and θ equals η/g. Comparing Eqs. (6.8) and (6.25), we find that

$$G(t - s) = g \exp\left[-\frac{(t - s)}{\theta}\right] \tag{6.26}$$

which when introduced into Eq. (6.5), reveals that

$$\tau(t) = g\gamma e^{-t/\theta} \tag{6.27}$$

and it is seen that in a stress relaxation experiment the stress decays over a time scale of the order of θ. Consequently, θ is called a *relaxation time*. Although a single relaxation time might fit data on monodisperse polymers, it will certainly not suffice for polydisperse materials. It is for this reason that, even for monodisperse samples, we modify the mechanical analog to make it consist of N Maxwell elements in parallel; each spring has a modulus g_i and each dashpot has a damping constant η_i. The stress relaxation modulus now becomes

$$G(t - s) = \sum_{i=1}^{N} g_i \exp\left[-\frac{(t - s)}{\theta_i}\right] \tag{6.28}$$

in which $\theta_i = \eta_i/g_i$, and the set of constants g_i and θ_i is called a *discrete relaxation time spectrum*.

In principle, the relaxation time spectrum can be obtained by fitting Eq. (6.28) to stress relaxation data. In practice, however, it is easier to measure the storage and loss moduli and to use these data to compute the relaxation time spectrum. This computation requires that we be able to interrelate one linear viscoelastic function to another, and available methods have been discussed by Ferry [1]. Here we employ the technique proposed by Baumgaertel and Winter [17], because this is particularly straightforward. If we introduce the stress relaxation modulus, [Eq. (6.28)], into Eqs. (6.12) and (6.13) and carry out the integrations, we get

$$G'(\omega) = \sum_{i=1}^{N} \frac{g_i(\omega\theta_i)^2}{1 + (\omega\theta_i)^2} \tag{6.29}$$

$$G''(\omega) = \sum_{i=1}^{N} \frac{g_i(\omega\theta_i)}{1 + (\omega\theta_i)^2} \tag{6.30}$$

The constants g_i and θ_i are found by simply fitting Eqs. (6.29) and (6.30) to measured G' and G'' data using a nonlinear least-squares procedure; typically we choose between one and two relaxation modes per decade of frequency [17]. Figure 6.9 shows a master curve of G' and G'' values in a temperature range of 130–250°C on an injection molding–grade sample of polystyrene; all the data have been combined by means of a horizontal shift using time–temperature superposition with a 150°C reference temperature. The stress relaxation modulus calculated in the manner of Baumgaertel and Winter is displayed in Figure 6.10 along with actual stress relaxation data at 150°C; the agreement is very good.

So far we have considered a discrete spectrum of relaxation times, i.e., the stress relaxation modulus was made up of a finite number of terms that traced their origin to an equal number of Maxwell elements in the mechanical analog of the polymer. By letting the number of Maxwell elements become infinite, we can define a continuous relaxation time spectrum $H(\theta)$ and say that $H\,d(\ln\theta)$ is

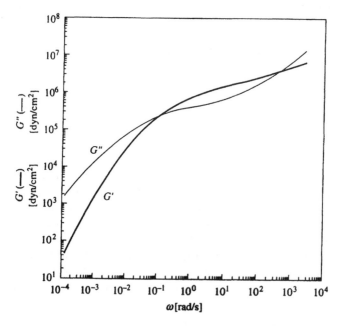

FIGURE 6.9   Master curve of storage and loss moduli of polystyrene. (From Ref. 15, with permission of The McGraw-Hill Companies.)

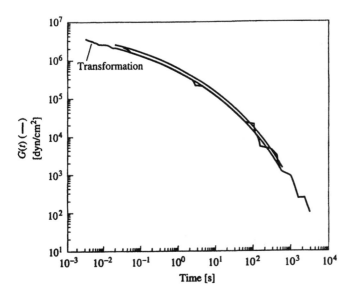

FIGURE 6.10 Stress relaxation modulus of polystyrene at 150°C. (From Ref. 15, with permission of The McGraw-Hill Companies.)

the contribution to the modulus from relaxation times that range in value from θ to θ + dθ. The stress relaxation modulus now takes the form [18]

$$G(t) = \int_{-\infty}^{\infty} H(\theta) \left[\exp\left(\frac{-t}{\theta}\right) \right] d(\ln \theta) \tag{6.31}$$

and dynamic mechanical data can again be used to compute $H(\theta)$; details are available in the literature [1,18–20].

We re-emphasize that once the stress relaxation modulus has been determined, whether in the form of Eq. (6.28) or Eq. (6.31), the Boltzmann superposition principle [Eq. (6.8)], can be employed to calculate the stress resulting from any specified deformation, provided that the total deformation is small or the rate of deformation is slow in order that linear behavior be exhibited.

V. DYNAMIC RHEOLOGICAL PROPERTIES OF SOLUTIONS

Typical dynamic rheological behavior of polymer solutions at several concentrations is shown in Figures 6.11a and 6.11b [21,21a]. Both the storage modulus and the loss modulus increase with concentration of polymer. At these low concentrations, few if any molecular entanglements are to be expected, but the solu-

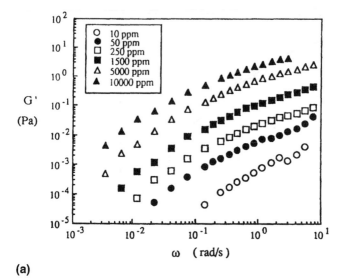

(a)

(b)

FIGURE 6.11 Dynamic data for solutions of Separan AP30 polyacrylamide in distilled water in a concentration range of 10–10,000 ppm: (a) storage modulus versus frequency, (b) loss modulus versus frequency. (From Ref. 21a. Copyright CRC Press, Boca Raton, Florida.)

tions still show elasticity. This elasticity, no doubt, results from distortion of the coiled molecules by the shearing field, with resultant orientation of some of the molecular segments. When the flow stops, the deformed molecules relax back to their normal, randomly coiled state. In concentrated solutions, entanglements can exist that greatly enhance the elasticity. Consequently, for high-molecular-weight polymers the storage modulus, at a fixed frequency, is a highly nonlinear function of concentration. The changes in the moduli, though, are much less at high frequencies than at low frequencies when the concentration is varied. Also, it is found that data taken at different constant temperatures and a fixed concentration obey the principle of time–temperature superposition [21,22].

It is common practice to represent the behavior of unentangled polymer solutions with the help of bead-spring theories (see Refs. 1, 15, 23, and 24 and sec. III of Chap. 4). Here each polymer molecule is modeled as a chain of $N + 1$ beads connected by N elastic springs. By making simplifying assumptions, such as the absence of any polymer–solvent interactions, we can carry out a force balance and determine the tension in each spring resulting from flow-induced extension and orientation of the molecule. The stress contributed by the polymer is then obtained by adding together the contributions from all the different molecules. In the Rouse theory, it is assumed that the molecule is free draining, i.e., that flow is not affected by the presence of the polymer. The resulting expressions for the storage and loss moduli are [23]:

$$G'(\omega) = \frac{cRT}{M} \sum_{i=1}^{N} \frac{\omega^2 \theta_i^2}{1 + \omega^2 \theta_i^2} \tag{6.32}$$

$$G''(\omega) = \omega \eta_s + \frac{cRT}{M} \sum_{i=1}^{N} \frac{\omega \theta_i}{1 + \omega^2 \theta_i^2} \tag{6.33}$$

in which c is the mass concentration of polymer, R is the universal gas constant, T is the absolute temperature, M is polymer molecular weight, and the relaxation times θ_i are given by

$$\theta_i = \frac{6(\eta_0 - \eta_s)M}{\pi^2 i^2 cRT} \tag{6.34}$$

wherein η_0 is the solution zero-shear viscosity and η_s is the solvent viscosity.

In the Zimm theory, hydrodynamic interactions are taken into account, but this does not change the functional form of G' and G''; only the magnitude of the various relaxation times changes [23]. In terms of data representation, therefore, Eqs. (6.32) and (6.33) suggest that we plot results in dimensionless form as reduced moduli versus reduced frequency, defined by [1]:

$$G'_R = \left(\frac{M}{cRT}\right) G' \tag{6.35}$$

$$(G'' - \omega\eta_s)_R = \left(\frac{M}{cRT}\right)(G'' - \omega\eta_s) \tag{6.36}$$

$$\omega_R = \left[\frac{(\eta_0 - \eta_s)M}{cRT}\right]\omega \tag{6.37}$$

If the theory is correct, this form of plotting the storage and loss moduli ought to lead to concentration-and-temperature-independent master curves. This is, indeed, found to be the case for aqueous polyacrylamide solutions, and results are displayed in Figure 6.12 [22]. As expected, the reduced storage modulus has a slope of 2 at low frequencies and the reduced loss modulus has a slope of unity. At high frequencies, the Rouse theory predicts a merger of the two curves, each with a slope of 1/2, whereas the Zimm theory says that the two curves become parallel, with a slope of 2/3. It appears that the data of Tam and Tiu [21,22] follow the trend predicted by the Zimm model.

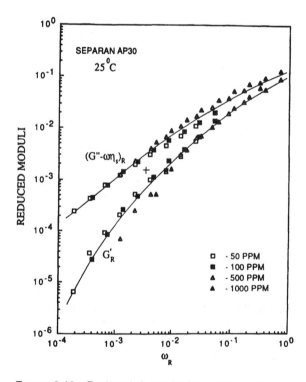

FIGURE 6.12 Reduced dynamic data of four different concentrations of Separan AP30 shown in Fig. 6.11. (From Ref. 22.)

In closing this section, we mention that Osaki [25] has examined the viscoelastic properties of extremely dilute polymer solutions and used the collected
data to carry out a detailed investigation of the appropriateness of different molecular theoretical explanations.

VI. RELEVANCE TO NONLINEAR VISCOELASTICITY

As mentioned earlier, the Boltzmann superposition principle is the correct constitutive equation for viscoelastic liquids provided that the total strain is small. Indeed, as Bird et al. demonstrate [26, p. 283], serious errors can arise if this theory
is used to predict stresses for large deformations. The problem is related to the
fact that the infinitesimal strain tensor appearing in Eq. (6.8) is not valid for large
deformations, and a finite measure of strain is required. Now, however, there is
a great choice in admissible strain measures that all reduce to the same quantity
for infinitesimal deformations [18,26]: picking a specific strain measure is usually

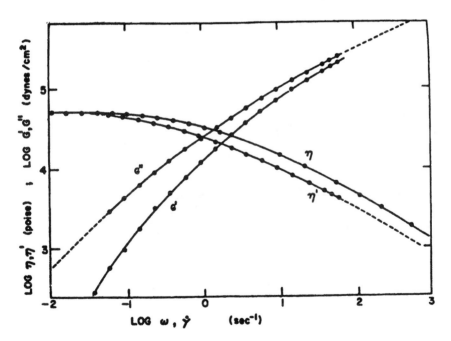

FIGURE 6.13 Steady-shear viscosity, dynamic viscosity, storage modulus, and
loss modulus as a function of either angular frequency or shear rate. The absolute
value of the complex viscosity is nearly identical to the shear viscosity. (From Ref.
29.)

justified by comparison with experimental data; and the Finger measure of strain has proven particularly popular [18]. With the use of this strain measure, nonlinear behavior can be described, but predicted stress values are usually higher than those measured experimentally. The problem appears to be related to the fact that the stress relaxation modulus becomes a function of strain for large strains. For many polymers, however, the nonlinear relaxation modulus $G(t,\gamma)$ can be written as a product of the linear relaxation modulus $G(t)$ and a function of strain, $h(\gamma)$, which is called the *damping function* and whose value is less than unity [27]. Different functional forms of the damping function result in different stress predictions, and the "correct" form is still to be decided. As a consequence, the development and testing of constitutive theories of nonlinear viscoelasticity is still an active area of research.

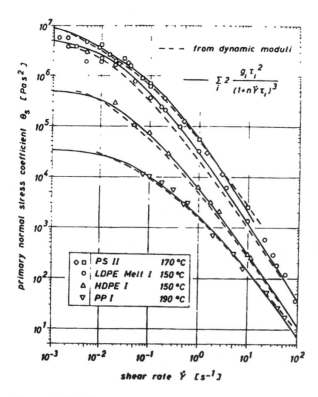

FIGURE 6.14   Primary normal stress coefficient versus shear rate. Symbols represent experimental data. Broken lines are the predictions of Eq. (6.39) (From Ref. 30.)

While the theoretical relationship between linear and nonlinear viscoelasticity is a matter of debate, for many polymers certain empirical relationships have been found to be true between material functions characteristic of small strains and those characteristic of large strains. In particular, the modulus of the complex viscosity η^*, defined as

$$|\eta^*| = [(\eta')^2 + (\eta'')^2]^{1/2} \tag{6.38}$$

where the dynamic viscosity $\eta' = G''/\omega$ and $\eta'' = G'/\omega$, when plotted versus frequency, often superposes with the steady-shear viscosity as a function of the shear rate [28]. This is known as the Cox–Merz rule, and it provides information about a nonlinear property from a measurement of a linear property. Figure 6.13 shows typical steady-shear and dynamic properties for a low-density polyethylene melt [29]; while the steady-shear viscosity equals the dynamic viscosity only at low deformation rates, the magnitude of the complex viscosity is essentially the same as the shear viscosity over the complete frequency or shear rate range.

Another useful empiricism, Laun's rule, relates the first normal stress coefficient to dynamic data [30]:

$$\psi_1 = \frac{2G'}{\omega^2}\left[1 + \left(\frac{G'}{G''}\right)^2 \right]^{0.7}_{\omega \doteq \gamma} \tag{6.39}$$

Figure 6.14 shows that for melts of polyethylene, polypropylene, and polystyrene, Eq. (6.39) does a good job of interrelating linear and nonlinear viscoelastic data.

VII. CONCLUDING REMARKS

Dynamic mechanical properties of polymer melts and polymer solutions are measured on a routine basis for polymer characterization purposes. Often these measurements are directly related to the shear viscosity or the first normal stress coefficient in shear. They also allow for the determination of the relaxation time spectrum, which, together with the Boltzmann superposition principle, gives information on any small-strain property of interest. As we shall see later, dynamic testing is also extremely useful for probing the structure and examining the behavior of crystalline polymers, crosslinking systems, solid-in-liquid suspensions, emulsions, polymer blends, and polymer composites. A particular advantage of small-amplitude deformations is that material structure is not altered during the course of measurement; this is especially important for multiphase systems. Finally, we note that complex waveforms are sometimes used instead of sinusoidal oscillations, and large-amplitude testing is sometimes done to investigate material response in the nonlinear region [31].

REFERENCES

1. J. D. Ferry. Viscoelastic Properties of Polymers. 3rd ed. Wiley, New York, 1980.
2. T. Murayama. Dynamic Mechanical Analysis of Polymeric Material. Elsevier, Amsterdam, 1978.
3. J. M. Dealy. Rheometers for Molten Plastics. Van Nostrand Reinhold, New York, 1982.
4. G. Marin. Oscillatory Rheometry. In: A. A. Collyer and D. W. Clegg, eds. Rheological Measurement. Elsevier, London, 1988, pp. 297–343.
5. R. W. Whorlow. Rheological Techniques. 2nd ed., Ellis Horwood, New York, 1992.
6. B. Maxwell and R. P. Chartoff. A polymer melt in an orthogonal rheometer. Trans. Soc. Rheol. 9:41–52 (1965).
7. R. K. Krishnaswamy and D. S. Kalika. Dynamic mechanical relaxation properties of poly(ether ether ketone). Polymer 35:1157–1165 (1994).
8. J. E. Mark, A. Eisenberg, W. W. Graessley, L. Mandelkern, and J. L. Koenig, Physical Properties of Polymers. American Chemical Society, Washington, DC, 1984, pp. 97–153.
9. S. Onogi, T. Masuda, and K. Kitagawa. Rheological properties of anionic polystyrenes. I. Dynamic viscoelasticity of narrow-distribution polystyrenes. Macromolecules 3:109–116 (1970).
10. L. R. G. Treloar. The Physics of Rubber Elasticity. 3rd ed., Clarendon, Oxford, UK, 1975.
11. T. Masuda, K. Kitagawa, T. Inoue, and S. Onogi. Rheological properties of anionic polystyrenes. II. Dynamic viscoelasticity of blends of narrow-distribution polystyrenes. Macromolecules 3:116–125 (1970).
12. T. Masuda, Y. Ohta, and S. Onogi. Rheological properties of anionic polystyrenes. III. Characterization and rheological properties of four-branch polystyrenes. Macromolecules 4:763–768 (1971).
13. R. A. Mendelson. Flow properties of polyethylene melts. Polym. Eng. Sci. 9:350–355 (1969).
14. D. W. Mead. Determination of molecular weight distributions of linear flexible polymers from linear viscoelastic material functions. J. Rheol. 38:1797–1827 (1994).
15. A. Kumar and R. K. Gupta. Fundamentals of Polymers, McGraw-Hill, New York, 1998.
15a. S. Wu. Dynamic rheology and molecular weight distribution of insoluble polymers: tetrafluoroethylene-hexafluoropropylene copolymers. Macromolecules 18:2023–2030 (1985).
16. G. Astarita and G. Marrucci. Principles of Non-Newtonian Fluid Mechanics. McGraw-Hill, London, 1974.
17. M. Baumgaertel and H. H. Winter. Determination of discrete relaxation and retardation time spectra from dynamic mechanical data. Rheol. Acta 28:511–519 (1989).
18. J. M. Dealy and K. F. Wissbrun. Melt Rheology and Its Role in Plastics Processing. Van Nostrand Reinhold, New York, 1990.
19. R. I. Tanner. Note on the iterative calculation of relaxation spectra. J. Appl. Polym. Sci. 12:1649–1652 (1968).

20. J. Honerkamp and J. Weese. A nonlinear regularization method for the calculation of relaxation spectra. Rheol. Acta 32:65–73 (1993).
21. K. C. Tam. Rheology and turbulent flow behavior of polymer solutions. PhD dissertation, Monash University, Australia, 1990.
21a. K. C. Tam, C. Tin. Water soluble polymers (rheological properties). In: J. C. Salamone, ed. Polymeric Materials Encyclopedia. CRC Press, Boca Raton, FL, Vol. 11, 1996, pp. 8655–8677.
22. K. C. Tam and C. Tiu. Steady and dynamic shear properties of aqueous polymer solutions. J. Rheol. 33:257–280 (1989).
23. W. W. Graessley. The entanglement concept in polymer rheology. Adv. Polym. Sci. 16:1–179 (1974).
24. R. B. Bird, C. F. Curtiss, R. C. Armstrong, and O. Hassager. Dynamics of Polymeric Liquids, 2nd ed. Vol. 2. Wiley, New York, 1987.
25. K. Osaki. Viscoelastic properties of dilute polymer solutions. Adv. Polym. Sci. 12: 1–64 (1973).
26. R. B. Bird, R. C. Armstrong, and O. Hassager. Dynamics of Polymeric Liquids. 2nd ed. Vol. 1. Wiley, New York, 1987.
27. M. H. Wagner. Analysis of time-dependent non-linear stress-growth data for shear and elongational flow of a low-density branched polyethylene melt. Rheol. Acta 15: 136–142 (1976).
28. W. P. Cox and E.H. Merz. Correlation of dynamic and steady flow viscosities. J. Polym. Sci. 28:619–622 (1958).
29. R. N. Shroff. Dynamic mechanical properties of polyethylene melts: calculation of relaxation spectrum from loss modulus. Trans. Soc. Rheol. 15:163–175 (1971).
30. H. M. Laun. Prediction of elastic strains of polymer melts in shear and elongation. J. Rheol. 30:459–501 (1986).
31. A. A. Collyer, ed. Techniques in Rheological Measurement. Chapman and Hall, London, 1993.

7

Extensional Viscosity

I. INTRODUCTION

As pointed out in Chap. 1, polymeric liquids can undergo tensile deformation in addition to the more familiar shear deformation. The importance of extensional flow has been recognized for some time now, and in the last 30 years a significant portion of the literature on rheology has been devoted to this topic alone [1–3 and references therein]. This is due to both practical and theoretical reasons: (1) Industrial polymer fabrication processes such as fiber spinning, film blowing, blow molding of bottles, and flat film extrusion involve a predominantly tensile deformation, and this mode of deformation is also encountered in the peel testing of pressure-sensitive adhesives [4]. (2) Due to the fact that polymer molecules uncoil and stretch during extensional flow, the resistance to deformation in extension is significantly greater than the corresponding resistance to shear flow. Consequently, the steady-state uniaxial extensional viscosity, especially at large values of the stretch rate, is hundreds of times greater than the shear viscosity, and, in general, one quantity cannot be obtained from a measurement of the other. Note that the ratio of the extensional viscosity to the zero-shear viscosity is generally called *Trouton's ratio*; for Newtonian liquids, Trouton's ratio has a value of 3.

If the stretch rate in a uniaxial extension experiment (that begins from rest at time zero) is small, we can use the theory of linear viscoelasticity, Eq. (6.8), to compute the extensional viscosity. Employing the Boltzmann superposition principle in combination with Eqs. (1.14), (1.17), and (1.18) we find that the tensile stress growth coefficient is given by

$$\eta_{\bar{E}}^{+}(t) = 3 \int_0^t G(t') \, dt' = 3\eta^{+}(t) \tag{7.1}$$

in which η^* is the shear stress growth coefficient defined as the ratio of the instantaneous shear stress to the shear rate during startup of shearing at a constant shear rate. At steady state, of course, the preceding result leads to Eq. (1.15). In other words, the steady-state behavior of polymeric fluids is Newtonian provided we are in the linear viscoelastic region. The predicted relationship between η_E^* and η^* is, indeed, observed, and it can be seen to hold at the lowest stretch rate of 0.001 sec^{-1} in Fig. 7.1, which shows tensile stress growth data on a low-density polyethylene sample [5]; at higher deformation rates, however, the data deviate from linear behavior, and the deviation occurs at smaller strain values with increasing stretch rates. The occurrence of stresses that are higher than those predicted by the theory of linear viscoelasticity is known as *strain hardening* or *extension thickening*. Note from Fig. 7.1 that a steady state in the stress is not attained at the highest stretch rate values.

Observed steady-state extensional viscosity behavior is compared with shear viscosity in Fig. 7.2. Three general types of tensile behavior have been reported in the past as a function of tensile stress (or rate of elongation). Tensile viscosity may be nearly independent of tensile stress for some polymers [6,7]. In other cases, extensional viscosity starts to increase with tensile stress at a value of stress about the same as where the shear viscosity starts to decrease [7]. In

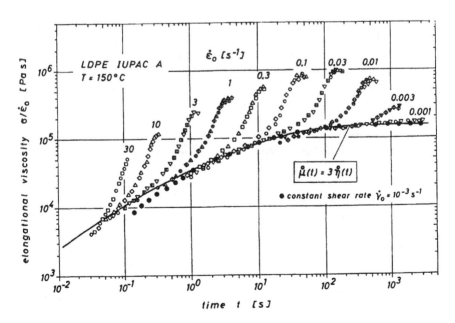

FIGURE 7.1 Tensile stress growth data for LDPE. (From Ref. 5. Used by permission of Steinkopff Publishers, Darmstadt, FRG.)

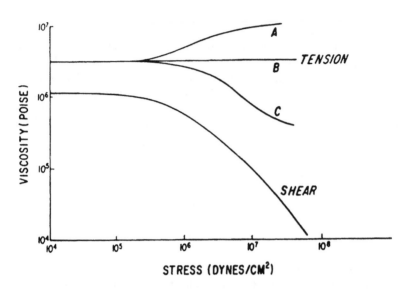

FIGURE 7.2   Three types of tensile viscosity behaviors compared to shear viscosity as a function of tensile or shear stress for polymeric fluids.

still other cases, particularly those involving suspensions, the elongational viscosity starts to decrease at about the same value of stress as where the shear viscosity decreases from the Newtonian value [7.8]. Note though that stress levels or stretch rates in industrial polymer processing operations are usually significantly higher than those indicated in Fig. 7.2; however, these flows are typically of short duration, and steady states in the stress are rarely, if ever, attained. Furthermore, tensile viscosity increases with decreasing temperature, and this increase often overshadows the change in tensile stress due to changes in the rate of deformation. At present, there is no single theory capable of quantitatively predicting this wide diversity of behavior found for extensional viscosity.

In some polymer processing operations, such as fiber spinning, it is desirable that for process stability the extensional viscosity increase with the rate of elongation. The reason for this is as follows: If a weak spot develops in the fiber, which results in a decrease in cross-sectional area, the rate of elongation at that spot increases. The increase in rate of elongation causes an increase in the tensile viscosity, which resists the further stretching of the thin section. In tubular film blowing, it is found that the response of a polymer melt to an extensional deformation, and this depends on the structure of the polymer, determines how easily the film thins and how stable it is against rupturing [9]. Thus, extensional viscosity measurements can be useful in selecting polymers that are easier to process.

Extensional flow measurements can also be made on polymer solutions, and these data are relevant for a number of industrial applications. It is found, for example, that, unlike Newtonian liquids, drops of viscoelastic liquids suspended in an air stream do not always disintegrate easily. Instead, depending on the tensile stresses generated, they form strands and big blobs, or they form long filaments that gradually break up into a very fine spray [10]. Matta and Tytus [11] showed that the addition of a small amount of polymer to a liquid could change the breakup mechanism from shear to extension, resulting in an order-of-magnitude increase in the average drop size; this phenomenon is clearly important in fuel atomization and the aerial spraying of insecticides. Similarly, in roll coating of paints, high-molecular-weight thickeners modify the surface-tension-induced breakup of roll fibers, and this can lead to increased spatter [12–14].

II. LOW-STRETCH-RATE BEHAVIOR OF EXTENSIONAL VISCOSITY

In order to understand and interpret extensional flow measurements properly, it is necessary to ensure that the results depend neither on the technique of stretching nor on any assumptions regarding the constitutive behavior of the material. This requires that data be obtained either under constant-stress or under constant-strain-rate conditions. This task has been accomplished only for relatively viscous polymer melts and a few polymer solutions, and even then for stretch rate values that are significantly lower than for typical industrial practice. The difficulty arises from the mobile nature of liquids and the impossibility of gripping them and making them stretch in a prescribed manner. Nonetheless, enough reliable data are now available on several polymers [see Ref. 2 for an extensive listing] that some general comments can be made regarding the nature of the extensional viscosity. Note that data on unfilled polymer melts and polymer solutions are qualitatively similar, although Trouton's ratio of polymer solutions can exceed Trouton's ratio of polymer melts by about two orders of magnitude [3].

A. Measurement Techniques for Polymer Melts

A conceptually simple way of making extensional viscosity measurements is to clamp or glue one end of a cylindrical sample of molten polymer to a force transducer and to move the other end outwards according to the regimen of Eq. (1.12) [15,16]. The polymer may be stretched horizontally or vertically, but it is generally necessary to immerse the sample in an inert liquid to prevent gravity from contributing to the deformation. An alternative to controlling the stretch rate is to impose a constant stress and to monitor the resulting strain as a function of time [15,17–19]; a constant extensional viscosity is achieved if the strain begins to increase linearly with time. Shown in Fig. 7.3 is the "creepmeter" developed by Munstedt [19]; for some time, this instrument was commercialized by the Gottfert and Rheometrics companies. A disadvantage of this design is that the duration of the experiment is limited to the time taken by the moving end of the sample to traverse the length of the liquid bath, which typically corresponds to about 3–4 strain units; the upper limit on the stretch rate is a few reciprocal seconds. Note that the same results are obtained regardless of whether a constant strain rate is imposed or whether the stress is held fixed, provided that care is taken to minimize temperature and stretch rate nonuniformities in the sample during the process of deformation. However, the total strain needed to achieve a steady state is found to be smaller if the instrument is operated in the constant-stress mode.

The maximum-strain limitation, imposed by the size of the apparatus, can be removed if one stretches a sample at constant length. This can be done in one of two ways: (1) the two ends are pulled outwards at the same constant speed, or (2) one end is kept stationary while the velocity of the other end is maintained constant. While details of the different designs may be found in the literature [2,3], the rotary clamp technique of Meisssner is particularly noteworthy [20].

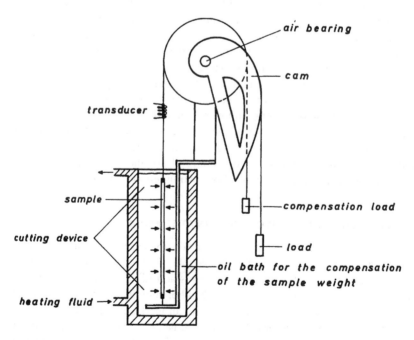

FIGURE 7.3. Schematic drawing of the tensile creep apparatus. (From Ref. 19. Used by permission of Steinkopff Publishers, Darmstadt, FRG.)

Meissner used two pairs of gears to grip and stretch a cylindrical specimen; Hencky strains as large as 7, equivalent to a thousandfold increase in specimen length, could be achieved. More recently, Meissner and Hostettler [21] pioneered the development of new clamps that make use of metal conveyor belts for stretching samples that are supported by an inert gas that is heated to the test temperature; the stretching force is monitored with the use of leaf-spring transducers. This design has now been commercialized by the Rheometrics Scientific Company, and a sketch of their elongational rheometer for melts is shown in Fig. 7.4. This instrument can stretch samples at rates up to 1 sec⁻¹, and has a maximum attainable strain of 7, operation up to 350°C, and a force measurement range of between 0.001 and 2 N.

B. Experimental Results and Their Constitutive Modeling

The first systematic attempt at obtaining consistent rheological data on chemically similar polymer melts was made by the IUPAC working party on the structure and properties of commercial polymers, which in 1975 reported work done on three LDPE samples [22]. One of these well-characterized samples, IUPAC A, was subjected to extensive elongational testing by Laun and Munstedt; Fig. 7.1

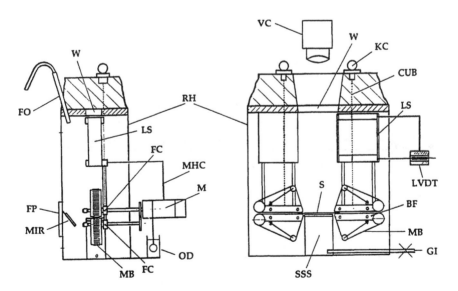

FIGURE 7.4 Rheometrics Scientific elongational rheometer for melts. RH, rheometer housing; MB, metal conveyor belt; BF, belt fixture; S, sample; LS, leaf spring; M, motor; GI, gas inlet; SSS, sample support system; VC, video camera. (From Ref. 21. Used by permission of Steinkopff Publishers, Darmstadt, FRG.)

shows some of their results. The elongational viscosity, calculated using the steady-state stress values in Fig. 7.1, is displayed in Fig. 7.5 as a function of the stretch rate [23]. The results span six decades of stretch rate and show that Trouton's ratio equals 3 at low stretch rates. As the stretch rate is increased, the extensional viscosity increases, goes through a maximum, and finally decreases to a value even below that of the zero-shear value. (Note that there has been some concern in the literature about whether or not a true steady state is actually achieved in experiments such as these [24], and this issue has still not been resolved in a completely satisfactory manner.)

These investigators also determined the temperature dependence of the extensional viscosity of IUPAC A, and their measurements are plotted in Fig. 7.6 as a function of tensile stress [5]. It is seen that the shape of the extensional viscosity curve is not altered by changing the temperature. In fact, all the data can be made to overlap by means of a vertical shift, demonstrating the validity of the time–temperature superposition principle; the temperature shift factors are identical to those obtained from shear data.

Variables other than temperature and stretch rate also influence the extensional flow behavior of polymer melts. These are material structural variables such as polymer molecular weight, molecular weight distribution, and chain branching. How these variables influence extensional viscosity appears to depend

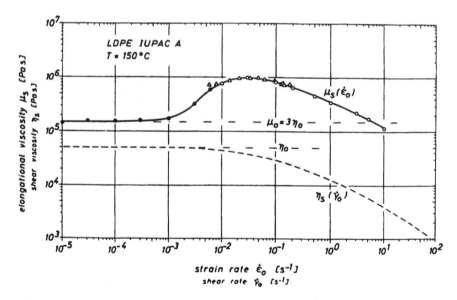

FIGURE 7.5 Shear and elongational viscosities of LDPE as a function of deformation rate calculated using the data of Fig. 7.1. (From Ref. 23. Used by permission of Steinkopff Publishers, Darmstadt, FRG.)

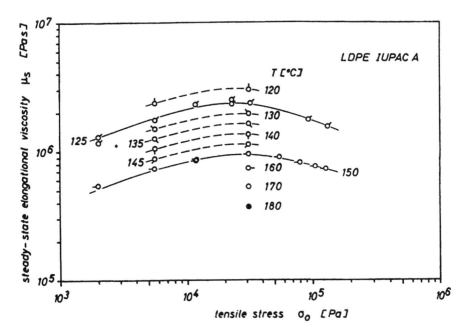

FIGURE 7.6.   Stress dependence of steady-state elongational viscosity at different
temperatures. (From Ref. 5. Used by permission of Steinkopff Publishers, Darmstadt, FRG.)

on whether one is in the linear region or in the nonlinear region. Data on several
polystyrenes [25,26] suggest that, at low deformation rates, the value of the extensional viscosity is a function of the molecular weight alone and is proportional
to the 3.4–3.6 power of the weight-average molecular weight. Representative
data are shown in Fig. 7.7 [25], and these are consistent with the observation
that the extensional viscosity of molten polymers is generally equal to three times
the shear viscosity at low stretch rates; recall from Chap. 3 that the zero-shear
viscosity of entangled polymer melts behaves in a similar manner with changes
in the weight-average molecular weight. At higher deformation rates, however,
the extensional viscosity (unlike the shear viscosity) increases, and it can reach
values that are several times larger than the Trouton value with a broadening of
the molecular weight distribution. This effect is especially noticeable in the presence of a high-molecular-weight tail, even when the highest-molecular-weight
components cannot be detected by the use of a gel permeation chromatograph.

The influence of chain branching on extensional viscosity has been explored in the literature, mainly by comparing and contrasting results obtained
using different members of the polyethylene family. Whereas HDPE is an essen-

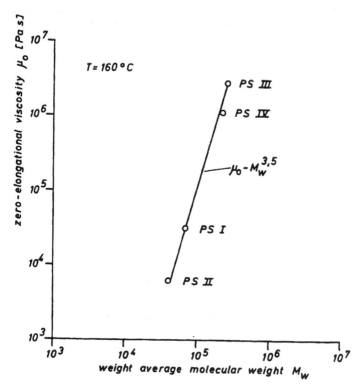

FIGURE 7.7   Molecular weight dependence of elongational viscosity. (From Ref. 25.)

tially linear molecule, LDPE and LLDPE are branched, but the branches of LLDPE are relatively short. In qualitative terms, it is found that it is easier to stretch branched polyethylenes in a uniform, homogeneous manner as compared to linear polyethylene [27], although polymers with extremely long branches may also be difficult to stretch [28]. In quantitative terms, the presence of long-chain branching shows up as strain hardening during uniaxial elongation at a constant stretch rate; the longer the branches, the greater is the extent of strain hardening. This is shown in Fig. 7.8, where the behavior of LLDPE is compared with that of two LDPE's having different extents of long-chain branching [9]. The weight-average molecular weight of LDPE2 exceeds that of both LDPE1 and LLDPE by at least a factor of 4, and it also has much longer branches as compared to LDPE1. Yet LDPE2 has the lowest values of tensile stress in the linear viscoelastic region, and this is consistent with the observation made in Sec. II of Chap. 6 that the introduction of long-chain branches in a polymer decreases the elasticity

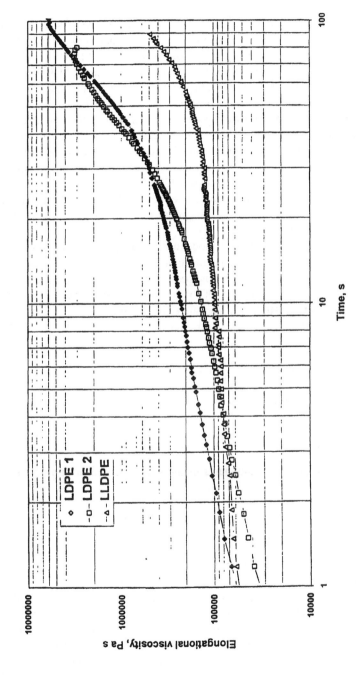

FIGURE 7.8. Tensile stress growth data for three different polyethylenes at 140°C. The stretch rate is 0.052 sec^{-1}. (From Ref. 9.)

as measured via the storage modulus. With increasing time of stretching, however, there is immense strain hardening in LDPE2 but very little in LLDPE. Indeed, it is in order to increase the extensional viscosity of LLDPE and to make the process of film blowing more stable that a small amount of LDPE is often blended with LLDPE [9]. At the other extreme, due to the absence of chain branching, the curve of extensional viscosity versus stretch rate for HDPE shows only a slight maximum; on lowering the molecular weight, the maximum disappears, and the steady-state behavior is Newtonian [29,30]. For all the different-molecular-weight HDPE polymers, though, further increases in the stretch rate result in decreases in the extensional viscosity, which can become significantly lower than thrice the zero-shear viscosity [29]. We caution, though, that exceptions to these generalizations can sometimes be found [31], but these may be due to phase separation resulting from the presence of two different kinds of molecules in the polymer.

In terms of the constitutive modeling of data of the kind shown in Figs. 7.1 and 7.5, the simplest approach is to employ a version of the Boltzmann superposition principle, but one that has been generalized to large deformations through the use of a finite measure of strain. As mentioned in Sec. VI of Chap. 6, the Finger measure of strain C_{ij}^{-1} appears to be the most appropriate one for this purpose. The resulting equation is known as the Lodge rubberlike liquid [32,33]:

$$\tau_{ij} = \frac{g}{\theta} \int_{-\infty}^{t} \exp\left[\frac{-(t-s)}{\theta}\right] C_{ij}^{-1} \, ds \qquad (7.2)$$

and its differential counterpart is the upper-convected Maxwell equation [33,34]:

$$\tau_{ij} + \theta \frac{\delta \tau_{ij}}{\delta \tau} = 2 \, g\theta D_{ij} \qquad (7.3)$$

in which the D_{ij} are the components of the strain rate tensor and $\delta\tau_{ij}/\delta t$ is the upper-convected derivative given by

$$\frac{\delta \tau_{ij}}{\delta \tau} = \frac{\partial \tau_{ij}}{\delta t} + \sum_{m=1}^{3} \left[v_m \frac{\partial \tau_{ij}}{\partial x_m} - \tau_{im} \frac{\partial v_j}{\partial x_m} - \tau_{mj} \frac{\partial v_i}{\partial x_m} \right] \qquad (7.4)$$

and the two constants g and θ are generally picked to give agreement with linear viscoelastic data; often a discrete relaxation spectrum is used.

Although the upper-convected Maxwell equation displays all the qualitative features of nonlinear viscoelastic behavior and it has been used extensively to model polymer processing operations [35], it has some rather severe shortcomings. In particular, this model predicts that the viscosity in shear is constant and the viscosity in extension becomes unbounded at a finite value of the stretch rate. In general, the use of this model results in stresses that are larger than those measured experimentally. Thus, at the very least, an additional parameter is needed to moderate the increase in stress with increasing deformation rate. For integral equations, this additional parameter could be part of the damping function introduced in Sec. VI of Chap. 6, whereas, for differential models it could be part of the time derivative. Alternatively, the two constants in Eqs. (7.2) and (7.3) could be made variables and could be made to depend on the invariants of the strain rate tensor. All these possibilities have been proposed, and specific choices have been justified based on physical or mathematical grounds; the resulting modified equations may be found in standard books [34–36]. An especially popular equation is the one due to Phan-Thien and Tanner [37]; when applied to extensional flow data, this equation is often written in the following form [38]:

$$\left[\frac{\alpha}{g} \, \text{tr} \, (\tau)\right] \tau_{ij} + \theta \, \frac{\delta \tau_{ij}}{\delta t} = 2 \, g \theta D_{ij} \qquad (7.5)$$

in which α is a new parameter and tr τ represents the trace or sum of the diagonal terms of the extra stress tensor.

Khan and Larson [34,38] examined the ability of several constitutive equations to portray polyethylene melt data in step shear, step biaxial extension, and constant-stretch-rate uniaxial extension. They found that Eq. (7.5) was one of the few simple equations that could fit data in all the three flow fields. However, for the IUPAC A polymer, a different value of the parameter α was required to fit data from each of the three types of deformations. In other words, parameter values that gave a good fit to the data shown in Fig. 7.5 did not fit the corresponding shear and biaxial extension data adequately. This kind of situation is encountered frequently in polymer rheology, and it is for this reason that the formulation and validation of rheological constitutive equations remains an area of active research.

C. Extensional Viscosity of Polymer Solutions

The constant-stretch-rate, uniform stretching of polymer solutions is a relatively recent development, whose impetus was provided by a series of international meetings in France between the years of 1988 and 1993. These conferences also led to a large body of published work on the shear and extensional viscoelastic properties of polymer solutions [39], including research on three extremely well-characterized polymer solutions: M1—a constant-viscosity elastic liquid composed of a 0.244% solution of polyisobutylene in a mixed solvent of 7% kerosene in polybutene [40]; A1—a highly elastic, but shear-thinning, liquid made up of 2% polyisobutylene in a mixture of *cis*- and *trans*-decalin [41]; and S1—a shear-thinning, viscoelastic solution of 2.5% polyisobutylene dissolved in a mixture of decalin and polybutene and having a relaxation time similar to that of M1 [42]. The participants in these workshop-type meetings used different techniques to make measurements on the same fluid, and not only did this make data comparison easy, it also made it possible rationally to discuss the merits and demerits of each proposed technique of extensional viscosity measurement.

The technique that ultimately led to the measurement of steady-state extensional viscosity was pioneered by Sridhar and coworkers [43,44]. These researchers inserted the fluid sample between two identical coaxial disks placed one below the other. The disk diameter ranged from 1 to 7 cm, and the sample thickness was about 2 mm. In the initial experiments [43], the fluid was stretched vertically by keeping the upper disk stationary and moving the lower disk downwards. Figure 7.9 shows a sequence of photographs of the stretching of a constant-viscosity, ideal elastic liquid at an average stretch rate of 0.52 sec^{-1}; except near the disk surfaces, the stretching is fairly uniform and the filament radius is independent of position over 95% of the filament length [43]. The setup was later refined to work in either the horizontal or the vertical configuration, and a schematic diagram of the apparatus is displayed in Fig. 10 [44]. In this version, both ends are moved outwards simultaneously, and the maximum final length of the filament is about 1 m; this produces large enough total strains for a steady state in the stress to

t(s)   1.2   2.9   4.8   7.6   9.6   11.5

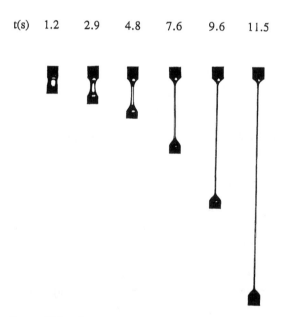

FIGURE 7.9.   Sequence of photographs showing the stretching of a filament of a 36 Pa-sec viscosity solution of 0.185% PIB in a solvent of kerosene and polybetene. The average stretch rate is 0.52 sec^{-1}, and the initial diameter and filament length are 3.03 mm and 1.8 mm, respectively. (From Ref. 43, with permission from Elsevier Science.)

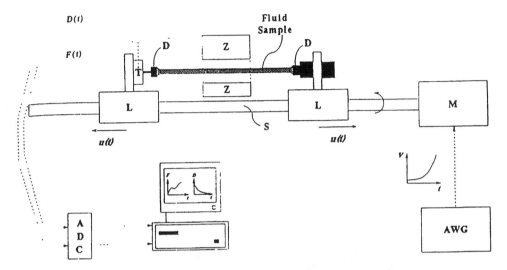

FIGURE 7.10 Schematic diagram of the filament-stretching rheometer: AWG = arbitrary waveform generator, M = servomotor, L = linear drive unit, S = shaft, D = disk, T = force transducer, Z = diameter measuring device, ADC = analog-to-digital converter. (From Ref. 44.)

be reached in many of the experiments. The force is measured using a transducer attached to one of the linear drive units; the filament diameter is determined with a high-resolution optical device. To compensate for end effects, the speed of the disks is continually adjusted to ensure that the filament diameter at the midpoint of the filament decreases exponentially with time, and additional refinements have been made to the apparatus since the time it was initially constructed. A version of this instrument, called the Filament Stretching Extensional Rheometer, is now available commercially from the Cambridge Polymer Group in Cambridge, MA. Data analysis is discussed by Szabo [44a].

Results for the growth of tensile stress during the constant-stretch-rate extension of a constant-viscosity, ideal elastic liquid at a variety of stretch rates are shown in Fig. 7.11 [44]. The similarity of these results to the LDPE melt data presented in Fig. 7.1 is unmistakable. In general, data for the different stretch-rate runs superpose at short times; during this time, the ratio of the stress growth coefficient to the shear viscosity attains a value of 3 and remains so for varying periods, depending on the extension rate imposed. This is then followed by a dramatic increase in the extensional stress as the fluid is stretched further; the deviation from linear viscoelastic behavior and commencement of strain hardening is noticeable at total strain values of about unity. The total strain in these

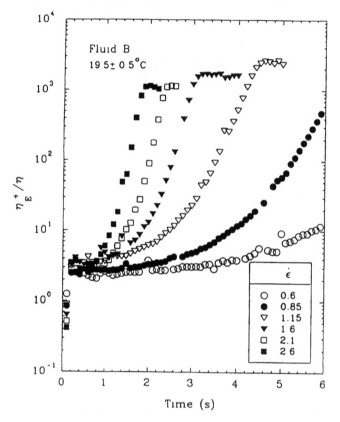

FIGURE 7.11 Tensile stress growth data for the constant-viscosity, elastic liquid of Fig. 7.9. (From Ref. 44.)

experiments is often as large as 7, a value comparable to that achieved in the melt stretching runs. Steady states are clearly achieved at high stretch rates, and the Trouton ratio approaches 10,000. Thus, the measured Trouton ratios for this polymer solution are easily about two orders of magnitude larger than those for the most extension-thickening polymer melts. Similar results are obtained on other constant-viscosity liquids, such as M1, and also on the shear-thinning solution A1 [44,45].

Regarding the use of rheological constitutive equations to describe the observed stress growth behavior of polymer solutions, we must remember to include the contribution to the stress made by the solvent. Although this contribution may not always be significant in extensional flow, it can have a major influence on shear behavior; if the solvent contribution is not included, errors will arise in the parameter values if these are obtained by fitting shear data [46]. As a consequence, the simplest constitutive equation for describing polymer solution viscoelasticity is a sum of Eq. (7.3) written for the polymer contribution and the Newtonian equation written for the solvent contribution; the resulting three-constant equation is commonly called the *Oldroyd model B* [46]. Tirtaatmadja and Sridhar [45] found that the Oldroyd B equation written using a four-element discrete relaxation time spectrum could describe the shear rheology of the constant-viscoity elastic liquids as well as the growth of tensile stress in uniaxial extension. However, owing to the inherent limitations of the Maxwell equation, this model cannot predict a steady state in the extensional stress. For the shear-thinning polymer solution A1, the use of the Oldroyd B model is clearly inappropriate, and here the White-Metzner equation was found to do a good job overall. This equation is the same as Eq. (7.3), but the two parameters are not constants; these are functions of the invariants of the rate-of-deformation tensor [47].

In closing this section we note that, at present, there are no published data on the influence of temperature, polymer concentration, molecular weight, molecular weight distribution, and chain branching on the extensional viscosity of polymer solutions. However, these will become available in the very near future [47a].

III. EXTENSIONAL VISCOSITY AT HIGH STRETCH RATES

The major limitation of the uniaxial extensiometers described so far is their inability to make measurements at stretch rates characteristic of polymer processing operations. Other techniques are therefore needed to stretch polymeric liquids at stretch rates of the order of 100 sec⁻¹. Invariably, these methods involve nonuniform stretching—neither the stress nor the stretch rate is constant in the Lagrangian sense. In addition, the fluid is rarely in a virgin, stress-free state to begin with. In this regard, fiber spinning and converging flow have proven to be especially popular. However, from stresses measured in these flow fields, we cannot obtain true extensional viscosity as defined by the equations given in Chap. 1. Nonetheless, data from properly designed nonuniform stretching experiments can, in principle, be analyzed with the help of rheological constitutive equations, i.e.; results can be employed to obtain the constants in a constitutive equation, which can, in turn, be used to predict the extensional viscosity. In addition, we can compute the ratio of instantaneous tensile stress to instantaneous stretch rate. This ratio is termed an *apparent extensional viscosity*, and it provides a simple measure of the resistance that polymeric fluids offer to extensional deformation.

A. Fiber Spinning

Polymer melt spinning is one of the major processes for the manufacture of synthetic fibers, and different aspects of this operation may be found in Chap. 15

of Ref. 33. A simplified fiber spinning setup is shown in Fig. 7.12, which displays a schematic of the commercially available Rheotens tensile tester for polymer melts, made by Goettfert in Germany [48]. A capillary viscometer or a laboratory extruder pumps molten polymer through a die of circular cross section, and the melt strand is pulled vertically downward at a constant velocity by two counterrotating, knurled wheels. For isothermal testing, the fiber may be enclosed in a constant-temperature oven, and the drawdown force is typically measured using a tensiometer. In the Rheotens instrument, the pulling speed can also be increased from rest at a constant rate until the filament breaks; the maximum velocity is 1.2 m/sec and the maximum acceleration is 12 cm/sec². Representative results on the stretching of a LDPE sample in room temperature air are shown in Fig. 7.12 [48]. The speed at fiber rupture is indicative of melt extensibility, and the fracture force is judged to be the melt strength.

For the purpose of data analysis in a constant-pulling-speed experiment, one assumes that the liquid is incompressible, the flow is steady and axisymmetric, and the velocity is uniform over the cross section. Under these conditions, the fluid velocity depends only on position down the spinline, and the velocity as well as the velocity gradient can be computed from a knowledge of the volumetric flow rate and the experimentally determined diameter profile. The visco-

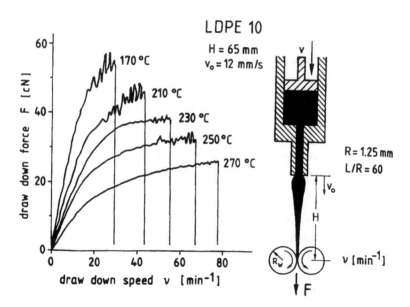

FIGURE 7.12 Representative tensile testing results obtained using the Rheotens instrument shown on the right of the figure. (From Ref. 48.)

elastic fluid stress at any location is then calculated from the measured pulling force through a momentum balance [1,3,33]. The ratio of the stress to the stretch rate at the same location is called the *apparent extensional viscosity*, and it is plotted either against the stretch rate or against position down the spinline. An example of such a graph is given in Fig. 7.13 for the isothermal spinning of a low-density polyethylene at three different temperatures [49]; the apparent elongational viscosity is plotted versus the stretch rate, which, in this case, increases with position in the flow direction. It is evident that (1) the stretch rates achieved are not higher than those in the filament stretching runs of Laun and Munstedt [5,23], and (2) although the Trouton ratios are similar to those shown in Figs. 7.5 and 7.6, the shape of the extensional viscosity curves is quite different. Concerning these observations, one generally has to go to nonisothermal spinning in order to attain high stretch rates, and the kind of data shown in Fig. 7.13 cannot be expected to agree quantitatively with rod-pulling results, because steady states are rarely, if ever, achieved in fiber spinning. It is only for Newtonian liquids that the apparent extensional viscosity equals the true extensional viscosity. Still, fiber spinning data are useful, and these may be utilized to determine the constants

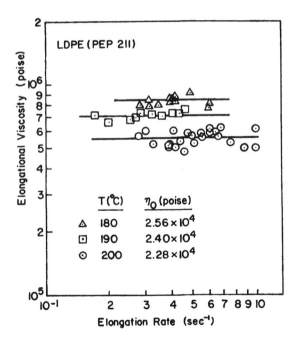

FIGURE 7.13   Apparent elongational viscosity versus elongation rate for the fiber spinning of an LDPE sample at three different temperatures. (From Ref. 49.)

in any candidate constitutive equation [50]; the equation can then be used for process analysis and simulation.

On going to nonisothermal fiber spinning, we find that the measured force is essentially the viscoelastic force in the fiber. Recognizing this fact, Laun and Schuch [48] simply divided the measured force by the fiber cross-sectional area at the end of the spinline to obtain the extensional stress there. They then computed the apparent (nonisothermal) extensional viscosity at that location and compared it with the true extensional viscosity measured at the temperature at which the polymer leaves the barrel of the capillary viscometer. Results on three IUPAC LDPE samples are shown in Fig. 7.14 [48]. These three materials are from the same batch, and their shear behavior is essentially identical. Differences in sample preparation, however, result in differences in extensional behavior. Not surprisingly, the nonisothermal Rheotens data do not superpose with the isothermal, homogeneous filament stretching data, but they appear to reveal the right trends. Also the order of magnitude of the apparent extensional viscosity is correct. This

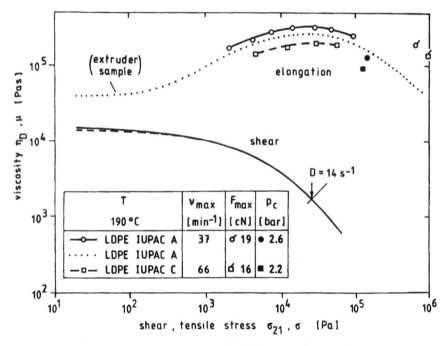

FIGURE 7.14 Comparison of different methods of extensional viscosity measurement. Unfilled symbols with a tick mark are Rheotens data; filled symbols are the results of converging-flow experiments. (From Ref. 48.)

simple test, therefore, seems to have practical utility. The constitutive modeling of Rheotens data may be found in Ref. 51.

In the fiber spinning of polymer solutions, as in the case of polymer melts, a fluid jet emerging from a nozzle is stretched and the resulting tensile force is measured. A variety of devices have been proposed over the years [3]; Fig. 7.15 shows the spinline viscometer of Ferguson and Hudson [52], which has been the basis of a commercial rheometer manufactured by Carri-Med (now TA Instruments) in the United Kingdom. In this instrument, fluid is stretched when it comes into contact with a rotating drum, and a variation in stretch rates is achieved by varying the speed of rotation of the drum. With increasing speed of rotation, though, there is slip between the drum and the liquid stream, and this places an upper limit on the achievable stretch rate. The force (up to 1 g) generated by stretching of the polymer solution is measured by the deflection of a thin-walled tube through which the fluid passes. The entire assembly is housed inside a heated environmental chamber, and this allows for operation to a temperature of 100°C. Data analysis for solutions is identical to that for melts except that the contributions to the measured force from fluid inertia, surface tension, gravity, and air drag can, in general, not be neglected.

A version of the spinline rheometer that can be used to make extensional flow measurements on extremely low-viscosity polymer solutions has been developed by Sridhar and Gupta, and this is shown in Figure 16 [53]. In this instrument,

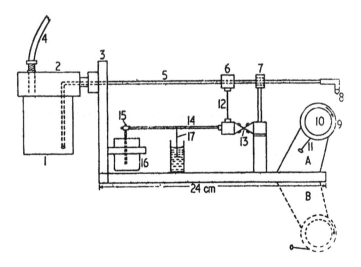

FIGURE 7.15 The spinline viscometer. 1, fluid reservoir; 5, metal tube; 8, nozzle assembly; 9, rotating drum; 11, scraper plate; 13, pivot assembly; 16, inductive displacement transducer; 17, damper. (From Ref. 52. Used by permission of Institute of Physics Publishing, Bristol, U.K.)

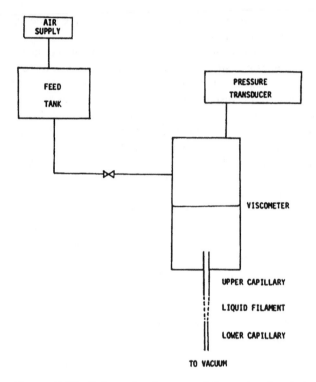

FIGURE 7.16 Schematic diagram of the extensional viscometer developed by Sridhar and Gupta. (From Ref. 53. Used by permission of Steinkopff Publishers, Darmstadt, FRG.)

liquid enters a tank at constant flow rate, compresses air to some steady-state value, and leaves through a capillary. When the liquid strand exiting the capillary is stretched, the air pressure in the tank drops to a new steady-state value, and this is measured using a pressure transducer. A simple force balance shows that the decrease in pressure equals the tensile stress in the liquid at the capillary outlet [53]. The tensile stress at other axial locations along the fiber is again obtained via a momentum balance. Since this viscometer has no moving parts, and since pressure measurements can be made with great ease and accuracy, apparent extensional viscosities of even very dilute aqueous polymer solutions can be determined [54]. Note that in this instrument, fluid stretching is accomplished by using a suction device in which the liquid emerging from the capillary is sucked into another capillary of still smaller diameter, producing a stretched jet in the air gap between the capillaries [55]; stretch rates in excess of 1000 sec^{-1} can be attained [54]. At these high stretch rates, however, it is very easy to cause polymer chain scission [56], and it is advisable not to recycle the solution in the process of extensional viscosity measurement.

Although measured apparent extensional viscosity data as a function of the stretch rate along the spinline can be profitably analyzed with the help of constitutive equations (see Ref. 57 for a successful illustration using data on fluid M1), Ferguson and Hudson [58] have suggested that we ought to represent all transient extensional viscosity results on a three-dimensional plot of elongational viscosity, strain, and time of elongation. An example of such a plot, utilizing data from different experiments, is displayed in Fig. 7.17 [58]. According to these authors, such a procedure yields a unique surface; different techniques and different runs from a given technique merely trace out different lines on this surface. If such a surface is available, the tensile stress growth function can be found by walking on this surface along a line that emanates radially from the origin. This is an

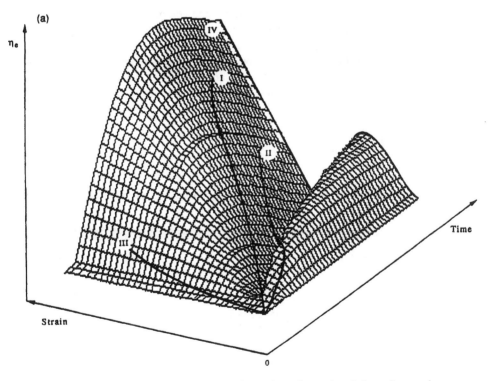

Figure 7.17 A three-dimensional plot of transient elongational viscosity as a function of time and total strain for fluid M1. Results from different experiments are shown as: I, spinline; II, falling drop; III, falling weight; and IV, equilibrium. (From Ref. 58, with permission from Elsevier Science.)

extremely appealing idea, but according to Petrie [59] the theoretical evidence to support the thesis of Ferguson and Hudson is not definitive.

B. Converging Flow

When a liquid flows through a conical channel, as shown in Fig. 7.18, there is obviously shearing of the fluid near the wall, but the average velocity increases monotonically as fluid elements move toward the apex. This increase of velocity is the result of decreasing cross-sectional area for flow, and it endows the flow field with a significant extensional character. The relative amounts of shear and extension depend on the channel geometry, the volumetric flow rate, fluid properties, and the presence of any low-viscosity lubricant near the walls. A similar situation arises in the entrance region leading from a reservoir to a capillary of radius R; this can be exploited for the measurement of the apparent extensional viscosity [60,61].

What we measure in this flow field are the volumetric flow rate Q and the entrance pressure drop, P_{ent}, defined as the total pressure drop across the abrupt contraction between two points where the flow is fully developed less the pressure drop for fully developed Poiseuille flow. According to Cogswell [60,61], P_{ent} can be separated into a shear component and an extensional component. The latter contrbution can be used to compute the average net tensile stress in the entry region. The result is:

$$\sigma_L = \left(\frac{3}{8}\right)(n + 1)P_{ent} \tag{7.6}$$

in which n is the power-law index for shear flow. The average stretch rate is calculated to be [62]:

$$\dot{\varepsilon} = \frac{4\tau\dot{\gamma}}{3(n + 1)P_{ent}} \tag{7.7}$$

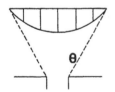
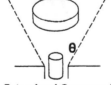

Simple Shear Component   **Extensional Component**

Figure 7.18 Converging flow into a conical channel. (From Ref. 60.)

where τ is the shear stress at a shear rate $\dot{\gamma}$ equal to $4Q/(\pi R^3)$. Dividing the average net tensile stress by the average stretch rate gives an average extensional viscosity.

Laun and Schuch [48] used the converging-flow method to measure the extensional-flow properties of two IUPAC LDPE samples, and the results are shown in Fig. 7.14. It appears that the technique not only ranks the polymer melts correctly but also gives data that agree with the true extensional viscosities. Indeed, Kwag and Vlachopoulos [63] used the data of Laun and Schuch in a finite element simulation and concluded that Cogswell's method was an acceptable approximation at high stretch rates. Although this method poses no experimental problems for work with either polymer melts or low-viscosity polymer solutions, it should be kept in mind that increasing the stretch rate by increasing the volumetric flow rate results in a reduction in the fluid residence time in the converging capillary; as a consequence and keeping in view the data of Fig. 7.1, the measured average-extensional-viscosity is likely to become progressively smaller than the true extensional viscosity with increasing stretch rate [64]. We also note that Binding [65] has carried out a very sophisticated analysis of contraction flows, and commercial capillary instruments such as the Rosand RH7 twin-bore viscometer (Rosand Precision, United Kingdom) are now available for making measurements as a function of both stretch rate and temperature. Most recently, this technique has been used to evaluate the influence of pressure on the elongational properties of polymer melts [65a].

C. Opposed-Nozzle Rheometer

Fuller et al. [66] have described an extensional viscometer for mobile liquids that uses two opposing nozzles attached to tubes immersed in a beaker of the test fluid. As shown in Fig. 7.19, one of the tubes is fixed, and the other is mounted on a knife edge and is free to swing under the influence of a torque that arises from the fluid forces between the nozzles. The force of interest is a tensile force generated when fluid is sucked into the nozzles by the application of vacuum, creating an approximate extensional flow, with a stretch rate as large as 10,000 sec^{-1}, in the small gap between the nozzles. This instrument is manufactured by Rheometrics Scientific under the name Fluids Extensional Analyzer, and it provides an average extensional viscosity; the tensile stress is calculated via the measured torque, and the ratio of the average fluid velocity in the nozzle to half the nozzle separation is taken to be the stretch rate. It is often found that measurements on Newtonian liquids give a value of the extensional viscosity that is slightly different from three times the shear viscosity. The flow field between the opposed nozzles has been thoroughly analyzed by Scriven and coworkers [67,68], who have concluded that "opposed nozzles should be considered as useful indexers rather than rheometers."

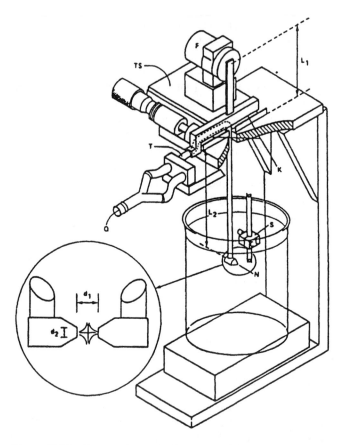

FIGURE 7.19 Opposed-nozzle rheometer. N, opposing nozzles; S, fixture defining nozzle separation; K, knife edge fulcrum; F, force transducer; L, L_2, lever arms. (From Ref. 66.)

IV. CONCLUDING REMARKS

It is now well established that when the mode of deformation in a polymer processing situation is predominantly extensional in character, it is the uniaxial extensional viscosity of the polymer, rather than the shear viscosity, that is the better predictor of observed behavior. This has been established, for example, in film blowing [69] and blow molding [70], even when the extensional viscosity measurements were made at relatively low stretch rates. These measurements are well understood in terms of their dependence on molecular parameters such as molecular weight, molecular weight distribution, and long-chain branching; data on well-characterized melts and solutions have also been used to test the validity of different proposed rheological constitutive equations. However, there is still a pressing need to develop techniques capable of measuring the true extensional viscosity of polymeric liquids at deformation rates more representative of industrial polymer processing. Even though in this chapter we have described only those measurement techniques that have led to commercial instruments, many ingenious approaches have been tried in the past [1–3,71–73]. We mention that, in addition to the uniaxial extension viscosity, we can also define an equal biaxial extension viscosity and a planar extension viscosity [71–73]. Popular methods for experimentally determining biaxial extension viscosity include bubble inflation [74], lubricated squeezing [75,76], and sheet stretching [77,78]. Although, as compared to uniaxial extension viscosity, the results of these techniques may be more relevant to processes such as film blowing and blow molding, the development of these methods has not been pursued with the same vigor as in the case of uniaxial elongation.

REFERENCES

1. C. J. S. Petrie. Elongational Flows. Pitman, London, 1979.
2. M. Rides, C. R. G. Allen, S. Chakravorty. Review of Extensional Viscoelasticity Measurement Techniques for Polymer Melts, Report #CMMT(A)44, National Physical Laboratory, Teddington, U.K., 1996.
3. R. K. Gupta, T. Sridhar. Elongational rheometers. In: A. A. Collyer, D. W. Clegg, eds. Rheological Measurement. 2nd ed. Chapman and Hall, London, 1998, pp. 516–549.
4. R. J. Good, R. K. Gupta. The coupling of interfacial, rheological, and thermal control mechanisms in polymer adhesion. In: L.-H. Lee, ed. Adhesive Bonding. Plenum Press, New York, 1991, pp. 47–73.
5. H. Munstedt, H. M. Laun. Elongational behavior of a low density polyethylene melt. II. Transient behavior in constant stretching rate and tensile creep experiments. Comparison with shear data. Temperature dependence of the elongational properties. Rheol. Acta 18:492–504 (1979).
6. R. L. Ballman. Extensional flow of polystyrene melt. Rheol. Acta 4:137–140 (1965).
7. F. N. Cogswell. Tensile deformations in molten polymers. Rheol. Acta 8:187–194 (1969).
8. L. Nicodemo, B. De Cindio, L. Nicolais. Elongational viscosity of microbead suspensions. Polym. Eng. Sci. 15:679–683 (1975).
9. P. Micic, S. N. Bhattacharya, G. Field. Rheological behavior of linear low density polyethylene/low density polyethylene blends under elongational deformation—the effect of molecular structure and melt morphology. Intern. Polym. Processing 12:110–115 (1997).
10. S. J. Armour, J. C. Muirhead, A. B. Metzner. Filament formation in viscoelastic fluids. Proc. IX Intl. Congress on Rheology, Acapulco, Mexico, 2/1984, pp. 143–151.

11. J. E. Matta, R. P. Tytus. Viscoelastic breakup in a high velocity airstream. J. Appl. Polym. Sci. 27:397–405 (1982).

12. J. E. Glass. Dynamics of roll spatter and tracking. Part III. Importance of extensional viscosities. J. Coatings Technol. 50 (641):56–71 (1978).

13. D. F. Massouda. Analysis and prediction of roll spatter from latex paints. J. Coatings Technol. 57(722):27–36 (1985).

14. D.A. Soules, R.H. Fernando, J. E. Glass. Dynamic uniaxial extensional viscosity (DUEV) effects in roll application I. Rib and web growth in commercial coatings. J. Rheol. 32:181–198 (1988).

15. J. Rhi-Sausi, J. Dealy. An extensiometer for molten plastics. Polym. Eng. Sci. 16: 799–802 (1976).

16. P. K. Agrawal, W. K. Lee, J. M. Lorntson, C. I. Richardson, K. F. Wissbrun, A. B. Metzner. Rheological behavior of molten polymers in shearing and in extensional flows. Trans. Soc. Rheol. 21:355–379 (1977).

17. F. N. Cogswell. The rheology of polymer melts under tension. Plast. Polym. 36: 109–111 (1968).

18. G. V. Vinogradov, V. D. Fikhman, B. V. Radushkevich. Uniaxial extension of polystyrene at true constant stress. Rheol. Acta 11:286–291 (1972).

19. H. Munstedt. Viscoelasticity of polystyrene melts in tensile creep experiments. Rheol. Acta 14:1077–1088 (1975).

20. J. Meissner. Rheometer zur Untersuchung der deformationsmechanischen Eigenschaften von Kunststoff-Schmelzen unter definierter Zugbeanspruchung. Rheol. Acta 8:78–88 (1969).

21. J. Meissner, J. Hostettler. A new elongational rheometer for polymer melts and other highly viscoelastic liquids. Rheol. Acta 33:1–21 (1994).

22. J. Meissner. Basic parameters, melt rheology, processing and end-use properties of three similar low density polyethylene samples. Pure Appl. Chem. 42:553–612 (1975).

23. H. M. Laun, H. Munstedt. Elongational behavior of a low density polyethylene melt. I. Strain rate and stress dependence of viscosity and recoverable strain in the steady-state. Comparison with shear data. Influence of interfacial tension. Rheol. Acta 17: 415–425 (1978).

24. T. Raible, A. Demarmels, J. Meissner. Stress and recovery maxima in LDPE melt elongation. Polymer Bull. 1:397–402 (1979).

25. H. Munstedt. Dependence of the elongational behavior of polystyrene melts on molecular weight and molecular weight distribution. J. Rheol. 24:847–867 (1980).

26. A. Franck, J. Meissner. The influence of blending polystyrenes of narrow molecular weight distribution on melt creep flow and creep recovery in elongation. Rheol. Acta 23:117–123 (1984).

27. J. Linster, J. Meissner. Melt elongation and structure of linear polyethylene (HDPE). Polymer Bull. 16:187–194 (1986).

28. V. S. Au-Yeung, C. W. Macosko, V. R. Raju. Extensional flow of linear and star-branched hydrogenated polybutadiene with narrow molecular weight distribution. J. Rheol. 25:445–452 (1981).

29. H. Munstedt, H. M. Laun. Elongational properties and molecular structure of polyethylene melts. Rheol. Acta 20:211–221 (1981).

30. F. P. La Mantia, A. Valenza, D. Acierno. Influence of long chain branching on the elongational behavior of different polyethylenes and their blends. Polymer Bull. 15: 381–387 (1986).

31. H. Munstedt, S. Kurzbeck, L. Egersdorfer. Influence of molecular structure on rheological properties of polyethylenes. Part II. Elongational behavior. Rheol. Acta 37: 21–29 (1998).

32. A. S. Lodge. Constitutive equations from molecular network theories for polymer solutions. Rheol. Acta 7:379–392 (1968).

33. A. Kumar, R. K. Gupta. Fundamentals of Polymers. McGraw-Hill, New York, 1998, Chap. 14.

34. R. G. Larson. Constitutive Equations for Polymer Melts and Solutions. Butterworths, Boston, 1988.

35. M. J. Crochet, A. R. Davies, K. Walters. Numerical Simulation of Non-Newtonian Flow. Elsevier, Amsterdam, 1984.

36. J. M. Dealy, K. F. Wissbrun. Melt Rheology and Its Role in Plastics Processing. Van Nostrand Reinhold, New York, 1990.

37. N. Phan-Thien, R. I. Tanner. A new constitutive equation derived from network theory. J. Non-Newt. Fluid Mech. 2:353–365 (1977).

38. S. A. Khan, R. G. Larson. Comparison of simple constitutive equations for polymer melts in shear and biaxial and uniaxial extensions. J. Rheol. 31:207–234 (1987).

39. Special issue of J. Non-Newt. Fluid Mech. 30:97–368 (1988).

40. T. Sridhar. An overview of the project M1. J. Non-Newt. Fluid Mech. 35:85–92 (1990).

41. N. E. Hudson, T. E. R. Jones. The A1 project—an overview. J. Non-Newt. Fluid Mech. 46:69–88 (1993).

42. Special issue of J. Non-Newt. Fluid Mech. 52:105–228 (1994).

43. T. Sridhar, V. Tirtaatmadja, D. A. Nguyen, R. K. Gupta. Measurement of Extensional Viscosity of Polymer Solutions. J. Non-Newt. Fluid Mech. 40:271–280 (1991).

44. V. Tirtaatmadja, T. Sridhar. A filament stretching device for measurement of extensional viscosity. J. Rheol. 37: 1081–1102 (1993).

44a. P. Szabo. Transient filament stretching rheometer. I: Force balance analysis. Rheol. Acta 36:277–284 (1994).

45. V. Tirtaatmadja, T. Sridhar. Comparison of constitutive equations for polymer solutions in uniaxial extension. J. Rheol. 39:1133–1160 (1995).

46. G. Prilutski, R. K. Gupta, T. Sridhar, M. E. Ryan. Model viscoelastic liquids. J. Non-Newt. Fluid Mech. 12:233–241 (1983).

47. J. L. White, A. B. Metzner. Development of constitutive equations for polymeric melts and solutions. J. Appl. Polym. Sci. 7:1867–1889 (1963).

47a. R. K. Gupta. Studies on uniaxial extension of dilute polymer solutions. Ph.D. dissertation. Monash University, Clayton, Australia, 1999.

48. H. M. Laun, H. Schuch. Transient elongational viscosities and drawability of polymer melts. J. Rheol. 33:119–175 (1989).

49. C. D. Han, R. R. Lamonte. Studies on melt spinning. Part I. Effect of molecular structure and molecular weight distribution on elongational viscosity. Trans. Soc. Rheol. 16:447–472 (1972).

50. J. A. Spearot, A. B. Metzner. Isothermal spinning of molten polyethylenes. Trans. Soc. Rheol. 16:495–518 (1972).

51. V. Rauschenberger, H. M. Laun. A recursive model for Rheotens tests. J. Rheol. 41:719–737 (1997).

52. J. Ferguson, N.E. Hudson. A new viscometer for the measurement of apparent elongational viscosity. J. Physics E: Scientific Instruments 8:526–530 (1975).

53. T. Sridhar, R. K. Gupta. A simple extensional viscometer. Rheol. Acta 24:207–209 (1985).

54. R. C. Chan, R. K. Gupta, T. Sridhar. Fiber spinning of very dilute solutions of polyacrylamide in water. J. Non-Newt. Fluid Mech. 30:267–283 (1988).

55. M. Khagram, R.K. Gupta, T. Sridhar. Extensional flow of xanthan gum solutions. J. Rheol. 29:191–207 (1985).

56. J. A. Odell, A. Keller, M. J. Miles. A method for studying flow-induced polymer degradation: verification of chain halving. Polym. Commun. 24:7–10 (1983).

57. D. A. Nguyen, R. K. Gupta, T. Sridhar. Experimental results and constitutive modelling of the extensional flow of M1. J. Non-Newt. Fluid Mech. 35:207–214 (1990).

58. J. Ferguson, N. E. Hudson. Transient elongational rheology of polymeric fluids. Eur. Polym. J. 29:141–147 (1993).

59. C. J. S. Petrie. Three-dimensional presentation of extensional flow data. J. Non-Newt. Fluid Mech. 70:205–218 (1997).

60. F. N. Cogswell. Measuring the extensional rheology of polymer melts. Trans. Soc. Rheol. 16:383–403 (1972).

61. F. N. Cogswell. Converging flow of polymer melts in extrusion dies. Polym. Eng. Sci. 12:64–73 (1972).

62. F. N. Cogswell. Converging flow and stretching flow: a compilation. J. Non-Newt. Fluid Mech. 4:23–38 (1978).

63. C. Kwag, J. Vlachopoulos. An assessment of Cogswell's method for measurement of extensional viscosity. Polym. Eng. Sci. 31:1015–1021 (1991).

64. R. K. Gupta, T. Sridhar. Viscoelastic effects in non-Newtonian flows through porous media. Rheol. Acta 24:148–151 (1985).

65. D. M. Binding. Contraction flows and new theories for estimating extensional viscosity. In A. A. Collyer, ed. Techniques in Rheological Measurement. Chapman and Hall, London, 1993, pp. 1–32.

65a. D. M. Binding, M. A. Couch, K. Walters. The pressure dependence of the shear and elongational properties of polymer melts. J. Non-Newtonian Fluid Mech. 79:137–155 (1998).

66. G. G. Fuller, C. A. Cathey, B. Hubbard, B. E. Zebrowski. Extensional viscosity measurements for low-viscosity fluids. J. Rheol. 31:235–249 (1987).

67. P. R. Schunk, J. M. de Santos, L. E. Scriven. Flow of Newtonian liquids in opposed-nozzles configuration. J. Rheol. 34:387–414 (1990).

68. P. Dontula, M. Pasquali, L. E. Scriven, C. W. Macosko. Can extensional viscosity be measured with opposed-nozzle devices? Rheol. Acta 36:429–448 (1997).

69. A. Ghaneh-Fard, P.J. Carreau, P. G. Lafleur. Study of kinematics of film blowing of different polyethylenes. Polym. Eng. Sci. 37:1148–1163 (1997).

70. D. H. Sebastian, J. R. Dearborn. Elongation rheology of polyolefins and its relation to processability. Polym. Eng. Sci. 23:572–575 (1983).

71. R. W. Whorlow. Rheological Techniques. 2nd ed. Ellis Horwood, New York, 1992, pp. 229–274.
72. J. M. Dealy. Extensional flow of non-Newtonian fluids—a review. Polym. Eng. Sci. 11:433–445 (1971).
73. J. M. Dealy. Extensional rheometers for molten polymers: a review. J. Non-Newt. Fluid Mech. 4:9–21 (1978).
74. D. D. Joye, G. W. Poehlein, C. D. Denson. A bubble inflation technique for the measurement of viscoelastic properties in equal biaxial extensional flow. Trans. Soc. Rheol. 16:421–445 (1972).
75. C. W. Macosko. Rheology: Principles, Measurements and Applications. VCH, New York, 1994, pp. 297–302.
76. Sh. Chatraei, C. W. Macosko, H. H. Winter. Lubricated squeezing flow: a new biaxial extensional rheometer. J. Rheol. 25:433–443 (1981).
77. J. Meissner. Experimental aspects in polymer melt elongational rheometry. Chem. Eng. Commun. 33:159–180 (1985).
78. J. Meissner. Polymer melt elongation—methods, results, and recent developments. Polym. Eng. Sci. 27:537–546 (1987).

8

Rigid-Rod and Liquid-Crystal Polymer Rheology

I. INTRODUCTION

Thus far, we have examined the flow behavior of flexible-chain polymers. If the polymers contain para-linked aromatic rings, such as those found in aromatic polyamides, polyesters, and polyazomethines, rotation about the backbone is inhibited, and this leads to a stiff, rigid, extended chain structure; chain rigidity typically results in a value of the Mark–Houwink exponent "a" that is greater than 0.8 (see Eq. 4.8). When such rodlike molecules, especially those having a large length-to-diameter or aspect ratio, are dissolved in a solvent, the resulting isotropic solution becomes quite viscous at low polymer concentrations. This happens because the polymer molecules in solution form an entangled matlike structure that refuses to accommodate additional molecules without forcing some of the dissolved molecules to bend. If chain flexibility is prohibited, a further increase in polymer concentration can occur only through the development of an anisotropic or ordered phase that involves parallelization of the polymer chains. Such a phase can possess long-range order similar to crystalline solids, and it is known as a *mesophase* or a *polymeric liquid crystal*. This ordering is accompanied by a sharp reduction in the solution viscosity. This is shown in Fig. 8.1 for a 50/50 copolymer of n-hexyl and n-propylisocyanate of 41,000 molecular weight dissolved in toluene at room temperature [1]. Fig. 8.1 clearly illustrates that it is easier for rods to slide past each other when they are oriented parallel to each other; in practical terms this implies ease of processing. Polymers that form a liquid crystalline phase in solution are known as *lyotropic*. Three different physical structures are found to occur with rodlike molecules; these are shown in Fig.

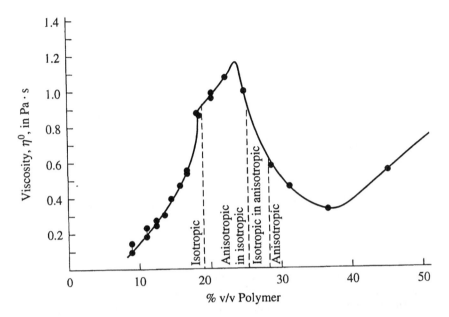

FIGURE 8.1 Viscosity as a function of polymer concentration for the system poly (50% n-hexyl + 50% n-propyl) isocyanate of M_w = 41,000 in toluene at 25°C. (From Ref. 1, with permission from Elsevier Science.)

8.2 [2]. In the *nematic* phase, there is no long-range order of positions, but there is a preferred direction, called the *director*, although a distribution of angles with respect to the director is observed. Liquid crystals in the *cholesteric* phase show an increase in order over the nematic phase, with the direction of the director varying helically along an axis perpendicular to the plane of the director. Finally, the *smectic* phase shows the most order, albeit in only one dimension.

The formation of a liquid crystalline phase within an isotropic system composed of monodisperse, rigid, rodlike particles in solution was predicted theoretically by Flory using concepts of statistical thermodynamics [3]. It was shown that at a critical volume fraction equal to $8/x$, where x is the aspect ratio of the rods, there would be phase separation; a nematic phase with a polymer concentration of $12.5/x$ was predicted to arise. For total polymer concentrations between $8/x$ and $12.5/x$, the isotropic and nematic phases would coexist, but there would only be a single anisotropic phase for a polymer volume fraction in excess of $12.5/x$. According to Wissbrun [2], this prediction has been verified, at least semiquantitatively, for a number of lyotropic systems. Furthermore, since the rod length is proportional to the polymer molecular weight, the critical concentration

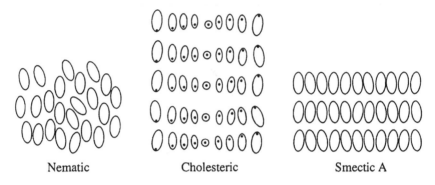

Nematic   Cholesteric   Smectic A

FIGURE 8.2   Schematic representation of mesophase types. (From Ref. 2.)

for transformation from the isotropic to the anisotropic state is inversely proportional to molecular weight. The best-known commercial example of a lyotropic liquid crystal polymer is poly-p-phenyleneterephthalamide (PPT) manufactured by Du Pont under the tradename Kevlar®. This is an extended-chain, para-oriented polyamide (or polyaramid) made by reacting p-phenylenediamine and terephthaloyol chloride. Another well-studied polymer is the (model) synthetic polypeptide poly-benzyl-L-glutamate (PBLG); when dissolved in a solvent such as m-cresol, it forms a helix that acts like a rigid rod. Some cellulosic polymers and biopolymers also exhibit liquid crystalline behavior. The impetus for the study of the rheological properties of these lyotropic polymeric liquid crystals has been the observation that fibers prepared from the anisotropic phase of materials such as Du Pont's Kevlar® have ultrahigh strength coupled with ultrahigh modulus (fibers made from the same polymer and using the same solvent but employing concentrations that give isotropic solutions have only about one-half the tensile strength [4]). These desirable properties are the consequence of preordered polymer domains that are not only preserved during fiber formation but also often enhanced by actual crystallization in an extended-chain conformation. The extended-chain morphology, together with molecular alignment, results in high strength, and the materials so formed are also chemically inert and thermally stable because of the highly aromatic chain chemistry. These high-strength and high-modulus fibers find application primarily as reinforcing members in polymer-matrix composites. A description of the chemistry of synthesis of commercially important liquid crystalline polymers (LCPs), their processing methods, resulting mechanical and optical properties, and current needs and future opportunities may be found in Ref. 5.

Because lyotropic liquid crystalline polymers cannot be extruded, injection molded, or blown into films, other polymers that can be melt processed have been developed. Commercialized in the 1980s [5], these so-called thermotropic liquid crystal polymers convert to a mesophase when the solid polymer is heated to a temperature above the crystalline melting point; crystalline melting points are typically in the 275–420°C range. Thus, these polymers show at least three thermal transitions. In increasing order of temperature, these are the glass transition temperature, the crystalline melting point, and the nematic-to-isotropic transition temperature. The most widely studied class of thermotropic polymers is aromatic polyesters. But since these materials melt with decomposition, they are copolymerized with flexible-chain polymers, which results in a lowered melting point.

An example of this approach is the copolyester of *p*-hydroxybenzoic acid (HBA) and polyethylene terephthalate (PET); the melt viscosity of this copolymer as a function of HBA content at different shear rates is shown in Fig. 8.3 [6]. It is seen that the viscosity behavior is similar to that shown in Fig. 8.1 insofar as the viscosity goes through a maximum at a particular HBA content, which is around 30 % in the present case. As might be anticipated, this is due to the formation of a mesophase.

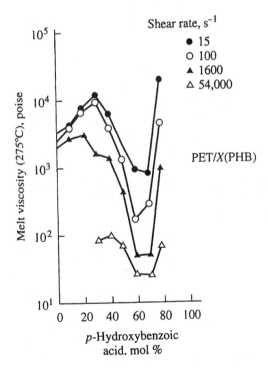

FIGURE 8.3 Melt viscosity of PET modified with HBA. (From Ref. 6. Copyright © 1976 by John Wiley & Sons, Inc. Reprinted by permission of John Wiley & Sons, Inc.)

A feature of the flow behavior of some thermotropic LCPs is that the transformation from the isotropic to the nematic phase, brought about by cooling, is accompanied by a drastic drop in the shear viscosity. This extremely low viscosity is used to advantage in injection molding, where complex, thin-walled molds can be filled quite rapidly. Short cycle times in injection molding also result from short cooling times due to small values of both the sensible heat and the latent heat; material consistency goes from free-flowing mobile liquid to rock solid over a narrow temperature range.

Note that, despite the incorporation of flexible spacers in the polymer backbone, commercially important thermotropic LCPs begin to degrade soon after reaching the nematic-to-isotropic transition temperature. Consequently, other aromatic polyesters have been synthesized as model compounds. An example is poly[(phenylsulfonyl)-p-phenylene 1,10-decamethylene-bis(4-oxybenzoate)], or PSHQ10, where the 10 refers to the number of methylene units present; this polymer has a melting point of 115°C, a nematic-to-isotropic transition temperature of 175°C, and a thermal degradation temperature of about 350°C [7]. The large intervals between these temperatures make it possible to investigate the rheological behavior of this polymer in both the isotropic and nematic states. A plot of the shear viscosity of PSHQ10 as a function of temperature is displayed in Fig. 8.4, where it is seen that, in the temperature range between the nematic and isotropic regions, viscosity increases sharply with increasing temperature [8].

The general rheological behavior of LCPs can be quite different and significantly more complicated as compared to the corresponding behavior of flexible-chain polymers observed in Chaps. 3–7. In particular, shear viscosity as a function of shear rate can show three distinct regions, as illustrated in Fig. 8.5 [9]. Quite surprisingly, there may be no zero-shear-rate limiting viscosity. Also, the measured viscosity may depend on the technique used for sample preparation as well as on the thermal and deformational history imposed on the polymer [8]; this is especially true in region I of Fig. 8.5. Another very unusual observation made with lyotropic LCPs has been the occurrence of negative first normal stress difference values during steady-shear flow. Data on poly(benzyl glutamate) solutions in m-cresol shown in Fig. 8.6 [10] reveal that, as expected, the first normal stress difference initially increases with increasing shear rate, but it soon reaches a maximum and then abruptly becomes negative. Equally abruptly, it becomes positive at still higher shear rates. Other unusual features of LCP rheology include little die swell, shear thickening during steady-shear flow, and long transients and stress overshoots during startup of shear flow [11].

In the rest of this chapter, we systematically examine the rheology of both lyotropic and thermotropic LCPs and compare and contrast this behavior with the corresponding behavior of isotropic polymers. In addition we examine the ability of different molecular theories to explain the observations made with LCPs. However, we restrict ourselves to rodlike, main-chain LCPs. In this con-

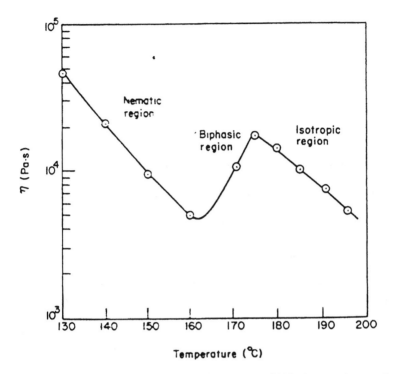

Figure 8.4 Viscosity versus temperature for PSHQ10 at a shear rate of 0.01 sec⁻¹. (From Ref. 8. Copyright © 1994 by John Wiley & Sons, Inc. Reprinted by permission of John Wiley & Sons, Inc.)

nection, we merely note that there also exist technologically important discotic LCPs, such as those derived from carbonaceous pitches [12] and side-chain LCPs that contain mesogenic groups as appendages on the polymer backbone [13].

II. VISCOSITY BEHAVIOR IN THE ISOTROPIC STATE

Ample reliable data are now available on both lyotropic and thermotropic LCPs in the isotropic state. At the low concentration end, Baird and Smith found that the intrinsic viscosity of a given molecular weight fraction of PPT depended strongly on sulfuric acid strength, exhibiting a maximum at an acid concentration of about 100% [14]. This suggests that polymer chain rigidity depends on the nature of the solvent. Not surprisingly, when the intrinsic viscosity of PPT is plotted on logarithmic coordinates as a function of molecular weight, different straight lines are obtained, depending on the strength of sulfuric acid used. This is shown in Fig. 8.7 [14]. The slope is a maximum, equal to 1.62, when the acid is

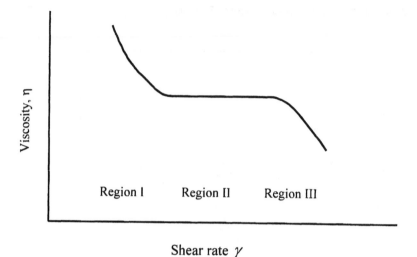

FIGURE 8.5 Shear rate dependence on apparent viscosity for liquid crystals.

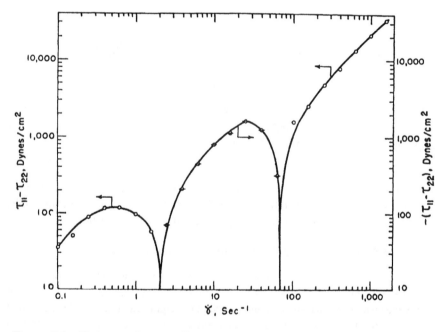

FIGURE 8.6 First normal stress difference versus shear rate for a 16.4 wt% racemic mixture of poly(benzyl-L-glutamate) and poly(benzyl-D-glutamate) in *m*-cresol. (From Ref. 10. Copyright © 1978 by John Wiley & Sons, Inc. Reprinted by permission of John Wiley & Sons, Inc.)

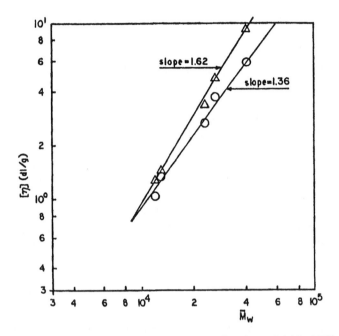

FIGURE 8.7 Intrinsic viscosity versus molecular weight for PPT dissolved in sulfuric acid. Symbols denote acid strength: triangles—100.2%, circles—96.6%. (From Ref. 14. Copyright © 1978 by John Wiley & Sons, Inc. Reprinted by permission of John Wiley & Sons, Inc.)

100% concentrated. According to the theory of Kirkwood and Auer, the intrinsic viscosity of monodisperse rigid rods is given by [15,16]:

$$[\eta] = \frac{2\pi N_A L^3}{45M \ln (L/D)} \tag{8.1}$$

in which N_A is Avogadro's number, M is polymer molecular weight, L is the length of the rod, and D is rod diameter. Since L is proportional to M, the dependence of intrinsic viscosity on M for rods of large aspect ratio becomes [16]:

$$[\eta] = \frac{CM^2}{\ln M} \tag{8.2}$$

where C is a constant independent of M. Clearly, the maximum measured value of the Mark–Houwink exponent of the PPT solutions is smaller than what would be predicted based on Eq. (8.2). More rigid molecules, such as poly(benzobisoxazole) dissolved in a variety of solvents, have a Mark–Houwink exponent of 1.85 [17]; this is close to the theoretical value. More flexible (but still rigid) molecules, such as the thermotropic polymer PSHQ10 dissolved in trichlorobenzene, have a Mark–Houwink exponent of 1.1.

In the limit of infinite dilution, polymer solution viscosity is linearly dependent on polymer concentration. This happens because the polymer molecules are so far apart that one molecule does not feel the presence of the other molecules. As a consequence, the solution viscosity is the product of the number of polymer molecules with the contribution of any one of the dissolved molecules. At a higher concentration, but still in the dilute region, the viscosity increases in a nonlinear manner with concentration due to the action of mutual hydrodynamic forces between molecules. The actual computation of the relative viscosity η_R is done by taking the ratio of the energy dissipated by shear flow per unit time per unit volume by the solution to the corresponding quantity for the solvent alone. The various methods of carrying out the viscosity calculation for solute particles of different shapes and sizes have been reviewed by Frisch and Simha [18], and the general result is that the concentration dependence of the viscosity can be written as:

$$\frac{\eta_R - 1}{\phi} = [\eta] + k_1[\eta]^2 \phi \tag{8.3}$$

in which ϕ is the volume fraction and k_1 is a shape-dependent constant; for rigid rods this constant has a value 0.73 [18].

The extension of the dilute solution result to still higher concentrations can be done in several ways. By adding solute sequentially and by considering each resulting solution to be the solvent for the solute that is added next, Brodnyan [19] derived the following result:

$$\eta_R = \exp\left[\frac{2.5\phi + 0.399(x - 1)^{1.48}\phi}{1 - k\phi}\right] \tag{8.4}$$

where x is the aspect ratio of the rigid rods and the constant k has a value of 1.91. Equation (8.4) has been shown to represent data on dilute poly-γ-benzyl-L-glutamate solutions [19].

Further increases in concentration result in polymer–polymer interactions. This has two consequences that are similar to those observed in the case of flexible-chain macromolecules: (1) the solution zero-shear viscosity increases very rapidly with polymer concentration or molecular weight, and (2) the viscosity becomes shear thinning. Fig. 8.8 shows the zero-shear viscosity of PPT solutions in 100% sulfuric acid as a function of the product of molecular weight and concentration [20]. Also shown are similar data on nylon 6,6 in the same solvent. While the behavior of the two solutions is qualitatively similar, there are important differences that are quite evident. At low concentrations, the viscosity of the rigid polymer is about one order of magnitude higher than that of the flexible

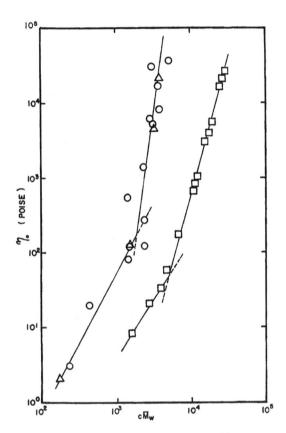

FIGURE 8.8 Zero-shear viscosity of polyamide solutions in 100% sulfuric acid at 24°C. Squares represent nylon 6,6 data; circles are PPT data. Triangles are PPT data at 60°C but shifted to 24°C. (From Ref. 20.)

polymer. Also, polymer–polymer interactions occur much earlier in the PPT solutions as compared to the nylon solutions. Beyond this point, the slope of the viscosity plot is an expected 3.43 for nylon but a rather large 6.8 for PPT. (For the undiluted thermotropic PSHQ10 polymer, the corresponding value of the slope is 6.5 [7]). Thus, chain stiffness promotes network formation. It is this network formation that makes the onset of shear thinning occur at a shear rate about one decade lower for the PPT solutions. Note that data on both solutions yield master curves when plotted in dimensionless form in the manner of Fig. 4.6.

The steady as well as transient behavior of the extra stress tensor (and not just the viscosity) of a dilute solution of rigid rods for any specified deformation history can be computed using the kinetic theory of rigid dumbbells [21]—a typical polymer molecule is idealized as two beads joined by a rigid rod. Under the influence of flow, the dumbbells translate and rotate and attain some average orientation. This orientation, coupled with a knowledge of the tension in the connector, gives the stress contributed by the polymer molecules. The complete stress tensor is then obtained by adding to this contribution the stress in the solvent; the solvent stress is taken to be Newtonian. On carrying out the calculations [21], we find, as expected, that the difference between the solution zero-shear viscosity and the viscosity of the solvent is proportional to polymer concentration. At high shear rates, the viscosity becomes non-Newtonian and decreases with the -2/3 power of the shear rate. An obvious modification of the rigid-dumbbell model is the use of N beads spaced evenly along a massless and inextensible rod [22]. Another modification is to make all these N beads touch each other; this is the wormlike chain [23]. The latter model is able to fit the PPT and nylon 6,6 viscosity versus shear rate data of Baird and Ballman [20], as shown in Fig. 8.9 [24]. Again, the high shear rate viscosity decreases with the $-2/3$ power of the shear rate.

At high polymer concentrations, in the entangled regime, the free rotation of each rod is restricted by the presence of the other rods. Also, translational motion is allowed only in the direction of the long axis. These twin facts are responsible for the observed nonlinear viscoelasticity and are incorporated into the molecular theory of Doi and Edwards [25], Kuzuu and Doi [26], and Doi

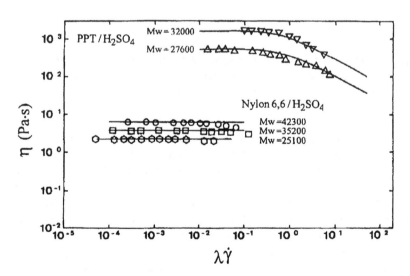

FIGURE 8.9 Wormlike model predictions fitted to the viscosity data of Baird and Ballman [20]. (From Ref. 24.)

[27]. For concentrated isotropic solutions, the theory predicts that the zero-shear viscosity is proportional to the sixth power of the molecular weight and the cube of the concentration [27]. Also, it is predicted that a universal curve should result if the zero-shear viscosity, made dimensionless with respect to the viscosity at the isotropic–nematic point, is plotted as a function of the volume fraction divided by the critical volume fraction for liquid crystal formation. At high shear rates, a power-law dependence on the shear rate is predicted [25]; this is in accord with experimental observations. Using this theory, other rheological functions can also be calculated for both shear and elongation, in steady as well as transient states; viscoelastic properties of rigid polymers are found to be qualitatively similar to those of flexible polymers [26].

III. CONSTANT-SHEAR-RATE BEHAVIOR IN THE NEMATIC STATE

Grizzuti et al. [28] have published cone-and-plate rheological data on anisotropic, aqueous solutions of hydroxypropyl cellulose (HPC) of 100,000 molecular weight at a polymer concentration of 50% by weight. Their experience in making these measurements revealed significant differences compared to the behavior of isotropic solutions. Firstly, it took as long as one month to properly dissolve the polymer. Secondly, very long-time transients in both the shear stress and the first normal stress difference were observed during initiation of flow at constant shear rate; representative results are shown in Fig. 8.10 [28]. Each of these two functions displays a large initial peak followed by a sequence of damped oscillations. Note that the first normal stress difference takes much longer to stabilize, and, at the experimental conditions of Fig. 8.10, its steady-state value is negative. Thirdly, even after following a strict experimental protocol, as many as 150 strain units were required for steady-state conditions to be attained. At a shear rate of 0.001 sec^{-1}, this translates into a wait of 2 days! When shearing was maintained for this long, the sample tended to dry, and consistent data could be obtained only by modifying the viscometer by creating a mercury seal around the edges; this is shown in Fig. 8.11. Data obtained show that the relaxation time of this lyotropic LCP is about 25 minutes.

The steady-shear viscosity of the HPC solution as a function of shear rate is shown in Fig. 8.12, and the general behavior is reminiscent of the three-region curve displayed in Fig. 8.5: at the lowest and highest shear rates, viscosity decreases with increasing shear rate. The average slope of the initial shear-thinning region is -0.7, and this rules out the occurrence of a yield stress; a yield stress would have necessitated a slope of -1. The average slope of the final shear-thinning region is -0.6, and these two shear-thinning regions are separated by a central region where the viscosity is a much weaker function of the shear rate.

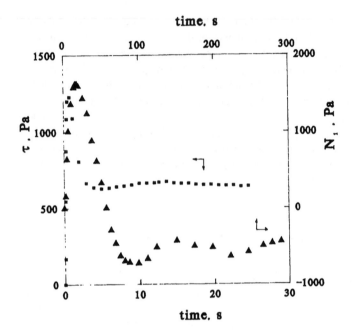

FIGURE 8.10 Startup of shear flow of an anisotropic HPC solution at a shear rate of 5 sec⁻¹. (From Ref. 28.)

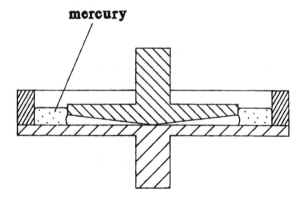

FIGURE 8.11 Modified cone-and-plate apparatus of Grizzuti et al. [28]. The mercury bath submerges the sample free edge.

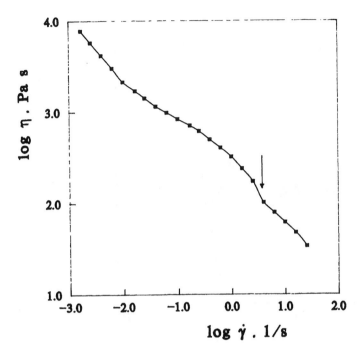

FIGURE 8.12 Steady-shear viscosity versus shear rate for an aqueous HPC solution in the liquid crystalline state. (From Ref. 28.)

The qualitative behavior of the first normal stress difference is similar to that shown in Fig. 8.6.

Compared to viscosity and first normal stress difference measurements, lyotropic LCP data on the second normal stress difference in shear are virtually nonexistent. One set of data that is available is on PBLG in m-cresol solutions and is due to Magda et al. [29]. These authors used a cone-and-plate viscometer for the simultaneous measurement of both normal stress differences (see Sec. IV.A of Chap. 5 for the theory), and results are shown in Fig. 8.13. A number of comments can be made by examining this figure. N_2 is an oscillatory rather than a monotonically decreasing function of the shear rate. Additionally, it is not always a negative quantity. At low shear rates, where N_1 is positive, N_2 is negative; but when N_1 changes sign with increasing shear rate, so does N_2; i.e., the second normal stress difference becomes positive! A second flip-flop in the sign occurs at still higher shear rates; then onward, the second normal stress difference decreases continuously with increasing shear rate. Another noteworthy point is that the ratio $|N_2/N_1|$ is above 0.5 for much of the first positive N_1 flow region.

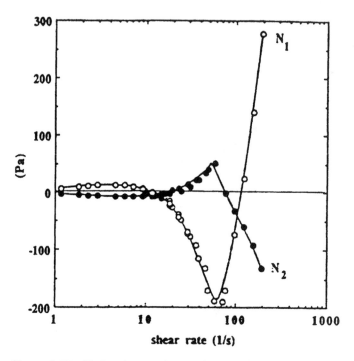

FIGURE 8.13 First and second normal stress differences measured for a fully liquid crystalline, 12.5% PBLG solution in *m*-cresol. (Reprinted with permission from Ref. 29. Copyright 1991 American Chemical Society.)

Finally, a progressive increase in the polymer concentration does not change the general shape of the curve of N_2 versus shear rate, but suppresses and ultimately eliminates the flow region in which N_2 is positive [29].

The steady-shear viscosity behavior of thermotropic LCPs is very similar to that of lyotropic LCPs. This is reported as a function of shear rate at six different temperatures in Fig. 8.14 [30]. The polymer is PSHQ6-12, a noncrystallizable cousin of PSHQ10 having a random distribution of flexible spacers of two different lengths—6 and 12 methylene groups—in the main chain. The nematic-to-isotropic change occurs at a temperature around 192°C. As expected, Fig. 8.14 shows the three-region viscosity curve in the nematic phase but Newtonian behavior with a high viscosity level in the isotropic phase. The corresponding first normal stress difference results as a function of the shear stress are given in Fig. 8.15 [30]. Only positive N_1 values were observed with this polymer over the entire shear rate range examined. When plotted in the manner of Fig. 8.15, a temperature-independent straight line of slope 2 was obtained for the isotropic data (compare to Figs. 5.7 and 5.8), but there was no data superposition for the nematic data. Note that negative N_1 values have been observed for thermotropic

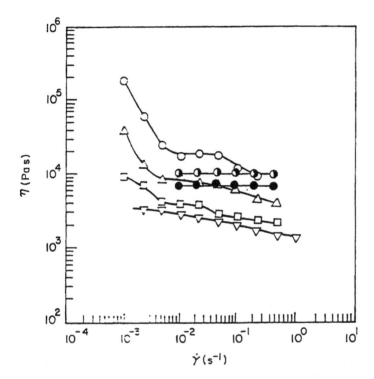

FIGURE 8.14 Viscosity versus shear rate for PSHQ6-12 samples at various temperatures. For the unfilled symbols, stress levels decrease with increasing temperature; measurement temperatures are 140, 150, 160, and 180°C. The temperature is 195°C for the half filled circles and 205°C for the filled circles. (Reprinted with permission from Ref. 30. Copyright 1997 American Chemical Society.)

LCPs as well [31], although such reports are relatively infrequent. Also note that shear thickening is sometimes observed with thermotropic LCPs in a limited range of temperatures and shear rates [32].

In terms of explaining and making sense of all these diverse observations, the theory that has proven to be the most successful has been authored by Doi [27,33] and modified by others. This theory for lyotropic LCPs is highly mathematical, but particularly lucid expositions have been provided by Marrucci and Greco [34] and Marrucci [35]. The essential ideas can be sketched as follows. In the nematic phase, the rodlike molecules are, on average, aligned in the direction of the director n, as shown in Fig. 8.16. However, the individual molecules show a distribution of orientations around u around n; here u is a unit vector aligned with the molecular axis. One can, therefore, define an orientation distribution function $f(u)$ such that $f(u)\, d\theta$ is the fraction of molecules whose molecular

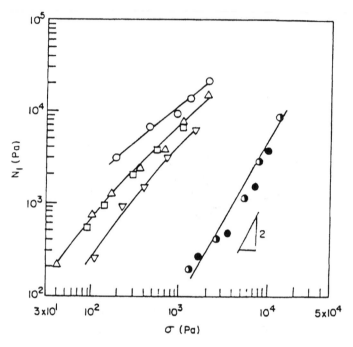

FIGURE 8.15 First normal stress difference versus shear stress for PSHQ6-16 at various temperatures. Symbols have the same meaning as in Fig. 8.14. (Reprinted with permission from Ref. 30. Copyright 1997 American Chemical Society.)

orientation falls in the neighborhood $d\theta$ of u. At equilibrium, the us are symmetrically distributed about n, but during flow f changes. The time-dependent value $f(u,t)$ is obtained by solving the diffusion equation for the rods, and this equation involves the rotational diffusivity and the imposed velocity gradient. While in the isotropic case the diffusion coefficient decreases with increasing concentra-

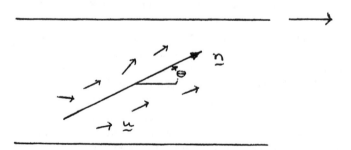

FIGURE 8.16 Flow director n and individual orientations u.

tion due to polymer–polymer interactions, in the anisotropic case its value depends on the extent of orientation and is larger than in the isotropic situation. It is this increase in diffusivity that causes the reduction in viscosity on the formation of the nematic phase. A correct formulation of the diffusion equation is at the heart of the Doi theory. On solving the diffusion equation, one gets the distribution function that can be employed to obtain the average value of any quantity of interest. In particular, one calculates the order parameter tensor $\underset{\sim}{S}$ given by

$$\underset{\sim}{S} = \langle \underset{\sim}{u}\underset{\sim}{u} \rangle = \int \underset{\sim}{u}\underset{\sim}{u}\, f(u,t)\, d\theta \tag{8.5}$$

which is the second-order moment of the orientational distribution. The utility of Eq. (8.5) is that it allows for the determination of the stress tensor $\underset{\sim}{\sigma}$, which can be written as

$$\underset{\sim}{\sigma} = 3ckT\underset{\sim}{S} + c\,\langle \underset{\sim}{u} \nabla V \rangle \tag{8.6}$$

in which c is the polymer concentration, k is Boltzmann's constant, and V is a potential that is experienced by a given molecule due to the presence of the other molecules. The functional form of V has to be specified before the stress can be calculated; this is part of the Doi theory.

On cessation of steady flow, the molecules possess uniaxial symmetry about the director, and the order parameter tensor takes the form

$$\underset{\sim}{S} = S\left(\underset{\sim}{n}\underset{\sim}{n} - \frac{\underset{\sim}{\delta}}{3} \right) + \frac{\underset{\sim}{\delta}}{3} \tag{8.7}$$

where the scalar S is called the order parameter and $\underset{\sim}{\delta}$ is the unit tensor. For perfect alignment, S is unity; for a random distribution of rods, S takes the value zero. The equilibrium value of S is obviously zero in the isotropic phase, but in the nematic phase it is given by

$$S = 0.25 + 0.75\left(1 - \frac{8\phi^*}{9\phi} \right)^{0.5} \tag{8.8}$$

in which ϕ is the polymer volume fraction and ϕ^* is the value of ϕ at which only the anisotropic phase is thermodynamically stable. Clearly, the lowest value of S for a fully liquid crystalline system is 0.5.

For shear flow at low shear rates, the Doi theory predicts that the zero-shear viscosity is

$$\frac{\eta}{\eta^*} = \left(\frac{\phi}{\phi^*} \right)^3 \frac{(1 - S)^4(1 + S)^2(1 + 2S)(1 + 1.5S)}{(1 + 0.5S)^2} \tag{8.9}$$

where η^* is the viscosity at ϕ^*; the value of S is specified by Eq. (8.8). The shape of the curve of zero-shear viscosity versus concentration given by Eq. (8.9) is very similar to actual observations, as displayed in Fig. 8.1. Note, however, that Eq. (8.9) predicts a monotonic decrease in viscosity with increasing concentration due to a monotonic increase in the value of the order parameter. Other predictions in steady-shear flow include a linear dependence of the first normal stress difference of anisotropic solutions on the shear rate rather than a quadratic one, as in the case of flexible polymer solutions; this seems to be reasonable [31,36]. Additionally, the minimum value of the low shear rate ratio of the first normal stress difference to the shear stress is calculated to be 1.5, and it is a monotonically increasing function of the order parameter [37]; this value of 1.5 is significantly larger than what is found experimentally [38]. Finally, the second normal stress difference is negative and, in magnitude, equals 14% of the first normal stress difference when S is 0.5. It is important to mention that the Doi theory originally predicted orientation in shear because of a closure approximation; removal of this approximation leads to a negative value of N_1.

For shear flow at high shear rates, the stresses predicted by the Doi theory, and these have to be determined numerically, tend to saturate and become constant. This is clearly unrealistic, and according to Doraiswamy and Metzner [39] and Maffetone et al. [39a] this results from a neglect of rod-solvent friction and the stress in the solvent itself. While the contribution of these two terms is negligible at low shear rates, it cannot be disregarded at high shear rates. This is evident from an examination of Fig. 8.17, which shows the flow curve for a 40% by weight solution of 100,000 molecular weight HPC in glacial acetic acid [39]. Note that the experimental data in this figure are a combination of capillary viscometer and cone-and-plate viscometer data, and there is no influence of system geometry or flow inhomogeneity (this data superposition is also observed for thermotropic LCPs [31]). Agreement between the modified theory and experimental data is perfect.

If one re-examines the data presented in Fig. 8.17, one will notice that the viscosity is constant at low shear rates. Indeed, the molecular theory of Doi [27,33] cannot predict a zero-shear rate viscosity that is non-Newtonian. How then does one explain the shear-thinning Region I that is often observed? Also how does one explain the two changes in the sign of the first normal stress difference with increasing shear rate? It turns out that the essential physics that leads to the occurrence of these two phenomena is not included in the Doi theory as formulated originally. A basic tenet of this theory is that flow causes alignment of the director $\underset{\sim}{n}$; the larger the shear rate, the smaller is the (time-independent) value of the angle θ in Fig. 8.16.

However, as explained by Marrucci and Maffettone [40], when the shear rate is lowered progressively while one is still in Region III of Fig. 8.5 the director begins to tumble; i.e., the director rotates indefinitely in the plane of shear even when the shear rate is held constant. In between the flow aligning and tumbling regimes, there may be a "wagging" regime where the director continually oscil-

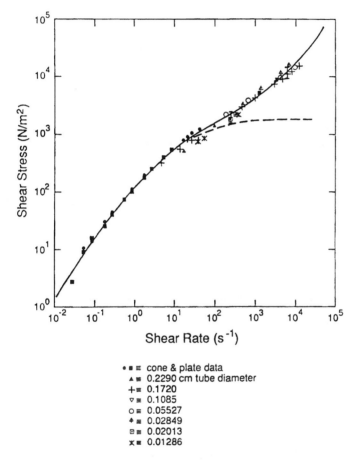

FIGURE 8.17 Comparison of data and theoretical correlation for the shearing flow of 40 wt% hydroxypropylcellulose in glacial acetic acid. The dashed curve is the original Doi theory. (From Ref. 39. Used by permission of Steinkopff Publishers, Darmstadt, FRG.)

lates between two limiting orientations [34,41]. The critical shear rate beyond which there is only flow alignment is located within the negative N_1 flow region, just past the minimum in the N_1 curve. Below the critical shear rate, the director either tumbles or wags and never achieves a steady orientation [29,40]. It is this director tumbling and wagging that is responsible for the observations of a negative first normal stress difference and a positive second normal stress difference. Furthermore, the tumbling of the director at low shear rates creates polydomains and leads to the formation of defects. These domains can be observed using polarized-light microscopy, and are shown in schematic form in Fig. 8.18. As explained originally by Onogi and Asada [9], only piled polydomains are present in the nematic phase at rest. On shearing, this structure gives way to small domains suspended in a continuous phase, and finally there is a transition to a monodomain continuous phase.

The Doi theory results presented in this section are valid only for the flow alignment region of homogeneous nematics where a single domain exists. Thus, to predict the viscosity behavior of Region I in Fig. 8.5, one needs to account for the presence of polydomains. Unfortunately, such a theory is lacking at present.

IV. EXTENSIONAL FLOW

Metzner and Prilutski [37] have carried out fiber spinning of a liquid crystalline, HPC-in-glacial-acetic-acid solution. They found that this material exhibited very

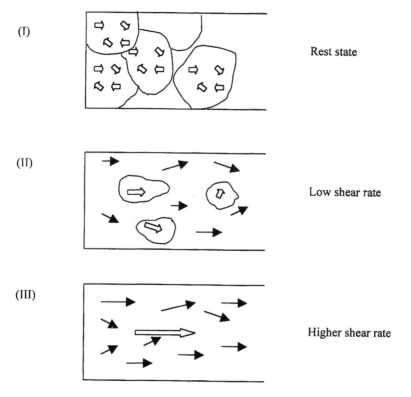

(I)   Rest state

(II)   Low shear rate

(III)   Higher shear rate

FIGURE 8.18   Schematic representation of the bulk structure of polymer liquid crystals with increasing shear rate.

little die swell—even less than the Newtonian value—despite the fact that the first normal stress difference in shear was fairly large. Recall from Sec. III.B of Chap. 5 that die swell and the primary normal stress difference are inextricably interrelated for flexible-chain polymer systems. Data for the tensile stress as a function of stretch rate are shown in Fig. 8.19, where it is seen that, within experimental error, all the data fall on the same curve [37,39]. The stress levels are low compared to those for flexible polymers, and it is obvious that there are no memory effects. In other words, any transients present are of short duration. This is contrary to shear behavior, where transients of very long duration are observed [37]. No doubt, the relative absence of transients is caused by the easy and complete alignment of the rodlike molecules under the influence of elongational flow. Figure 8.19 also reveals that at low stretch rates the stress is directly proportional to the stretch rate. Thus, the extensional viscosity is constant. However, the Trouton limit is not observed, and the extensional viscosity is slightly larger than thrice the zero-shear viscosity. Also, there is no strain hardening, and, as observed

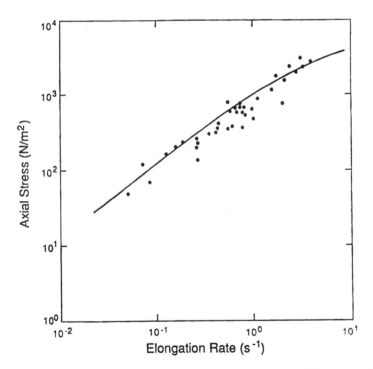

FIGURE 8.19 Comparison of the extensional-flow data of Metzner and Prilutski [37] with predictions of the Doi theory. (From Ref. 39. Used by permission of Steinkopff Publishers, Darmstadt, FRG.)

by Metzner and Prilutski, this bodes ill for the stability of polymer processing operations such as fiber spinning and film blowing. Indeed, other researchers have experienced difficulties is stretching lyotropic [42] as well as thermotropic [43,44] LCPs. Although one might argue that a fiber spinning experiment does not give the true extensional viscosity in general [43], the lack of memory effects means that transient and equilibrium values of the extensional viscosity are one and the same. In the present case, this was confirmed by Ooi and Sridhar [42], who measured the steady-state extensional viscosity of the same HPC solution with their filament stretching viscometer; there was very good overlap with the data shown in Figure 8.19. The solid line shown in this figure is the prediction of the Doi theory using the same parameter values as those employed for the data comparison in Figure 8.17. Clearly, the theory fits the data very well.

In closing, we note that Wilson and Baird [43] have measured the tensile stress growth coefficient for two thermotropic LCPs using a rotary clamp extensional viscometer. However, they were not able to obtain steady-state stress values due to sample necking or sample failure. Their transient data showed linear behavior at small strains and slight strain hardening at larger strains. They speculated that the strain hardening might be the result of residual crystallinity in their samples.

V. POLYDOMAINS AND TRANSIENT FLOW

So far in this chapter, we have focused either on steady-state flow or on flow at a constant deformation rate. Liquid crystals, however, are unusual in that they exhibit a virtual cornucopia of intriguing time-dependent stress responses. These transient data have been utilized for the purpose of probing the textural structural characteristics of LCPs in Regions I and II of the flow curve of Fig. 8.5 [45]. Besides the measurement of dynamic mechanical data and the stress response either after initiation of steady-shearing flow or upon cessation of such flow, experiments that have been conducted include stepwise changes in shear rate, flow reversal, and intermittent shear. It is found that the equation of linear viscoelasticity is not obeyed despite the fact that the strain rates are low enough that the viscosity is Newtonian [46]. These data, though, are useful for verifying scaling relations predicted based on dimensional arguments [34].

The preceding ideas can be illustrated using a flow-reversal experiment in which material is initially sheared at a constant shear rate. Upon attainment of a steady state, the flow direction (but not the magnitude of the shear rate) is changed, and the stresses are monitored as a function of time. If the instantaneous stress is made dimensionless with respect to the steady-state stress, dimensional analysis predicts that

$$\sigma_{red} = \frac{\sigma}{\sigma_{ss}} = f(\dot{\gamma}t) \tag{8.10}$$

and this scaling is valid provided that the material does not possess an internal time scale [47]. The truth of Eq. (8.10) is shown in Fig. 8.20 using data on a polybenzylglutamate solution in *m*-cresol [47]; a single, shear-rate-independent curve is obtained. According to Larson and Doi [46], simple scaling laws, such as Eq. (8.10), are a manifestation of the influence of texture on the rheology of LCPs, and they can be built into a phenomenological constitutive equation that might be useful for predicting other rheological phenomena. By using the Leslie–

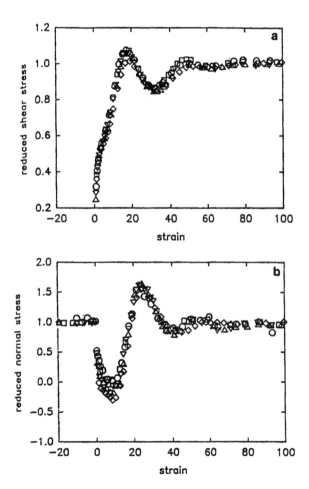

FIGURE 8.20 Reduced stresses (a: shear stress; b: first normal stress difference), after flow reversal for a PBG solution at 293 K at shear rates ranging from 0.3 to 6 sec⁻¹. (From Ref. 47, with permission from Elsevier Science.)

Ericksen continuum theory as the point of departure, these authors have formulated a model that involves averaging the predictions of this equation over a region that is large compared to a domain but small compared to bulk dimensions [47]. Results appear to be in qualitative agreement with experimental observations. Refinements of this idea and the formulation of new ideas for the description of rheological behavior of LCPs in the presence of a domain structure are topics of current research.

REFERENCES

1. S.M. Aharoni. Rigid backbone polymers: XVII. Solution viscosity of polydisperse systems. Polymer 21:1413–1422 (1980).
2. K. F. Wissbrun. Rheology of rod-like polymers in the liquid crystalline state. J. Rheol. 25:619–662 (1981).
3. P.J. Flory. Phase equilibria in solutions of rod-like particles. Proc. Roy. Soc. A 234: 73–89 (1956).
4. J. Preston. High-strength/high-modulus fibers from aromatic polymers. J. Chem. Education 58:935–937 (1981).
5. National Materials Advisory Board. Liquid Crystalline Polymers. National Academy Press, Washington, DC, 1990.
6. W.J. Jackson Jr., H. Kuhfuss. Liquid crystal polymers: I. Preparation and properties of p-hydroxybenzoic acid copolyesters. J. Polym. Sci. Polym. Chem. Ed. 14:2043–2058 (1976).
7. S.S. Kim, C.D. Han. Effect of molecular weight on the rheological behavior of thermotropic liquid-crystalline polymer. Macromolecules 26:6633–6642 (1993).
8. S.S. Kim, C.D. Han. Effect of shear history on the steady shear flow behavior of a thermotropic liquid-crystalline polymer. J. Polym. Sci.: Part B: Polym. Phys. 32: 371–381 (1994).
9. S. Onogi, T. Asada. Rheology and rheo-optics of polymer liquid crystals. Proc. VIIIth Int. Congress on Rheol., Naples, Italy, 1:127–147 (1980).
10. G. Kiss, R.S. Porter. Rheology of concentrated solutions of poly(γ-benzyl-glutamate). J. Polym. Sci.: Polym. Symp. 65:193–211 (1978).
11. J.M. Dealy, K.F. Wissbrun. Melt Rheology and Its Role in Plastics Processing. Van Nostrand Reinhold, New York, 1990, pp. 424–440.
12. H. Marsh, R. Menendez. Carbons from pyrolysis of pitches, coals and their blends. Fuel Processing Technol. 20:269–296 (1988).
13. G.S. Attard, G. Williams. Liquid-crystalline side-chain polymers. Chem. Britain 22: 919–924 (1986).
14. D.G. Baird, J.K. Smith. Dilute solution properties of poly(1,4-phenylene terephthalamide) in sulfuric acid. J. Polym. Sci.: Polym. Chem. Ed. 16:61–70 (1978).
15. J.G. Kirkwood, P.L. Auer. The visco-elastic properties of solutions of rod-like macromolecules. J. Chem. Phys. 19:281–283 (1951).
16. H. Yamakawa. Modern Theory of Polymer Solutions. Harper & Row, New York, 1971, pp. 324–328.

17. C.-P. Wong, H. Ohnuma, G.C. Berry. Properties of some rodlike polymers in solution. J. Polym. Sci.: Polym. Symp. 65: 173–192 (1978).

18. H.L. Frisch, R. Simha. The viscosity of colloidal suspensions and macromolecular solutions. In: F.R. Eirich, ed. Rheology. Vol. 1. Academic Press, New York, 1956, pp. 525–613.

19. J.G. Brodnyan. The concentration dependence of the Newtonian viscosity of prolate ellipsoids. Trans. Soc. Rheol. 3:61–68 (1959).

20. D.G. Baird, R.L. Ballman. Comparison of the rheological properties of concentrated solutions of a rodlike and a flexible chain polyamide. J. Rheol. 23:505–524 (1979).

21. R.B. Bird, H.R. Warner Jr., D.C. Evans. Kinetic theory and rheology of dumbbell suspensions with Brownian motion. Adv. Polym. Sci. 8:1–90 (1971).

22. R.B. Bird, C.F. Curtiss. Kinetic theory and rheology of solutions of rigid rodlike macromolecules. J. Non-Newt. Fluid Mech. 14:85–101 (1984).

23. P.J. Carreau, D. De Kee, R.P. Chhabra. Rheology of Polymeric Systems. Hanser, Munich, 1997, pp.312–320.

24. P.J. Carreau, M. Grmela, A. Rollin. Rheological models for rigid and wormlike macromolecules. In: D. De Kee, P.N. Kaloni, eds. Recent Developments in Structured Continua. Longman Scientific and Technical, New York, 1990, Chap. 6.

25. M. Doi, S.F. Edwards. Dynamics of rod-like macromolecules in concentrated solution. Part 2. J. Chem. Soc. Faraday Trans. II 74:918–932 (1978).

26. N.Y. Kuzuu, M. Doi. Nonlinear viscoelasticity of concentrated solution of rod-like polymers. Polym. J. 12:883–890 (1980).

27. M. Doi. Molecular dynamics and rheological properties of concentrated solutions of rodlike polymers in isotropic and liquid crystalline phases. J. Polym. Sci.: Polym. Phys. Ed. 19:229–243 (1981).

28. N. Grizzuti, S. Cavella, P. Cicarelli. Transient and steady-state rheology of a liquid crystalline hydroxypropylcellulose solution. J. Rheol. 34:1293–1310 (1990).

29. J.J. Magda, S.-G. Baek, K.L. DeVries, R.G. Larson. Shear flows of liquid crystal polymers: measurements of the second normal stress difference and the Doi molecular theory. Macromolecules 24:4460–4468 (1991).

30. S. Chang, C.D. Han. A thermotropic main-chain random copolyester containing flexible spacers of differing lengths. 2. Rheological Behavior. Macromolecules 30: 1656–1669 (1997).

31. A.D. Gotsis, D.G. Baird. Rheological properties of liquid crystalline copolyester melts. II. Comparison of capillary and rotary rheometer results. J. Rheol. 29:539–556 (1985).

32. F.N. Cogswell, K.F. Wissbrun. Rheology and processing of liquid crystal polymer melts. In: D. Acierno, A.A. Collyer, eds. Rheology and Processing of Liquid Crystal Polymers. Chapman & Hall, London, 1996, pp. 86–134.

33. M. Doi. Rheological properties of rodlike polymers in isotropic and liquid crystalline phases. Ferroelectrics 30:247–254 (1980).

34. G. Marrucci, F. Greco. Flow behavior of liquid crystalline polymers. Adv. Chem. Phys. 86:331–404 (1993).

35. G. Marrucci. Theoretical aspects of the flow of liquid crystal polymers. In: D. Acierno, A.A. Collyer, eds. Rheology and Processing of Liquid Crystal Polymers. Chapman & Hall, London, 1996, pp. 30–48.

36. P. Moldenaers, J. Mewis. Transient behavior of liquid crystalline solutions of poly (benzylglutamate). J. Rheol. 30:567–584 (1986).

37. A.B. Metzner, G. Prilutski. Rheological properties of polymeric liquid crystals. J. Rheol. 30:661–691 (1986).

38. P. Moldenaers, J. Mewis. Relaxational phenomena and anisotropy in lyotropic polymeric liquid crystals. J. Non-Newt. Fluid Mech. 34:359–374 (1990).

39. D. Doraiswamy, A.B. Metzner. The rheology of polymeric liquid crystals. Rheol. Acta 25:580–587 (1986).

39a. P.L. Maffetone, G. Marrucci, M. Mortier, P. Moldenaers, J. Mewis. Dynamic characterization of liquid crystalline polymers under flow-aligning shear conditions. J. Chem. Phys. 100:7736–7743 (1994).

40. G. Marrucci, P.L. Maffettone. Description of the liquid-crystalline phase of rodlike polymers at high shear rates. Macromolecules 22:4076–4082 (1989).

41. J. Mewis, M. Mortier, J. Vermant, P. Moldenaers. Experimental evidence for the existence of a wagging regime in polymeric liquid crystals. Macromolecules 30: 1323–1328 (1997).

42. Y.W. Ooi, T. Sridhar. Uniaxial extension of a lyotropic liquid crystalline polymer solution. Ind. Eng. Chem. Res. 33:2368–2373 (1994).

43. T.S. Wilson, D.G. Baird. Transient elongational flow behavior of thermotropic liquid crystalline polymers. J. Non-Newt. Fluid Mech. 44:85–112 (1992).

44. D.D. Edie, M.G. Dunham. Melt spinning pitch-based carbon fibers. Carbon 27:647–655 (1989).

45. P. Moldenaers. Time-dependent effects in lyotropic systems. In: D. Acierno, A.A. Collyer, eds. Rheology and Processing of Liquid Crystal Polymers. Chapman & Hall, London, 1996, pp. 251–287.

46. R.G. Larson, M. Doi. Mesoscopic domain theory for textured liquid crystalline polymers. J. Rheol. 35:539–563 (1991).

47. P. Moldenaers, M. Mortier, J. Mewis. Transient normal stresses in lyotropic liquid crystalline polymers. Chem. Eng. Sci. 49:699–707 (1994).

9

Yield Stress, Wall Slip, Particle Migration, and Other Observations with Multiphase Systems

I. INTRODUCTION

So far in this book we have examined the rheology of single-phase polymer melts and polymer solutions. In the remainder of the book we discuss the flow of two-phase systems, such as suspensions, emulsions, and foams. The introduction of a dispersed phase within the matrix of a continuous polymer phase not only leads to a consideration of new variables but also results in new complications. The new variables include, among others, the relative amounts of the two phases and the shape, size, and size distribution of the dispersed phase. The complications arise from the very nature of a two-phase system and the need to measure true material properties uninfluenced by the nature of the flow field, the kind of instrument chosen, or the fixture geometry employed. If the densities of the two phases are not matched, composition nonuniformities can arise in the material over the time scale of measurement, due either to sedimentation or creaming of the dispersed phase; in an extreme situation, there can be complete separation of the two phases. Concentrated two-phase systems are structured fluids, and this structure can endow the fluid with a yield stress. If the magnitude of the yield stress is larger than the force of adhesion between the fluid and the viscometer wall, the application of a stress that is smaller in value than the yield stress can result in slip of the fluid at a solid surface rather than deformation in the bulk of the material. If adhesion is restored rapidly, and this happens commonly in the flow of gels, the phenomenon of stick-slip may be witnessed [1]. Here the motion of a solid boundary causes fluid that is in contact with it to deform elastically, building up stress. The stress overcomes the force of adhesion, resulting in loss of contact between the fluid and the solid surface. This, in turn, makes the stress in the fluid relax and restores fluid–solid contact and adhesion. Once this happens, the cycle begins anew. If adhesion is not restored quickly, there can be complete separation at the solid boundary. This is illustrated in Fig. 9.1 for the torsional flow of a food emulsion in a parallel-plate viscometer [2]. A thin, vertical line is painted on the outer surface of the sample, and it makes contact with both the plates. On rotating the lower plate, there is a progressive decrease in the slope of the marker line because material adheres to both solid surfaces. At a later time, however, there is a separation of the marker at the upper boundary; the emulsion still sticks to the moving lower plate but slips on the upper one. The loss of adhesion is accompanied by a reduction in strain, with a concomitant increase in the slope of the marker line. Upon increasing the speed of rotation, material slips at the lower boundary as well and merely rotates as a solid body. This same phenomenon also manifests itself in concentric cylinder instruments, where the

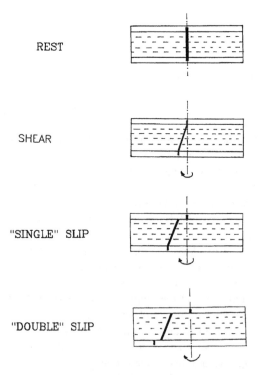

REST

SHEAR

"SINGLE" SLIP

"DOUBLE" SLIP

FIGURE 9.1   Macroscopic slip of a food emulsion in a parallel-plate viscometer.

member that is rotated simply cuts a hole through the sample. Note that slip and detachment can occur in single-phase fluids as well. Fig. 9.2 shows this for the stretching of a polyacrylamide solution emerging from a capillary; increasing the tensile stress makes the point of detachment move upstream into the capillary [3]. Note also that slip at the capillary wall has been proposed as the cause of the melt fracture instability in polymeric fluids [4]; this is examined in Chapter 17.

In addition to the macroscopic wall slip just described, we can also have microscopic or apparent wall slip resulting from wall depletion effects [5]. Since the region near the fixture walls must necessarily be depleted of the dispersed phase, whether particulates, droplets, or macromolecules, one has a low-viscosity fluid layer in contact with the fixture surface. This layer of the continuous phase acts as a lubricant and facilitates flow. The net result is a lowering of the measured viscosity, and one says that fluid slips at the wall. As shown later in this chapter, it is possible to correct for apparent slip provided that the dispersed-phase concentration remains uniform away from the wall. For flow through a tube, however, particles and drops that are excluded from the region in the immediate vicinity of the wall enter a faster-moving stream of liquid and are convected downstream

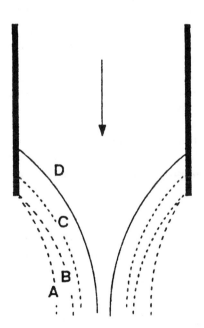

FIGURE 9.2 Illustration of the fluid detachment phenomenon. The take-up velocity increases as one goes from profile A to profile D. (From Ref. 3, with permission from Elsevier Science.)

more rapidly than the average of the suspending liquid, resulting in an axial solids concentration gradient. Particle migration can also occur due to forces arising from fluid inertia and fluid elasticity [6]. Furthermore, if the deformation rate varies with position in the flow field, the number of collisions experienced by a particle per unit time will also vary with position. This spatial variation of collision frequency can again cause particles, especially in concentrated suspension, to move across streamlines from a region of high shear rate to a region of low shear rate [7]. All this complicates the process of data analysis. Other complications can arise due to composition changes on account of polymer adsorption on the viscometer wall, sample drying, air entrapment within the sample, and the appearance of a new phase due to fluid deformation—examples include crystallization of polymer melts and the phase separation of polymer solutions. Flow can also result in segregation by size if the dispersed phase has a size distribution. Unless one is observant and works to actively suppress these phenomena, they can individually and collectively conspire to frustrate the investigator and the investigation!

In the chapters that follow, we examine the rheological properties of particulate and short-fiber suspensions, emulsions, foams, and granular powders. In these chapters, we assume that viscometric measurements yield true material data not corrupted by macroscopic wall slip or concentration inhomogeneities. We, however, allow for the possibility of thixotropy, or reversible, flow-induced microstructural changes [8]; the microstructure may be characterized by specifying the floc sizes, droplet sizes, fiber orientations, or the polymer entanglement density. It is assumed that flow determines the microstructure and that the microstructure determines the rheology. It is also assumed that transient measurements reveal information about microstructure evolution. Indeed, obtaining nonequilibrium structures is often the goal during the processing of two-phase systems such as immiscible polymer blends. In the present chapter, though, we seek to understand the causes of some of the unsettling effects just listed and how we might eliminate them or properly account for them in the data analysis.

II. UNSTABLE SUSPENSIONS

The flow of particulate suspensions is encountered in a myriad of applications involving materials such as filled polymers, paints, mineral slurries, fuels, and building materials. In order that rheological measurements made on these materials be reproducible in time and be relatable to the microstructure, it is necessary that particles neither settle nor agglomerate during the process of measurement. For colloidal particles, settling is not a concern unless flocculation takes place under the action of the London–van der Waals forces of attraction due to the large surface–area–to–volume ratio. Particle–particle collisions resulting from Brownian motion as well as any imposed flow field typically lead to the formation of flocs that can settle as a soft, loosely packed sediment under the influence of gravity. Such a suspension can be made stable against flocculation by either electrostatic or steric means [9]. However, if the densities of the two phases are not the same, the suspension will still be unstable against sedimentation; the deflocculated particles will settle in the form of a close-packed sediment that is difficult to redisperse. For large particles, settling is the major concern, since surface forces can safely be neglected; the rate of settling can be calculated theoretically [10], and settling can often be retarded by the use of a viscous suspending medium. However, even in the case of large particles, shear-induced collisions can result in agglomeration if binding mechanisms such as liquid bridges or mechanical interlocking are present [11]. Clearly, the ideal situation is one where the suspension is stable against both flocculation as well as sedimentation. This can be ensured in model studies but is obviously not always practical in the case of industrial materials.

For settling slurries that are stable against flocculation, proper rheological measurements can be made if the viscometer employed promotes fluid mixing. This situation can be achieved either by modifying standard instruments or by devising novel solutions. One possible design involving viscometer modification and taken from the work of Overend et al. [12] is shown in Fig. 9.3. This is a Couette viscometer in which an outer bowl surrounds a fixed cylinder in the form of a slotted sleeve having open ends. The motion of the bowl keeps the slurry from settling while viscosity is determined in the usual manner by measuring the torque resulting from shearing of the sample between the slotted sleeve and an inner rotating cylinder. The instrument is calibrated using Newtonian liquids of

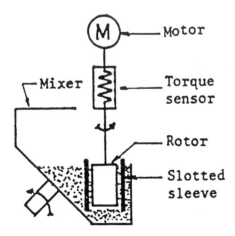

FIGURE 9.3   A modified Couette viscometer for use with settling slurries. (From Ref. 12.)

different viscosity levels, and there is provision for changing the temperature of measurement.

Kraynik and coworkers [13,14] have proposed using a single flighted, single-screw extruder of uniform channel depth and operated at no net discharge for measuring the flow curve of concentrated suspensions that tend to phase separate due to sedimentation; fluid recirculation resulting from relative motion between the extruder barrel and the screw keeps the solid particles in suspension. This instrument, called the helical-screw rheometer (HSR) and sketched in Fig. 9.4, can operate at high pressures, at elevated temperatures, and under conditions where chemical reactions cause large and rapid changes in the viscosity. If one assumes a specific rheological constitutive equation, such as the Bingham plastic or power-law model, for describing the flow behavior of the settling slurry, measurements of the pressure drop along the central portion of the extruder barrel as a function of the speed of screw rotation can yield the model parameters; this process has been applied to hydraulic fracturing fluids that are mixtures of inert particulates in chemically reactive polymeric liquids by Lord [15]. Note that, for unstable suspensions, Lombardi [16] suggests modifying the Bingham plastic equation by the addition of a term arising from Coulomb friction effects. A general analysis of the HSR that does not require assuming a particular constitutive model has also been published by Kraynik and Romero [14].

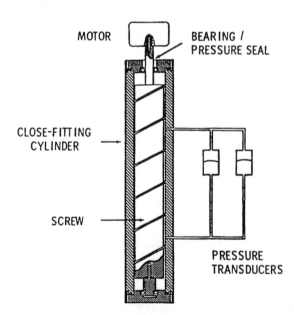

FIGURE 9.4 Diagram of the helical-screw rheometer showing parallel pressure tap arrangement. (From Ref. 13.)

When a concentrated suspension is not stable against flocculation, the particles may form a network structure that is so large that it entraps all available liquid. In this case, a large yield stress will develop, and there will be no visible sedimentation or phase separation. The resulting material will behave as a Bingham plastic liquid or as a power-law fluid with a yield stress (called a Herschel–Bulkley fluid). Since the suspension will not flow easily, standard viscometers cannot be employed to measure the fluid rheology. The helical-screw rheometer could be used, but the extruder will probably destroy the network structure and not give the (low shear rate) properties characteristic of the material at rest. The simplest approach in this situation is to utilize the squeezing-flow viscometer described in Sec. VII.B of Chapter 2. Representative data, for a 20% by volume suspension of 0.5 micron average-sized alumina in paraffin oil, plotted as the squeezing force versus plate separation for a constant squeezing velocity are shown in Fig. 9.5 [17]. The yield stress estimated by fitting these data to the Herschel–Bulkley model (the solid line) is almost 2,000 Pa; theory may be found in Covey and Stanmore [18]. The stress levels decrease rapidly as increasing

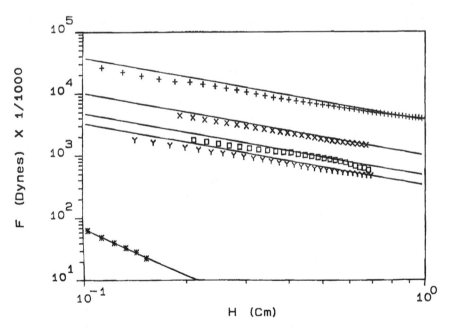

FIGURE 9.5 Squeezing force versus plate separation for an alumina-in-paraffin oil suspension at a plate velocity of 1 cm/min. Force values are lowered progressively as the amount of dispersant is increased in steps from nothing to 1, 1.5, 1.7, and 2% by weight of solids in the suspension. The solid lines are fits to the Herschel–Bulkley fluid model. (From Ref. 17.)

amounts of a dispersant, which acts as a steric stabilizer, is added to the suspension; here the dispersant is ICl's KD3 hypermer, and it has a polar anchoring group that attaches to the particulate surface while the polymer chain extends into the solvent. At the 2% dispersant level, the suspension behaves as a power-law liquid, and rheological properties can be measured with a cone-and-plate viscometer [19]. For this flocculated suspension, the Herschel–Bulkley model does a good job of data representation. This and other models that incorporate a yield stress are described in detail in Sec. VI of Chapter 10.

III. YIELD STRESS MEASUREMENT

The yield stress is defined as that stress below which a material is an elastic solid; this stress must be exceeded before flow can take place. A striking visual demonstration of this phenomenon was provided by Hartnett and Hu [20] by placing a nylon sphere having a density of 1.156 g/cm³ inside a graduated cylinder filled with an aqueous Carbopol (carboxylpolymethylene) solution of 1.008 g/cm³ density; due to the presence of a yield stress resulting from polymer network formation, the nylon ball remained immobile for several months! More commonly, though, yield stresses are observed in concentrated suspensions and emulsions, and their presence is detected by the tendency of the curve of shear stress versus shear rate to approach a finite, limiting stress value on reducing the shear rate to zero. The numerical value of the yield stress, however, appears to depend on the shear rate range of the viscometer and the technique employed, whether graphical or based on a rheological model, to extrapolate data to zero-shear rate. Even in the absence of thixotropy or time-dependent effects, difficulties can arise; they are illustrated in Fig. 9.6 [21]. It is evident that widely differing yield stress values can be computed. This observation prompted Barnes and Walters [22] to claim that the yield stress was a myth and that, even for structured fluids, the viscosity, although large, was always finite.

Extrapolation of the flow curve to vanishingly low shear rates involves approaching the yield stress from above, and this procedure gives the dynamic yield stress. The alternative, shown in Fig. 9.7, is to approach the yield stress from below [21]. In this method, which gives the static yield stress, the stress applied initially is lower than the yield stress. Under the influence of the applied stress, the material creeps, but the strain eventually becomes constant. For a higher value of the applied stress, though, the strain continues to increase, and ultimately the strain rate becomes constant. The yield stress is determined by progressively narrowing the stress interval between these two different kinds of behavior. Difficulties again arise due to the subjective nature of decision making, and a determination of whether a constant strain is attained or not depends on the time scale of observation [21]. In general, the static yield stress exceeds the dynamic yield stress in value. Several authors have reviewed the different meth-

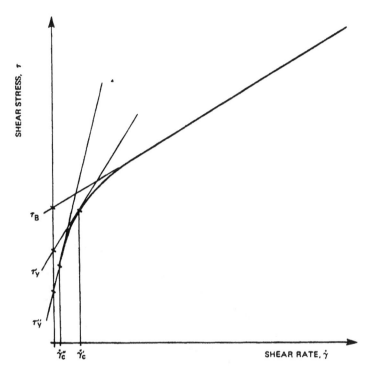

FIGURE 9.6   Extrapolation of the equilibrium-flow curve. (From Ref. 21. Used by permission of Steinkopff Publishers, Darmstadt, FRG.)

ods of yield stress measurement [23–26], and the techniques that have proven more popular than others are descibed next.

A.   Fluid Stability on an Inclined Plane [27]

An extremely simple method of measuring the static yield stress τ_y of a fluid is to slowly increase the angle of inclination θ (with the horizontal) of a plane that is covered by the fluid to depth H and exposed to the atmosphere. In this situation, the shear stress is zero at the air–liquid interface, and it increases linearly with depth perpendicular to the plane, becoming a maximum at the liquid–solid interface. The fluid remains stationary so long as the shear stress at the solid surface does not exceed the yield stress of the fluid; flow, indicated by the sudden motion of tracer particles on the liquid surface, takes place when the following condition is satisfied:

$$\tau_y = \rho g H \sin \theta \tag{9.1}$$

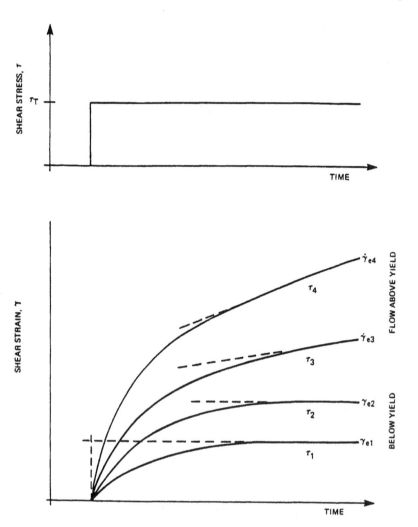

FIGURE 9.7 Creep response after application of a constant shear stress. (From Ref. 21. Used by permission of Steinkopff Publishers, Darmstadt, FRG.)

where g is the acceleration due to gravity and ρ is the liquid density. Thus, a single point measurement of the fluid depth and critical angle of inclination gives the yield stress. For better accuracy, we can plot the liquid depth as a function of the reciprocal of the sine of the critical value of the angle of inclination. The yield stress then follows from the slope of the resulting straight-line graph. According to Uhlherr et al. [27], this method can detect a yield stress as low as 3 mPa.

B. The Gun Rheometer

As shown in Fig. 9.8, this instrument is a pressure vessel to which is connected a tube containing the sample in the form of a cylindrical plug of diameter D and length L [21]. The pressure in the vessel is gradually increased until the sample is discharged. The yield stress is computed from a force balance and a knowledge of the minimum pressure drop Δp applied to eject the sample. The result is:

$$\tau_y = \frac{D}{4} \frac{\Delta p}{L} \qquad (9.2)$$

This is a simple technique suitable for liquids that are more viscous than those that might be employed with the inclined plane described previously. The instrument was developed at the Warren Spring Laboratory in the United Kingdom and is commercially available.

C. Vane Method

If a force or torque is applied to a solid object that is immersed in the fluid whose yield stress is to be determined, the stress that acts on the solid surface at the

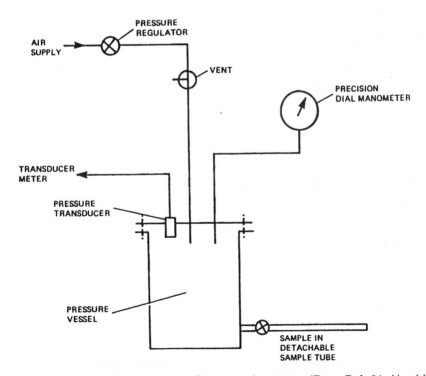

FIGURE 9.8 Schematic diagram of the gun rheometer. (From Ref. 21. Used by permission of Steinkopff Publishers, Darmstadt, FRG.)

point of incipient relative motion will equal the yield stress. As a consequence, the yield stress can be determined either from a force balance on a plate [25] or from a torque balance on a vane [26,28]. In the vane method, which is decidedly the more popular method, a four-bladed vane, similar to the one shown in Fig. 9.9 [28], is rotated very slowly at a constant rotational speed. The torque needed to maintain the motion is recorded as a function of time, and it is found that the torque goes through a maximum. Representative data for an alumina-in-paraffin oil suspension are displayed in Fig. 9.10 [17]. The initial, linear region in this figure represents elastic deformation of the material; the maximum in the torque, T_m, represents the point of yielding. The subsequent decrease in the torque is due to a combination of structure breakdown and material slip at the vane surface. Since the material between the blades moves along with the vane, a torque bal-

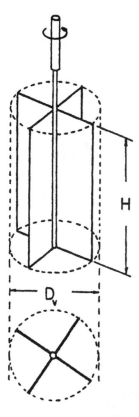

FIGURE 9.9 Schematic diagram of the vane used for yield stress measurement. (From Ref. 28.)

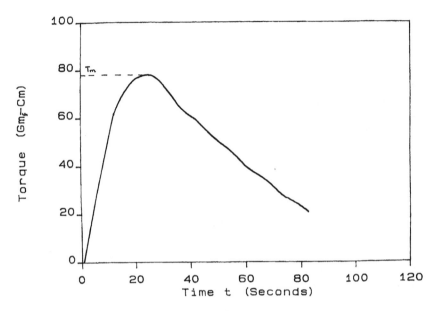

FIGURE 9.10 Typical plot of torque versus time for the rotation at 0.08 rpm of a vane in a 20% by volume suspension of alumina in paraffin oil. (From Ref. 17.)

ance, assuming that the stress on the curved surface as well as on the ends of the vane equals the yield stress, gives:

$$T_m = \frac{\pi D_v^3}{2} \left(\frac{H}{D_1} + \frac{1}{3} \right) \tau_1 \qquad (9.3)$$

from which the yield stress can be computed. In Eq. (9.3), D_1 is the vane diameter and H is the vane height. Note that the contribution to the torque from the stress acting on the surface of the shaft has been neglected here. The final results obtained ought to be independent of the vane diameter and the speed of rotation, and arriving at the proper experimental conditions may require some trial and error. An obvious variation of the procedure outlined here is to apply a constant and increasing torque to the vane and note the smallest torque value that results in continuous rotation [24]; this torque value is then inserted into the left side of Eq. (9.3).

D. Staged Squeezing

If a material possesses a yield stress, then upon cessation of steady shearing flow in a rotational viscometer, the torque signal does not relax to zero. Instead, it tends to a constant, residual value with increasing time. The fluid stress calculated from this residual torque is the yield stress. While this is an attractive technique in principle, it often does not give consistent results in practice, especially with highly flocculated systems that exhibit slip at the viscometer walls. The remedy is to use squeezing flow [17,18]. If one conducts constant-velocity squeezing, a residual force F remains when motion is stopped. If the Herschel–Bulkley model is used to describe the fluid rheology and if the yield stress is large, a theoretical analysis predicts that

$$\tau_v = \frac{3FH}{2\pi R^3} \tag{9.4}$$

where R is the plate radius and H is the plate spacing at the time of stopping the plate motion. By carrying out staged squeezing, i.e., stopping the plate motion intermittently and measuring both H and F, the yield stress can be determined from the slope of the straight-line plot of F versus the reciprocal of H. Rajaiah [17] found that this technique gave very good agreement with results obtained with the use of the vane method.

IV. MEASUREMENT OF THE WALL SLIP VELOCITY

If slip between the fluid sample and the viscometer wall is as large and as evident as that illustrated in Fig. 9.1 (see also Refs. 29 and 30), the slip velocity can be measured simply by flow visualization using photographic techniques and a marker line painted on the sample surface. This was done by Kalyon et al. [31] for the torsional flow of a very concentrated suspension containing 76.5% solids by volume, and the results are displayed in Fig. 9.11. It is seen that the slip velocity depends uniquely on the shear stress at the wall. By examining the residual stress after cessation of steady-shear flow, these authors determined that this suspension had a yield stress of between 4 and 20 Pa. When a stress that was lower than the yield stress was applied to the suspension in a parallel-plate viscometer, it was found that the material slipped at both surfaces and merely rotated as a solid body. This clearly demonstrated that fluid structure was responsible for the yield stress, which, in turn, led to the observed slip at the wall. Although the causes of wall slip can be many [5], in this particular case wall slip occurred because there was a thin, solid-free layer of liquid at the wall; under the applied stress, the bulk of the material could not deform and shearing took place only in this layer of fluid resulting in slip of the suspension. When the applied stress exceeds the yield stress, bulk deformation and wall slip occur simultaneously. Unless data are corrected for slip effects, true material properties cannot be calculated. If wall slip is ignored, the viscosity, in particular, will appear to decrease with decreasing fixture size—tube radius, gap separation, and cone angle [5].

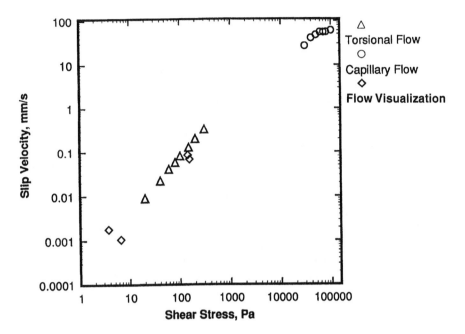

FIGURE 9.11 Slip velocity versus wall shear stress in parallel-disk torsional and capillary flows. (From Ref. 31.)

Although slip velocities can sometimes be determined by visual methods when working with suspensions, emulsions, foams and gels, the technique is certainly not relevant for unfilled polymer melts and polymer solutions. In general, therefore, one uses standard rotational as well as capillary instruments and makes measurements by varying the fixture size, which has the effect of varying the surface-area-to-volume ratio. These analyses are described next.

A. Measuring Slip in a Capillary Viscometer

This analysis goes back to the work of Mooney in 1931 [32] and is similar to the Rabinowitsch analysis, which leads to the expression for the wall shear rate, Eq. (2.2); the only difference is in the boundary condition at the capillary wall. Fluid velocity at the wall is now taken to be V, instead of zero, and this results in the following expression [32,33]:

$$8V_s = \frac{\partial(32Q/\pi D^3)}{\partial(1/D)}\bigg|_{\tau_w} \qquad (9.5)$$

where Q is the volumetric flow rate, D is the capillary diameter, and the differentiation in Eq. (9.5) is carried out at a constant value of the wall shear stress.

Representative capillary viscometer data, for capillaries having very large L/D values, plotted as shear stress versus shear rate, assuming no slip, for a 1% aqueous Separan AP-30 solution are shown in Fig. 9.12 [33]; tube diameters range from 0.01906 to 0.1097 cm. The curve marked "No Slip" is for cone-and-plate viscometer data, and it is clear that the different data sets do not superpose. For a given wall shear stress, decreasing tube diameter leads to progressively higher apparent shear rates due to the presence of wall slip. Fig. 9.13 quantifies the apparent slip phenomenon according to Eq. (9.5), and straight lines result on plotting $32Q/\pi D^3$ (the apparent wall shear rate) as a function of $1/D$ while keeping the wall shear stress fixed. The slip velocity calculated from the slope of these straight lines increases with increasing wall shear stress and can be represented as a power-law in the shear stress. Finally, Fig. 9.14 shows the fractional contribution of slip flow to the total flow rate, and this can be substantial for small diameter tubes, especially at low values of the wall shear stress. It is sometimes possible for the flow rate to be composed almost entirely of slip flow. In this situation, the material flows essentially like a plug. The decrease in the

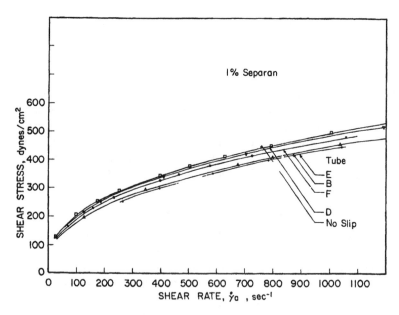

FIGURE 9.12 Influence of capillary diameter on the flow curve of a 1% aqueous polyacrylamide solution. (From Ref. 33.)

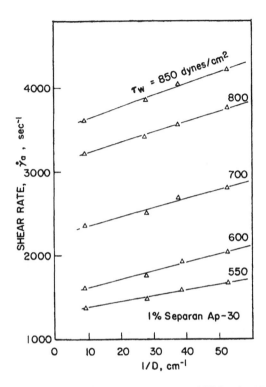

FIGURE 9.13   Shear stress versus $1/D$ for the data of Fig. 9.12. (From Ref. 33.)

slip flow contribution with increasing wall shear stress witnessed in Fig. 9.14 is the result of shear thinning.

   That the Mooney analysis yields reasonable and accurate results can be confirmed by flow visualization experiments. This is demonstrated in Fig. 9.11 where it is found that the slip velocity of concentrated suspensions calculated according to Eq. (9.5) is the same as that measured using a marker line method. Recently nuclear magnetic resonance (NMR) imaging techniques have become available for the experimental determination of the velocity profile of polymer solutions very close to a solid surface. Rofe et al. [34] have made such measurements on aqueous xanthan gum solutions and shown that NMR imaging gives results that are identical to those obtained with the help of Eq. (9.5). Before leaving the topic of apparent wall slip of polymer solutions, we want to reassure readers that this phenomenon does not influence viscometric data obtained on polymer solutions in the usual manner. In particular, there is no slip in a cone-

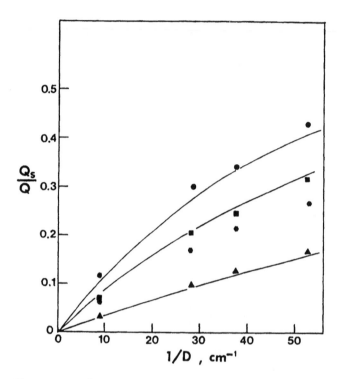

FIGURE 9.14 Contribution of slip to the total flow rate as a function of tube diameter and wall shear stress using the data of Figs. 9.12 and 9.13. The shear stress at the wall equals 250 dynes/cm² for the uppermost curve and 800 dynes/cm² for the lowermost curve. (From Ref. 33.)

and-plate viscometer and in a small-gap Couette viscometer because wall slip occurs only in situations where the shear rate varies with position within the flow field; in these situations, it is postulated that macromolecules move from regions of high shear stress to regions of low shear stress [35a]. Even when a capillary viscometer is used, wall slip is unlikely to be a problem because the shear rates are usually very high, and the contribution of slip flow is likely to be quite small (see Fig. 9.14). However, capillaries with large L/D values may favor slip [35b,c].

B. Wall Slip in Parallel-Plate Viscometers

If one assumes that material slips at both surfaces in a parallel-plate viscometer and that the slip velocity V_r, which depends on the wall shear stress, is the same

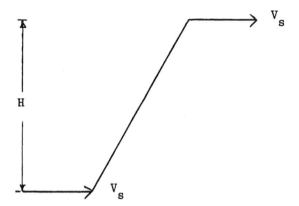

Figure 9.15 Circumferential velocity profile in the presence of slip at any radial position in a parallel-plate viscometer.

at both surfaces, then the circumferential velocity profile at any radial position will be as shown in Fig. 9.15. If the lower disk is stationary and the upper disk moves at a velocity V, the true deformation rate $\dot{\gamma}$ at any value of r is $(V-2V_s)/H$ and the apparent deformation rate $\dot{\gamma}_a$ (ignoring slip) is V/H, where V equals the product of r with the angular velocity. Thus,

$$\dot{\gamma}_a = \dot{\gamma}(\tau) + \frac{2V_s(\tau)}{H} \tag{9.6}$$

To obtain the slip velocity from Eq. (9.6), we need a relationship between the true and apparent shear rates. This is developed by writing the torque M as an integral of the shear stress over the surface of the upper disk, changing variables from r to $\dot{\gamma}_a$, and differentiating the result with respect to the apparent shear rate at the disk periphery, $\dot{\gamma}_{aR}$, to get the following expression for the shear stress at the edge of the disk [36]:

$$\tau_R = \frac{M}{2\pi R^3}\left[3 + \frac{\partial \ln M}{\partial \ln \dot{\gamma}_{aR}}\right] \tag{9.7}$$

Equation (9.7) allows us to relate, by experiment, the apparent shear rate at the edge of the disk to the shear stress at the same location for any chosen value of H. Equation (9.6) then says that a plot of the apparent shear rate versus $1/H$ at a fixed value of the shear stress should be a straight line of slope $2V_s$.

Figure 9.11 shows that, as expected, the foregoing procedure gives the same results as those obtained from the Mooney analysis. Once the slip velocity is known, the true shear rate is also known, and the true flow curve can be determined. Yoshimura and Prud'homme used this procedure to obtain corrected data on a paraffin-oil-in-water emulsion with a 0.923 volume fraction of the dispersed phase. These data are presented in Fig. 9.16, and these clearly show the presence of a yield stress [36]. Had corrections not been made, it would have been easy not to notice the yield stress.

Wall slip can often be prevented or reduced in a number of ways [5]. The standard technique is to roughen the fixture surfaces so that they tend to grip the sample better. Surface roughening might be accomplished by sand-blasting or by affixing an abrasive foil to the plate surface [2,30]. Alternately, fixture surfaces can be serrated or radially grooved, with adjacent grooves being close to each other [30]. Sometimes, the sample is actually glued on to the plate surface [29,37]. While these measures can help promote the bulk deformation of the fluid so that true material properties might be extracted from the data after making corrections for wall slip, they do not always work; the sample can fracture in the middle in a cohesive manner [29].

Analyses are also available for making wall slip corrections to data obtained from Couette viscometers [32,36], but the effort involved in implementing the procedure is significantly greater than that with the use of a parallel-plate viscometer. As a consequence there is little incentive to use a Couette viscometer when a sample is known to slip at the fixture wall.

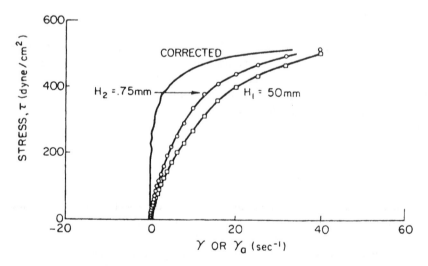

FIGURE 9.16 Stress versus apparent shear rate for an emulsion measured on parallel disks with two different gap separations. Also shown is the flow curve corrected to eliminate wall slip. (From Ref. 36.)

V. MIGRATION OF PARTICLES ACROSS STREAMLINES

We have seen that density differences and the presence of walls can result in the development of concentration nonuniformities during the flow of a dispersed phase in viscometers. Over the years it has been found that particles of a suspension or droplets of an emulsion can also move transverse to a streamline due to such causes as fluid inertia, fluid viscoelasticity, and particle deformability, and a large literature, both experimental and theoretical, now exists on the topic [38]. In the flow of an initially uniform, dilute suspension through a tube of radius R, for example, Segre and Silberberg [39,40] found that neutrally buoyant spheres in a Newtonian liquid collected into a thin annular region that was located at a radial distance about $0.6R$ away from the tube axis. Particle migration became more pronounced on increasing the tube length, the average flow velocity, and the ratio of the sphere radius to the tube radius. Figure 9.17 shows typical concen-

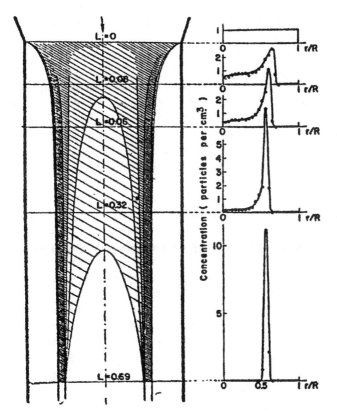

FIGURE 9.17   Particle concentration as a function of radial and longitudinal position during the tube flow of a dilute suspension. (Reprinted by permission from *Nature* (Ref. 39), Copyright 1961 Macmillan Magazines Limited.)

tration profiles at different axial positions, and it is seen that the effect can be quite severe [39]. Ho and Leal [41] theoretically showed that, in this case, particle migration was due to the influence of fluid inertia, and they were able to correctly predict the radial dependence of particle concentration. Note that neutrally buoyant spheres do not migrate in the inertialess flow of a Newtonian liquid [42]. Regarding the influence of inertia on non-neutrally buoyant spheres, it is found that, in tube flow, particles that are denser than the suspending liquid migrate toward the wall in downward flow and toward the centerline in upward flow [43]. Furthermore, as the particle Reynolds number increases, the approach toward the centerline becomes oscillatory. The review by Leal [38] may be consulted for the influence of fluid inertia on the migration of both spheres and axisymmetric, but nonspherical, particles in other flow fields, whether steady or pulsatile.

When the flow of a dilute suspension of spheres in a non-Newtonian medium is examined, the direction of particle migration appears to depend on the relative influence of elasticity and shear thinning [44,45]. Under the influence of fluid elasticity, particles move from regions of high shear rate to regions of low shear rate; this happens because a gradient in the first normal stress difference results in a ''lift force'' perpendicular to the streamline direction. Thus, in tube flow, spheres tend to accumulate along the axis, often associating with each other in the form of a necklace. The effect of shear thinning is opposite to that of fluid elasticity, and results in migration in a direction of increasing shear rate. When a viscoelastic fluid is also highly shear thinning, particles move toward the wall in Poiseuille flow but do not actually hit the wall [45]. The reason why shear thinning makes particles move across streamlines is not clearly understood.

At higher particle concentrations, particle motion is governed as much by particle–particle interactions as it is by fluid inertia and fluid rheology, and multiparticle interactions can often restrict particle mobility. An example is the pumping of a concentrated suspension of neutrally buoyant spheres from a reservoir of large diameter to a tube of small diameter. It is found that the volume-averaged solid concentration in the tube is less than that in the reservoir, and the measured concentration decreases with decreasing tube diameter [46]. In an extreme situation, there can be complete blockage of the tube [47]. Another unexpected observation made with concentrated suspensions that tend to migrate toward the axis in tube flow is the occurrence of an axial concentration gradient. This is illustrated in Fig. 9.18 [48], and it arises because particles that move away from the tube wall enter a faster-moving stream of liquid and are convected downstream more rapidly than the average of the suspending liquid. Kubat and coworkers [48,49] demonstrated this phenomenon by injecting polyethylene and nylon melts loaded with 5–25% glass beads of different diameters into a spiral mold. A gradient in glass concentration and a marked fractionation effect with respect to the size of

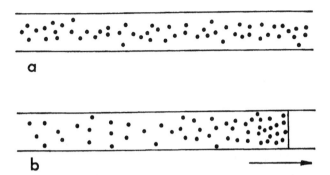

FIGURE 9.18 (a) Effect of particle migration during the flow of a concentrated suspension through a tube. (b) Increase in particle concentration at the front of a suspension during the filling of a tube. (From Ref. 48.)

the spheres were observed along the length of the spiral. It was found that the separation effect increased with increasing sphere size, and for large spheres the glass concentration at the tip of the spiral for the 2-m-long mold was six times the inlet concentration [49]. This effect is much more pronounced with spheres than it is with fibers [48,50].

Whenever there is a velocity gradient, the particle velocity varies with position, and faster-moving particles tend to collide with slower-moving particles whose trajectories are along adjacent streamlines. If, however, the shear rate varies with position in the flow field, the number of collisions experienced by a typical particle per unit time also varies with position. This gradient in the collision frequency across a particle makes the particles move from regions of high shear rate to regions of low shear rate [7]. Thus, in wide-gap, annular Couette flow, particles migrate from the high-shear-rate region near the inner rotating cylinder to the low-shear-rate region at the outer wall [51]. Similarly, in tube flow, migration is toward the centerline, but solid volume fractions in excess of 0.1 are needed to observe concentration differences [52]. The resulting gradient in the suspension viscosity, though, opposes this migration, and the interplay of these two forces can lead to the establishment of an equilibrium concentration profile; these profiles have been measured using nuclear magnetic resonance imaging [51,52].

In closing this section, we mention that deformable particles, i.e., droplets, also migrate across streamlines for many of the same reasons that cause the migration of particles [53]. However, the motion of droplets is much more complex than that of solid spheres, and droplet migration is observed even in the absence of inertial, gravitational, and non-Newtonian effects [54]. For concentrated emulsions, one also has to contend with drop coalescence; coalescence is considered in a later chapter.

VI. CONCLUDING REMARKS

The material presented in this chapter clearly demonstrates that making rheological measurements on multiphase systems requires much greater care than does making measurements on unfilled polymer melts and solutions. If proper care is not exercised, data may not be reproducible or they may lack internal consistency. Additionally, one may not be able to distinguish between time-dependent effects due to reversible structural changes (considered in the next chapter) and slip and migration effects discussed in this chapter. While we can systematically correct for phenomena such as wall slip and seek to prevent annoying phenomena such as air entrainment (which can also influence measurements [55]), we sometimes have to accept large data scatter when working with disperse systems [47]. This is a characteristic feature of the flow behavior of concentrated suspensions and emulsions.

REFERENCES

1. E. Rabinowicz. Stick and slip. Sci. Am. 194(5):109-118 (1956).
2. J. Plucinski, R.K. Gupta, S. Chakrabarti. Wall slip of mayonnaises in viscometers. Rheol. Acta 37: 256–269 (1998).
3. T. Sridhar, R.K. Gupta. Fluid detachment and slip in extensional flows. J. Non-Newt. Fluid Mech. 30:285–302 (1988).
4. A.V. Ramamurthy. Wall slip in viscous fluids and influence of materials of construction. J. Rheol. 30:337–357 (1986).
5. H.A. Barnes. A Review of the slip (wall depletion) of polymer solutions, emulsions and particle suspensions in viscometers: its cause, character, and cure. J. Non-Newt. Fluid Mech. 56:221–251 (1995).
6. R.K. Gupta. Particulate suspensions. In: S.G. Advani, ed. Flow and Rheology in Polymer Composites Manufacturing. Elsevier, Amsterdam, 1994, pp. 9–51.
7. D. Leighton, A. Acrivos. The shear-induced migration of particles in concentrated suspension. J. Fluid Mech. 181:415–439 (1987).
8. H.A. Barnes. Thixotropy—A review. J. Non-Newt. Fluid Mech. 70:1–33 (1997).
9. D.H. Napper. Polymeric Stabilization of Colloidal Dispersions. Academic Press, London, 1983.
10. J.M. Coulson, J.F. Richardson. Chemical Engineering, 3rd ed. Vol 2. Pergamon Press, Oxford, 1978, pp. 172–229.
11. S. Chimmili, D. Doraiswamy, R.K. Gupta. Shear induced agglomeration of particulate suspensions. Ind. Eng. Chem. Res. 37:2073–2077 (1998).
12. I.J. Overend, R.R.Horsley, R.L. Jones, R.K. Vinycomb. A new method for the mea-

surement of rheological properties of settling slurries. Proc. IX Intl. Congress on Rheol., Acapulco, Mexico, 1984, Vol. 2, pp. 583–590.

13. A.M. Kraynik, J.H. Aubert, R.N. Chapman. The helical screw rheometer. Proc. IX Intl. Congress on Rheol., Acapulco, Mexico, 1984, Vol. 4, pp. 77–84.

14. A. Kraynik, L.A. Romero. A general analysis for obtaining the viscosity function from viscometric measurements: a case study for the helical screw rheometer. Proc. Xth Intl. Congress on Rheol. Sydney, Australia, 1988, Vol. 2, pp. 43–45.

15. D.L. Lord. Helical screw rheometer: a new tool for stimulation fluid evaluation. Paper SPE 18213 presented at the 1988 Soc. Pet. Eng. Ann. Tech. Conf. and Exhibition, Houston, TX, October 2–5, pp. 341–347.

16. G. Lombardi. The role of cohesion in cement grouting of rock. Quinzieme Congress des Grands Barrages, Lausanne, Switzerland, 1985, Q58- R.13, pp. 235–261.

17. J. Rajaiah. Rheology and thermal conductivity of concentrated suspensions. Ph.D. dissertation, State University of New York, Buffalo, 1990.

18. G.H. Covey, B.R. Stanmore. Use of the parallel-plate plastometer for the characterization of viscous fluids with a yield stress. J. Non-Newt. Fluid Mech. 8:249–260 (1981).

19. J. Rajaiah, E. Ruckenstein, G.F. Andrews, E.O. Forster, R.K. Gupta. Rheology of sterically stabilized ceramic suspensions. Ind. Eng. Chem. Res. 33:2336–2340 (1994).

20. J.P. Hartnett, R.Y.Z. Hu. The yield stress—an engineering reality. J. Rheol. 33:671–679 (1989).

21. D. C-H. Cheng. Yield stress: a time-dependent property and how to measure it. Rheol. Acta 25:542–554 (1986).

22. H.A. Barnes, K. Walters. The yield stress myth? Rheol. Acta 24:323–326 (1985).

23. J.J. Vocadlo, M.E. Charles. Measurement of yield stress of fluid-like viscoplastic substances. Can. J. Chem. Eng. 49:576-582 (1971).

24. A.S. Yoshimura, R.K. Prud'homme, H.M. Princen, A.D. Kiss. A comparison of techniques for measuring yield stresses. J. Rheol. 31:699–710 (1987).

25. D. De Kee, C.J. Durning. Rheology of materials with a yield stress. In: A.A. Collyer, L.A. Utracki, eds. Polymer Rheology and Processing. Elsevier, London, 1990, pp. 177–203.

26. Q.D. Nguyen, D.V. Boger. Measuring the flow properties of yield stress fluids. Annu. Rev. Fluid Mech. 24:47–88 (1992).

27. P.H.T. Uhlherr, K.H. Park, C. Tiu, J.R.G. Andrews. Yield stress from fluid behavior on an inclined plane. Proc. IX Intl. Congress on Rheology, Acapulco, Mexico, 1984, Vol. 2, pp. 183–190.

28. N.Q. Dzuy, D.V. Boger. Yield stress measurement for concentrated suspensions. J. Rheol. 27:321–349 (1983).

29. A. Magnin, J.M. Piau. Shear rheometry of fluids with a yield stress. J. Non-Newt. Fluid Mech. 23:91–106 (1987).

30. A. Magnin, J.M. Piau. Cone-and-plate rheometry of yield stress fluids. Study of an aqueous gel. J. Non-Newt. Fluid Mech. 36:85–108 (1990).

31. D.M. Kalyon, P. Yaras, B. Aral, U. Yilmazer. Rheological behavior of a concentrated suspension: a solid rocket fuel simulant, J. Rheol. 37:35–53 (1993).

32. M. Mooney. Explicit formulas for slip and fluidity. J. Rheol. 2:210–222 (1931).

33. Y. Cohen, A.B. Metzner. Apparent slip flow of polymer solutions. J. Rheol. 29:67–102 (1985).
34. C.J. Rofe, L. de Vargas, J. Perez-Gonzalez, R.K. Lambert, P.T. Callaghan. Nuclear magnetic resonance imaging of apparent slip effects in xanthan solutions. J. Rheol. 40:1115–1128 (1996).
35a. U.S. Agarwal, A. Dutta, R.A. Mashelkar. Migration of macromolecules under flow: the physical origin and engineering implications. Chem. Eng. Sci. 49:1693–1717 (1994).
35b. L. de Vargas, J. Perez-Gonzalez, J. de J. Romero-Barenque. Experimental evidence of slip development in capillaries and a method to correct for end effects in the flow of xanthan solutions. J. Rheol. 37:867–878(1993).
35c. A.F. Mendez-Sanchez, J. Perez-Gonzalez, L. de Vargas, J. Tejero. Two-fluid model of the apparent slip phenomenon in Poiseuille flow. Rev. Mex. Fis. 45:26–30 (1999).
36. A. Yoshimura, R.K. Prud'homme. Wall slip corrections for couette and parallel disk viscometers. J. Rheol. 32: 53-67 (1988).
37. L.L. Navickis, E.B. Bagley. Yield stresses in concentrated dispersions of closely packed, deformable gel particles. J. Rheol. 27:519–536 (1983).
38. L.G. Leal. Particle motions in a viscous fluid. Ann. Rev. Fluid Mech. 12:435–476 (1980).
39. G. Segre, A. Silberberg. Radial particle displacements in Poiseuille flow of suspensions. Nature 189:209–210 (1961).
40. G. Segre, A. Silberberg. Behavior of macroscopic rigid spheres in Poiseuille flow. Part 2. Experimental results and interpretation. J. Fluid Mech. 14:136–157 (1962).
41. B.P. Ho, L.G. Leal. Inertial migration of rigid spheres in two-dimensional unidirectional flows. J. Fluid Mech. 65:365–400 (1974).
42. R. Clift, J.R. Grace, M.E. Weber. Bubbles, Drops, and Particles. Academic Press, New York, 1978, p. 229.
43. C.D. Denson, E.B. Christiansen, and D.L. Salt. Particle migration in shear fields. AIChE J. 12:589–595 (1966).
44. F. Gauthier, H.L. Goldsmith, S.G. Mason. Particle motions in non-Newtonian Media. II. Poiseuille flow. Trans. Soc. Rheol. 15:297–330 (1971).
45. M.A. Jefri, A.H. Zahed. Elastic and viscous effects on particle migration in plane-Poiseuille flow. J. Rheol. 33:691-708 (1989).
46. V. Seshadri, S.P. Sutera. Apparent viscosity of coarse, concentrated suspensions in tube flow. Trans. Soc. Rheol. 14:351–373 (1970).
47. D.C.-H. Cheng. Further observations on the rheological behavior of dense suspensions. Powder Technol. 37:255–273 (1984).
48. J. Kubat, A. Szalanczi. Polymer–glass separation in the spiral mold test. Polym. Eng. Sci. 14:873–877 (1974).
49. L. Borocz, J. Kubat. Phase separation effects in glass-sphere-containing polymer melts. Plast. Rubber Process. 4:82–86 (1979).
50. R.P. Hegler, G. Mennig. Phase separation effects in processing of glass-bead and glass-fiber-filled thermoplastics by injection molding. Polym. Eng. Sci. 25:395–405 (1985).
51. J.R. Abbott, N. Tetlow, A.L. Graham, S.A. Altobelli, E. Fukushima, L.A. Mondy,

T.S. Stephens. Experimental observations of particle migration in concentrated suspensions: Couette flow. J. Rheol. 35:773–795 (1991).

52. R.E. Hampton, A.A. Mammoli, A.L. Graham, N. Tetlow, S.A. Altobelli. Migration of particles undergoing pressure-driven flow in a circular conduit. J. Rheol. 41:621–640 (1997).

53. P.C.-H. Chan, L.G. Leal. The motion of a deformable drop in a second-order fluid. J. Fluid Mech. 92:131–170 (1979).

54. S. Haber, G. Hetsroni. The dynamics of a deformable drop suspended in an unbounded Stokes flow. J. Fluid Mech. 49:257–277 (1971).

55. D.M. Kalyon, R. Yazici, C. Jacob, B. Aral. Effects of air entrainment on the rheology of concentrated suspensions during continuous processing. Polym. Eng. Sci. 31:1386–1392 (1991).

10*

Solid-in-Liquid Suspensions

I. INTRODUCTION

The rheology of suspensions containing rigid fillers is important in many areas of polymer technology. Composite materials containing filler weight fractions in the range of 0.4–0.65 are not uncommon, and the fillers may act either as reinforcements or as diluents. A common diluent added to both thermoplastics and thermosets is calcium carbonate, often coated with a stearate. Talc is added to many thermoplastics to increase stiffness and high-temperature creep resistance. Most rubber formulations contain carbon black or silica for mechanical property enhancement, while rubber is added to polystyrene for the purpose of increasing the impact strength. Concentrated suspensions in polymeric liquids are also encountered in the injection molding of metal powders as well as ceramic powders; here the polymer merely acts as a binder. Additional examples of solid-in-liquid suspensions of technological interest include paints and building materials, coal-water fuels, drilling and fracturing fluids, thermal grease, and dental adhesives. Lattices are suspensions of rigid polymer particles in water, and plastisols are suspensions of polymer particles in a liquid plasticizer. The diversity of applications of suspensions, pastes, and slurries is truly immense, and this is matched only by the complexity of the rheology exhibited by these materials. Even in cases where the suspending liquid is Newtonian in behavior, the presence of a filler produces profound effects on the rheological behavior of the suspension. The rheology becomes even more complex if the liquid phase is non-Newtonian.

---

* Coauthored with Deepak Doraiswamy, E.I. du Pont de Nemours and Company, Wilmington, Delaware.

The simplest situation that we can consider, both theoretically and experimentally, is the shear viscosity versus shear rate behavior of a nonsettling suspension of spheres. This is illustrated in Fig. 10.1 [1]. If a small volume fraction of particles (typically <0.01) is suspended in a Newtonian liquid whose own viscosity is given by curve a, the suspension viscosity is uniformly raised to curve b. On further increasing the amount of solids, we find that the viscosity continues to increase but becomes shear thinning, although Newtonian behavior is still observed in the limits of both low and high shear rates (curve c). If, however, the suspending liquid is itself a polymeric fluid, its own viscosity is likely to be non-Newtonian and similar to curve c. Adding solids simply shifts the curve for the suspension upward to that labeled d. Additional increases in particulate content, whether in Newtonian or non-Newtonian media, can result in the appearance of an apparent yield stress (curve e), indicated by the disappearance of the Newtonian plateau and the approach to a slope of -1 at low shear rates. Finally, at solid contents approaching the maximum packing value (curve f), one gets shear thickening, particularly with uniformly sized particulates.

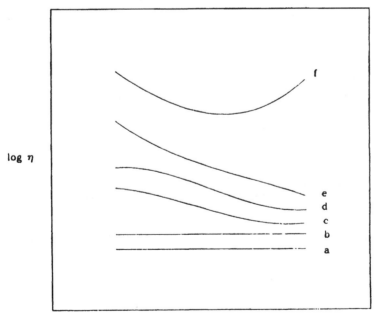

FIGURE 10.1   Qualitative representation of the influence of increasing filler volume fraction on suspension viscosity. (From Ref. 1, with permission from Elsevier Science.)

Einstein [2] first predicted the effect of a rigid filler on the viscosity of a Newtonian liquid. His simple equation for the creeping flow of dilute suspensions is

$$\eta = \eta_0(1 + k_E\phi) \tag{10.1}$$

The viscosity of the mixture is η and that of the suspending liquid is η_0. The volume fraction of filler is Φ, and k_E is the Einstein coefficient. For particles of spherical shape, k_E is 2.5. The magnitude of the Einstein coefficient is determined by the degree to which the particles disturb the streamlines in a flowing system. Some particle shapes, such as rods, disturb the streamlines more than do spheres and have correspondingly larger Einstein coefficients; thus the Einstein coefficient is 3.1 for cubes, and for uniaxially oriented fibers parallel to the tensile stress component, the Einstein coefficient is $2L/D$, where L/D is the fiber aspect ratio. Results for prolate and oblate spheroids of different elipticity values have been derived by Jeffrey [3,4]. Although the Einstein equation is valid only for very low concentrations of particles ($<1\%$), it is amazingly simple. The equation implies that the relative viscosity of the suspension is independent of the size and nature of particles. This equation also says that the intrinsic viscosity of a suspension is a dimensionless quantity that equals the Einstein constant when the concentration is expressed in volume fraction units. As the solids concentration is increased, particles begin to interact with each other, and the viscosity increases more than linearly with the volume fraction. Here it should be recognized that the Einstein equation incorporates just the first term in a more general power-series expansion, such as the Huggins equation introduced in Chap. 4. The coefficient of the quadratic term can be calculated, but its numerical value appears to depend on the assumptions made and the method of calculation used. The results have been summarized by Happel and Brenner [4].

II. SHEAR VISCOSITY OF CONCENTRATED SUSPENSIONS OF NONCOLLOIDAL PARTICULATES

The literature on the shear viscosity of particulate suspensions is truly immense, and it has been reviewed at fairly regular intervals [1,5–9]. This reflects both the theoretical and the practical importance of this topic. We have cited only a few, relatively recent, representative review articles, but a fairly extensive list of review articles can be compiled by consulting these references. While some of the reviews have a theoretical bias, others emphasize experimental aspects, and each may contain several hundred references!

A very large number of equations has been proposed for estimating the viscosity of a Newtonian liquid containing spherical particles up to moderate concentrations. One of the more successful equations, for monodisperse spheres, is due to Frankel and Acrivos [10], who assumed that the increase in viscosity on adding particulates was due to energy dissipation in the thin liquid film between neighboring spheres as they moved past each other. These authors further assumed that dissipation due to squeezing motion was dominant and that due to the sliding motion was negligible. By averaging the energy dissipated by all possible pairs of particles in the suspension, they obtained the following expression for the relative viscosity:

$$\eta_R = \frac{9}{8}\left[\frac{(\phi/\phi_m)^{1/3}}{1 - (\phi/\phi_m)^{1/3}}\right] \tag{10.2}$$

where Φ_m is the maximum possible solids volume concentration and is the value of Φ at which the suspension viscosity becomes infinitely large. Equation (10.2) depicts data correctly at large values of Φ, but it does *not* reduce to Eq. (10.1) as $\Phi \to 0$.

Several successful semi-empirical equations have emerged from the realization that a unique curve can be obtained by plotting the relative viscosity as a function of Φ/Φ_m. Thus, Chong et al. [11] found

$$\eta_R = \left[1 + 0.75\left(\frac{\phi/\phi_m}{1 - \phi/\phi_m}\right)\right]^2 \tag{10.3}$$

which *does* reduce to Eq. (10.1) at low values of Φ, provided that Φ_m equals 0.6.

The simplest one-parameter equation, however, is the one based on the work of Maron and Pierce [11a] and evaluated by Kataoka and coworkers [12,13]:

$$\eta_R = \left(1 - \frac{\phi}{\phi_m}\right)^{-2} \tag{10.4}$$

This was tested extensively by Poslinski et al. [14,15] by making room-temperature measurements on different concentrations of narrow-size-distribution glass beads suspended in four different (Newtonian) polybutene matrices. Some of these shear viscosity data as a function of shear rate are shown in Fig. 10.2. The average diameter of the glass beads was 15 microns, and the value of Φ_m determined by liquid displacement experiments was 0.62. As seen from Fig. 10.2, the suspension viscosity remains Newtonian at least till a sphere volume fraction of 0.5. Beyond this concentration, a yield stress appears to develop. Figure 10.3 displays the suspension relative viscosity as a function of Φ/Φ_m for all the data in the concentration range in which the viscosity is Newtonian. Also shown are predictions of Eqs. (10.2)–(10.4), and Eq. (10.4) appears to do the most reasonable job of representing the data. Note that the fits to the data were not appreciably improved when the maximum packing fraction was treated as an adjustable parameter [1].

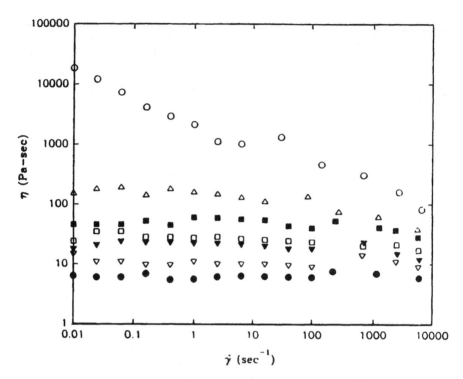

FIGURE 10.2 Shear viscosity of uniformly sized, glass sphere suspensions in polybutene. The various symbols are for solid volume fractions of 0, 0.1, 0.2, 0.3, 0.4, 0.5, and 0.6; the viscosity increases with increasing filler volume fraction. (From Ref. 1, with permission from Elsevier Science.)

When one goes to a distribution of sphere sizes, one finds little effect of the size distribution on the viscosity, provided that the particulate volume fraction is less than 0.2 [6]. On exceeding this value, however, one can get a dramatic lowering in viscosity, relative to the level for a unimodal suspension, by a proper selection of sphere sizes [11]. To predict this effect, especially if the size distribution is very wide, we can take the "effective medium" approach [16]. We begin with the suspending liquid alone, and add the solids in stages. We add the finest solids first, followed by the next coarser ones and so on. At each stage, the suspension viscosity is assumed to be given by Eq. (10.1) except that η_0 is taken to be the same as the viscosity of the suspension at the end of the previous stage. This leads to the following result:

$$d\eta = k_E \eta \, d\phi \qquad (10.5)$$

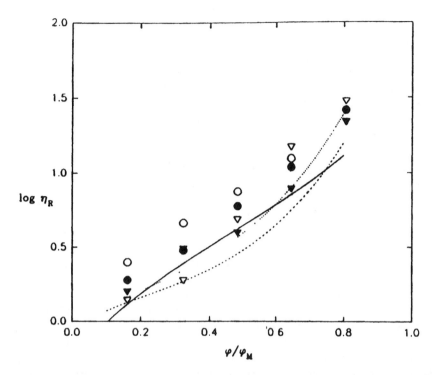

FIGURE 10.3 Relative viscosity of glass sphere suspensions in Newtonian media of different viscosity values. The solid line is Eq. (10.2); --- is Eq. (10.3); . . . is Eq. (10.4). (From Ref. 1, with permission from Elsevier Science.)

If, due to the presence of particles already in suspension, the free volume available to the added particles is reduced by a factor of $(1 - K\Phi)$, an integration of Eq. (10.5) gives

$$\eta = \eta_0(1 - K\phi)^{-k_E/K} \tag{10.6}$$

and K must equal the reciprocal of Φ_m because we expect the suspension viscosity to become infinitely large as $\Phi \rightarrow \Phi_m$. Also, since the Einstein coefficient equals the intrinsic viscosity, we have

$$\eta = \eta_0\left(1 - \frac{\phi}{\phi_m}\right)^{-[\eta]\phi_m} \tag{10.7}$$

which is the Krieger–Dougherty equation [17]. For nonspherical particles, both the intrinsic viscosity and the maximum packing fraction have to be determined experimentally. Since Eq. (10.7) is based on the Einstein equation, it properly reduces to Eq. (10.1) as the solids volume fraction approaches zero.

An alternative approach is to apply Eq. (10.4) to multimodal suspensions regardless of particle shape but with the maximum packing fraction corresponding to the actual suspension. Mixtures containing both large and small particles can pack more densely than particles of a uniform size. As the packing density increases, Φ_m increases and the relative viscosity decreases. While Φ_m can be calculated theoretically in some cases for different types of packing [18,19], experimentally it can be estimated from sedimentation volumes or even from the volume occupied by a given weight of a powder. Poslinski et al. [15] have published data on the viscosity of 15-micron- and 78-micron-diameter glass spheres mixed in various proportions and suspended in low-molecular-weight polybutenes. These mixtures had a shear viscosity that was independent of the shear rate; results are displayed in Fig. 10.4. The solid line in Fig. 10.4 is Eq. (10.4) utilizing experimentally measured values of the maximum packing fraction. Not only do all the data fall on a single curve, but the curve itself is well represented by Eq. (10.4).

Most polymeric liquids are non-Newtonian and, in particular, shear thinning. One finds that the addition of the same amount of particulates tends to raise the viscosity proportionately more at low shear rates than at high shear rates. Thus, the relative viscosity depends on shear rate as well as on the solids concentration. In order to predict this effect of shear rate, Jarzebski [20] assumed that the polymeric fluid could be represented as a power-law liquid with power-law index n. He then repeated the analysis of Frankel and Acrivos [10] for uniformly sized spheres and obtained the following result:

$$(\eta_R)_\gamma = \frac{9}{8}\left[\frac{(\phi/\phi_m)^{1/3}}{1 - (\phi/\phi_m)^{1/3}}\right]^n \tag{10.8}$$

This reduces to Eq. (10.2) when n equals unity. The subscript $\dot{\gamma}$ in this equation indicates that in the calculation of the relative viscosity, the suspension and suspending medium viscosities are measured at the same shear rate. Note that since the shear rate does not appear explicitly in Eq. (10.8), the flow curve for the composite has the same slope on logarithmic coordinates as the flow curve of the suspending liquid, and one curve is situated directly above the other. In order to determine the extent by which the composite viscosity exceeds the suspending liquid viscosity, one separately has to relate the power-law index to the shear rate.

In order to test Eq. (10.8), Poslinski and coworkers [14] measured the viscosity as a function of shear rate of 15-micron-diameter glass beads suspended in a shear-thinning thermoplastic polymer at three different temperatures. The flow curves of the polymer are shown in Fig. 10.5a. At each temperature, the

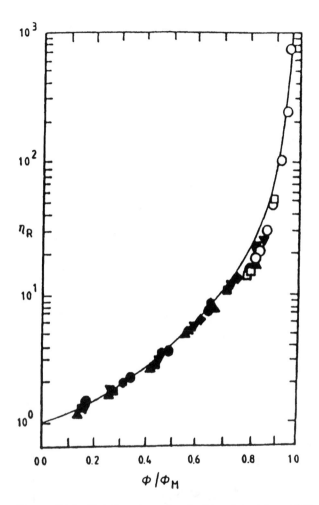

FIGURE 10.4 Relative viscosity of a bimodal mixture of glass spheres suspended in a Newtonian polybutene versus the reduced volume fraction. The different symbols represent different relative amounts of the two kinds of spheres; ϕ_m values range from 0.61 to 0.75. (From Ref. 15.)

neat polymer is Newtonian below a shear rate of 1 sec^{-1}, but the power-law index reduces to 0.75 at 10 sec^{-1} and finally to 0.5 at shear rates in excess of 10 sec^{-1}. The corresponding viscosity behavior of the glass sphere suspension at 150°C is shown in Fig. 10.5b as a function of shear rate at different loading levels. The power-law index of the suspension is the same as that of the unfilled polymer,

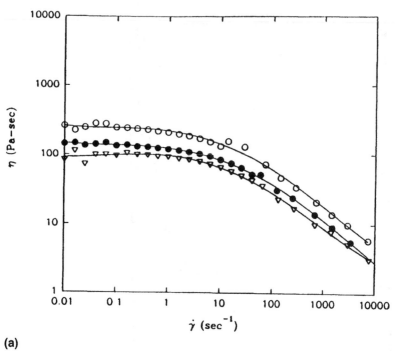

(a)

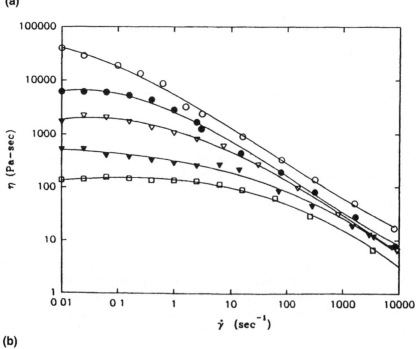

(b)

and also, as expected, the viscosity increase at low shear rates is greater than that at high shear rates. However, the point of onset of shear thinning moves to lower shear rates with increasing particulate concentration. The influence of shear rate on the relative viscosity is clearly visible in Fig. 10.6, where data at all three temperatures are presented and found to superpose. The different curves are for different constant-shear rates or power-law indices and represent Eq. (10.8); the maximum packing fraction is obtained experimentally and equals 0.62. Note that implicit in Eq. (10.8) is the assumption that flow activation energies of the filled and unfilled polymers are equal at a given shear rate, although these may vary with shear rate; the activation energy typically decreases with increasing shear rate [21]. As Fig. 10.6 demonstrates, agreement between data and theory is very good.

An alternate way of representing data was suggested by Kitano et al. [13]. These authors found that the relative viscosity, $(\eta_R)_\tau$, determined at the same value of the shear stress rather than at the same value of the shear rate, was independent of the shear stress. This is, indeed, found to be the case when the data of Fig. 10.5 are replotted in this manner in Fig. 10.7. Relative viscosities so obtained also superpose well with data presented in Fig. 10.3 and are, therefore, representable by Eq. (10.4). Thus, one may employ either Eq. (10.8) or the appropriately defined form of Eq. (10.4) for describing the viscosity of uniformly sized sphere suspensions in shear-thinning liquids. For a distribution of sphere sizes or for nonspherical particles, however, the only available approach that is likely to be satisfactory is to use Eq. (10.4) in conjunction with an experimentally determined value of Φ_m.

In closing this section, we mention that if the dispersed phase forms permanent, rigid aggregates, the apparent volume fraction Φ_a of the particulates will be greater than the true volume fraction Φ, because some liquid becomes immobilized within the aggregates and in the cusps where the particles touch each other. Thus, the apparent volume of a spherical aggregate is larger than the true volume occupied by the sum of the individual spheres making up the aggregate. The extent by which the former quantity exceeds the latter quantity will, in general, depend both upon the number of particles in the aggregate and upon the type of packing within the aggregate [22]. A further consequence of loss of the liquid

---

FIGURE 10.5 (a) Shear viscosity versus shear rate at temperatures of 130, 150, and 170°C for the thermoplastic polymer used by Poslinski et al. [14]. (b) Flow curves at 150°C for 15-micron-diameter glass spheres suspended in the thermoplastic polymer [14]. The various curves are for solid volume fractions of 0.13, 0.26, 0.35, 0.45, and 0.6; the viscosity increases with increasing filler volume fraction. (From Ref. 1, with permission from Elsevier Science.)

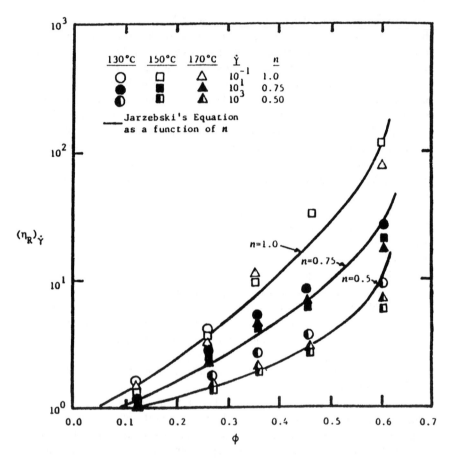

FIGURE 10.6 Data of Fig. 10.5b and similar data at 130°C and 170°C replotted as relative viscosity versus filler volume fraction. Solid lines represent the predictions of Eq. (10.8). (From Ref. 1, with permission from Elsevier Science.)

phase to the aggregates is that the maximum packing fraction appears to decrease. If aggregates pack in the same manner as the individual particles do in making up the aggregates, then the maximum packing fraction becomes the square of the maximum packing fraction in the absence of aggregate formation. It is, therefore, obvious that good engineering judgement needs to be exercised in the selection of any of the available equations in view of their specificity with respect to concentration range and nature of the medium.

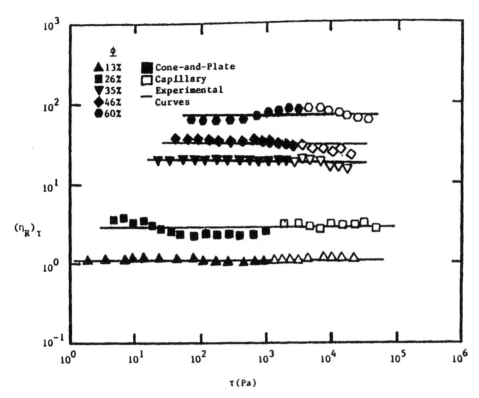

FIGURE 10.7 Data of Fig. 10.5 replotted as the relative viscosity at a constant shear stress versus the shear stress. (From Ref. 1, with permission from Elsevier Science.)

III. SHEAR VISCOSITY OF COLLOIDAL HARD SPHERES

Thus far we have considered the behavior of large particles suspended in high-viscosity liquids. In this situation, the suspension viscosity is governed by hydro-dynamic forces alone. On reducing the particle size to the colloidal range, we find that Brownian motion becomes important, and it tends to impart an isotropic structure to the suspension at rest. If this structure is disturbed by flow, shear thinning can be observed. Normally the reduction in particle size is accompanied by growth in the magnitude of colloidal forces. However, these forces can be almost totally neutralized with the use of proper techniques that impart colloidal stability (against flocculation) to the suspension; viscosity is then determined by the interplay between hydrodynamic and Brownian motion forces. A study of colloidal hard spheres, i.e., spherical particles stabilized in this manner, is instruc-tive if one wants to formulate low-viscosity suspensions. This is because hard sphere behavior represents the low-viscosity limit of a colloidal suspension, and the inclusion of electrostatic, steric, or van der Waals forces serves only to increase the suspension viscosity.

We have seen that in the presence of hydrodynamic forces alone, the relative viscosity of a suspension of spheres in a Newtonain liquid is a unique function of the reduced volume fraction. When Brownian motion is also important, its influence shows up in the form of an additional dimensionless group in any equation relating the relative viscosity to the other dimensionless variables. Most commonly, this group is the dimensionless shear stress [17]

$$\tau_r = \frac{\tau R^3}{kT} \tag{10.9}$$

in which R is the sphere radius, k is Boltzmann's constant, and T is the absolute temperature. The actual relationship between the various dimensionless groups, though, has to be determined by experiment. This was accomplished by Krieger and coworkers [17,23] using monodisperse polystyrene and polyvinyltoluene lattices in a diameter range of 0.19–1.12 micrometers. (Recall that lattices are suspensions of polymer spheres of 0.01–1-micrometer diameter in an aqueous medium.) The latex spheres are generally charged, and the presence of an electrical charge results in an increase in the viscosity. This happens because the electrical double layer makes the particles appear larger than they actually are, and the electrical charges bring about repulsion of particles. It is difficult for one charged particle to penetrate within the electrical double layer of another. The viscosity depends upon electrolyte concentration and upon pH, since the thickness of the electrical double layer depends upon concentration. The viscosity can be larger than expected for such a suspension by as much as ten to a hundred times at low ionic concentration [23]. As the concentration of ions increases to the order of 0.1 mol/L for univalent electrolytes, the viscosity goes through a minimum. Krieger and coworkers found that this electrolyte concentration was independent of the dimensionless shear stress. Since coulombic interactions can only increase the viscosity, the minimum in viscosity has to be representative of hard sphere behavior. Thus, Woods and Krieger [23] added the appropriate amount of electrolyte to all their lattices. An additional complication is the presence of a high concentration of surfactant, which arises due to the fact that these lattices are made by the process of emulsion polymerization. The surfactant not only influences the viscosity but also adsorbs on the sphere surface. This adsorbed layer makes the total particle diameter larger than the diameter of the pure polymer particle. With the help of surface tension measurements, Woods and Krieger ensured that there was just enough surfactant for monolayer coverage, and they took the increase in diameter into account while analyzing their viscosity results.

The data of Woods and Krieger as suspension relative viscosity versus the dimensionless shear stress at a single value of the solids content are shown as the solid line in Fig. 10.8. The symbols represent similar data of Papir and Krieger [24] on crosslinked polystyrene lattices in benzyl alcohol and in m-cresol; the use of organic suspending media allows one to get away from charge effects and also affords a convenient method of varying the suspending medium viscosity. As expected, all the data superpose properly. The general trend of the data on varying the dimensionless shear stress is sketched in Fig. 10.9a. An equivalent representation in terms of the particulate volume fraction is shown in Fig. 10.9b. Evidently, everything else being the same, a suspension of small particles has a higher viscosity compared to a suspension of large particles. This is especially true when the sphere volume fraction exceeds 0.2 [17].

In order to predict some of these results, Batchelor [25] used a pair interaction theory to calculate the relative viscosity in the limit of low shear rates. He determined that

$$\eta_R = 1 + 2.5\phi + 6.2\phi^2 \tag{10.10}$$

and found that Eq. (10.10) fitted Krieger's data up to a volume fraction of about 0.2. Thus Eq. (10.10) represents the initial part of the uppermost curve in Fig. 10.9b. At the other extreme of shear rates, Brownian motion is negligible, and

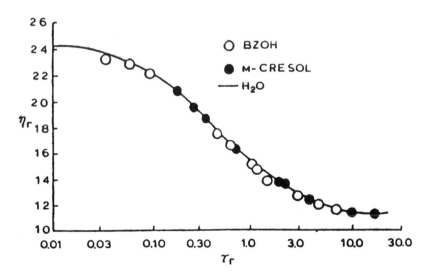

FIGURE 10.8 Relative viscosities in various media of 50% dispersions of spherical particles of different diameters, graphed versus reduced shear stress. (From Ref. 24.)

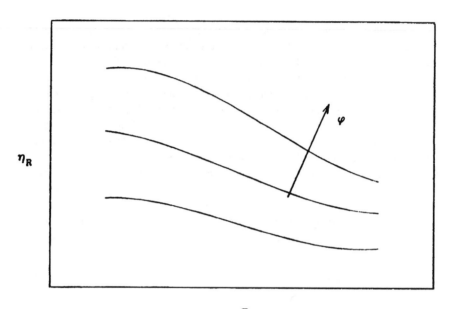

(a)

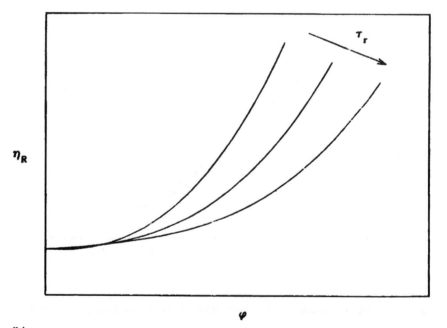

(b)

one again has the situation considered in the previous section. Indeed, all the high-shear-rate data (the lowermost curve in Fig. 10.9b) can be predicted using either Eq. (10.4) [26] or Eq. (10.7) [24]. Furthermore, if one takes the view that the maximum packing fraction should depend on the shear rate, then one finds that the relative viscosity is a unique function of the reduced volume fraction $\Phi / \Phi_m(\tau)$ [27]. This means that all the curves sketched in Fig. 10.9b should collapse into a single master curve that is identical to the curve seen in Fig. 10.4.

An alternate way of representing data at intermediate values of the dimensionless shear stress and a fixed value of the solids volume fraction (see Fig. 10.8 or Fig. 10.9a) is through the use of an empirical Ellis or Cross-type model [24]

$$\frac{\eta - \eta(\dot{\gamma} \to \infty)}{\eta(\dot{\gamma} \to o) - \eta(\dot{\gamma} \to \infty)} = \frac{1}{1 + b\tau,} \tag{10.11}$$

where b is a constant independent of the volume fraction.

In closing this section, it is worth reiterating the need for extra caution in the rheometry of particulate suspensions which are especially prone to equipment gap-size effects and slip. A good rule of thumb is that the gap size should be larger than 10 times the maximum particle size to minimize bridging effects and distortion of the flow field. Various analyses for incorporating slip effects, which are almost invariably present with low viscosity matrix fluids, have been discussed in Chap. 9. We also draw the attention of readers to falling-ball viscometry [28], a simple and inexpensive method for measuring the zero-shear viscosity of polymeric fluids, including suspensions at solids concentrations as high as 45% by volume [29]. Here we measure the terminal velocity of a sphere falling through an unbounded expanse of liquid whose viscosity is to be determined; the liquid does not need to be optically clear [29]. The viscosity is computed from a balance of forces acting on the sphere. The application of this technique to inelastic, shear-thinning liquids has been discussed in detail by Chhabra [30].

IV. SHEAR THICKENING IN COLLOIDAL HARD SPHERE SUSPENSIONS

Even though colloidal hard sphere suspensions always have a viscosity that is lower than that of the corresponding flocculated suspensions, they suffer from the generally undesirable phenomenon of shear thickening. This can usually be

---

FIGURE 10.9 Qualitative representation of the variation of suspension relative viscosity with (a) dimensionless shear stress and (b) filler volume fraction. (From Ref. 1.)

observed with unimodal suspensions having a solids volume fraction in the neighborhood of 0.5. Shear thickening is seen in Fig. 10.10 for PVC suspensions in
dioctyl phthalate [31]. At volume fractions exceeding 0.5, the shear thickening
is so severe that one actually observes a discontinuity in the flow curve; the
behavior, however, is entirely reversible on lowering the shear rate. Note also
that even suspensions of noncolloidal particles exhibit shear thickening. Indeed,
according to Barnes [32], shear thickening is so common that all suspensions of
solid particles can be made to show shear thickening given the right circumstances. In general, though, shear thickening rarely lasts for more than one decade
of shear rate, and the viscosity subsequently levels off or even decreases with
further increases in shear rate. The shear rate at which shear thickening is first
observed is a strong function of Φ/Φ_m; the critical shear rate decreases as Φ/Φ_m
increases toward unity. One way to ameliorate the viscosity increase is to use a
particle size distribution, for this results in an increase in Φ_m. It is also found
that the critical shear rate is lower for the same volume fraction of large spheres
than it is for small spheres. In addition, the severity of shear thickening goes
down as the particle diameter decreases. It is for this reason that the particle size
distribution should be broadened with the help of small spheres rather than large
spheres.

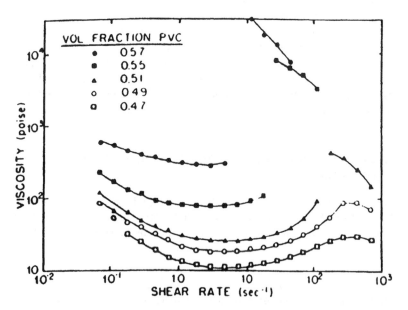

FIGURE 10.10 Effect of volume fraction and shear rate on the viscosity of 1.25-
micrometer PVC suspensions in dioctyl phthalate. (From Ref. 31.)

In terms of understanding why shear thickening takes place, Hoffman [33] postulated that under the influence of hydrodynamic forces the dispersed particles arrange themselves in ordered layers, and the layered structure is maintained due to the presence of electrostatic or steric repulsion forces. This was confirmed not only by Hoffman's own diffraction experiments but also by light-scattering experiments of others [34]. With increasing shear rate, a point is reached where the shear stress couple exceeds the forces that hold the spheres in layers, and the resulting disorder manifests itself as shear thickening. When measurements are made with a controlled stress viscometer, hysterisis can sometimes be observed, and the viscosity (or shear rate) can jump back and forth between two limiting cases corresponding to two different particle arrangements [35]. The influence of particle size, particle concentration, and suspending medium viscosity on the critical shear rate for the onset of shear thickening has been explored experimentally by Frith et al. [36], while Boersma et al. have used simple force balances to theoretically predict the critical shear rate [37]. Alternative theories of shear thickening have also been proposed, and the issue is not completely resolved [38].

V. ELASTIC EFFECTS

As in the case of other fluids, elastic effects in the flow of solid-in-liquid suspensions manifest themselves in the form of unequal normal stresses in steady shearing flow, a finite storage modulus in oscillatory flow, a non-Newtonian extensional viscosity in stretching flow, enhanced entrance pressure losses in flow through a capillary, and the presence of die swell on extrusion out of a tube. As discussed in Chap. 5, the diameter of a polymeric extrudate leaving a capillary is significantly larger than the diameter of the capillary itself, and the ratio of these two diameters is often correlated with the first normal stress difference in shear. Numerous studies have shown that extrudate swell reduces monotonically with addition of increasing amounts of particulates [39]; this reduction is greater than what might be expected based simply on the dilution of the polymer by the filler [40]. Since normal stresses are synonymous with elasticity, this reduction in die swell implies that filled polymers are less elastic than their "neat" counterparts. A similar conclusion is reached on examining data for entrance pressure drops in capillary flow. When a viscoelastic liquid flows from a reservoir into a capillary through a converging channel, it undergoes a pressure drop that is much larger than that encountered in the case of Newtonian liquids of similar viscosity; this excess pressure loss is ascribed to fluid elasticity [41]. It is found that the addition of particulate fillers reduces the entrance pressure drop, and hence elasticity, of polymer melts (see, for example, Ref. 42). More quantitative conclusions can be drawn by examining data on the first normal stress difference, the storage modulus, and extensional viscosity. This is done in the following sections.

A. First Normal Stress Difference

When solid particulates are added to a polymer melt, the first normal stress difference at a given shear rate generally increases when compared to the unfilled
polymer [43,44]. At a fixed concentration, straight lines having a negative slope
are obtained when the first normal stress coefficient is plotted against the shear
rate on logarithmic coordinates; suspension data lie above the data for the polymer alone. However, when normal stress data are plotted against the shear stress,
the filled system has the lower first normal stress difference [44,45]. This trend
reversal happens because a filled polymer achieves the same value of the shear
stress at a much lower value of the shear rate. Since normal stresses increase
with shear rate for both the filled and unfilled polymers, it is not surprising that the
primary normal stress difference for the neat polymer exceeds the corresponding
quantity for the suspension at a lower shear rate. Thus it is debatable whether
polymer melt elasticity, as measured by N_1, is increased or decreased with the
addition of fillers.

In terms of quantitative representation of data, Minagawa and White [43]
showed that the ratio of N_1 of the suspension to that of the polymer at the same
shear rate was independent of the shear rate and was an increasing function of
the volume fraction of suspended solids. By obtaining data on model suspensions
of glass beads in both constant-viscosity polybutene oils and shear-thinning polymer melts, Poslinski et al. [14] found that all the normal stress results could be
represented by an analogous form of Eq. (10.4):

$$\frac{N_1(\phi, \dot{\gamma})}{N_1(0, \dot{\gamma})} = \left[1 - \frac{\phi}{\phi_m} \right]^{-2} \tag{10.12}$$

where ϕ was as large as 0.6 and ϕ_m was 0.62. The fit of Eq. (10.12) to all the
reduced normal stress data is shown in Fig. 10.11, and there is good data superposition.

B. Storage Modulus

As with the first normal stress difference, it appears that the addition of fillers
to a polymeric fluid results in an increase in both the storage and loss moduli at
a given frequency, although reliable data are not very plentiful. Aral and Kalyon
[46] suspended neutrally buoyant glass beads of 12-μm diameter in a shear-thinning and viscoelastic silicone oil. They found that the suspending liquid itself
had a large value of G' and a large linear viscoelastic region. The strain amplitude
for linear behavior, though, shrank progressively with the incorporation of glass
beads, and the dynamic mechanical properties simultaneously increased in magnitude; this latter fact can be seen in Fig. 10.12, taken from their work [46].
Furthermore, the filler volume fraction had to be kept below 0.3 to ensure the

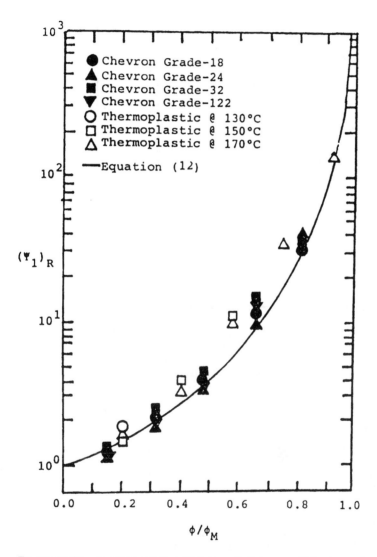

Figure 10.11 Average relative first normal stress coefficient as a function of the reduced volume fraction of glass beads for the suspensions of Fig. 10.2 and 10.5. (From Ref. 14.)

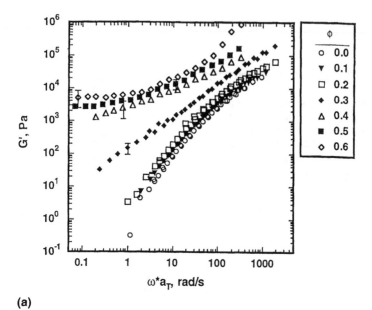

(a)

(b)

FIGURE 10.12 Dynamic mechanical properties of a suspension of neutrally buoyant glass beads in silicone oil at a reference temperature of 273 K: (a) storage
modulus and (b) loss modulus. (From Ref. 46.)

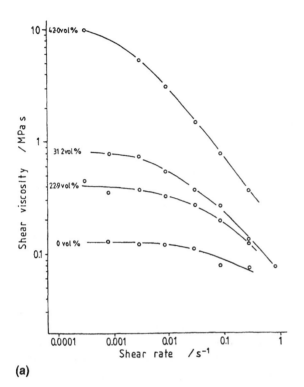

(a)

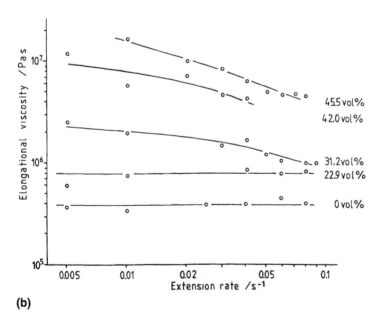

(b)

FIGURE 10.13 Viscosity of filled and unfilled PIB as a function of deformation rate at 294 K: (a) shear viscosity and (b) elongational viscosity. (From Ref. 47.)

absence of structure formation in the suspension. Under these conditions, the G' and G'' versus frequency data at different temperatures could be superposed by means of a horizontal shift, and the Cox and Merz rule was found to be obeyed. Additionally, G'' was greater than G' over the investigated frequency range, indicating that the viscous component of the modulus was dominant over the elastic counterpart. Faulkner and Schmidt [42] made storage modulus as well as loss modulus measurements on glass bead–filled polypropylene melts as a function of solids volume fraction at a single frequency. They found that G' of the suspension divided by the corresponding quantity for the suspending liquid was greater than unity and increased linearly with filler concentration. The relative loss modulus, however, was always larger than the relative storage modulus and increased with concentration at a faster rate. Results of Poslinski et al., however, indicate that elastic forces increase more rapidly than viscous forces [14].

C. Extensional Viscosity

Measurements of the steady extensional viscosity of particulate-filled polymers are rather limited. However, it appears that, at a given stretch rate, the steady extensional viscosity of liquids containing colloidal or noncolloidal particles increases with increasing volume fraction of solids. This is shown in Fig. 10.13 for suspensions of colloidal alumina powder in polyisobutylene (PIB) stretched at a constant extension rate [47]. Both the shear and extensional viscosities exhibit zero-shear-rate values (at least up to a particle volume fraction of about 0.3) that give way to monotonically decreasing viscosities with increasing deformation rate beyond a critical value of the deformation rate; this critical value decreases with increasing solids concentration. When these data are used to determine the ratio of the extensional to shear viscosities at low deformation rates, a Trouton ratio of approximately 3 is calculated; thus, Newtonian behavior is observed. As will be seen in the next chapter, a very different situation arises for fiber-filled suspensions.

VI. FLOCCULATED SUSPENSIONS

When fine particles are dispersed in low-viscosity liquids, the observed rheological behavior is the result of interplay between hydrodynamic forces and van der Waals, coulombic, steric and Brownian motion forces [48]. For concentrated systems, Brownian motion is relatively weak, and the rest state is characterized by the formation of structures—flocs or a lattice. This structure formation is the result of particle–particle interactions owing to surface forces and is different from network formation resulting from polymer–polymer interactions (entanglements). When suspensions are prepared in organic or nonpolar media, coulombic effects are unimportant and the van der Waals forces of attraction make the suspended particulates flocculate, trapping liquid and forming a gel. The result is a high viscosity and the appearance of an apparent yield stress with decreasing shear rate. This is shown in Fig. 10.14, taken from the work of Poslinski et al. [14]. The floc sizes can be reduced by shearing, and the result is the severe shear thinning observed in Fig. 10.14. If shearing is stopped, the structure can reform; and if the time scale for recovery is large, time-dependent effects or thixotropy can arise; the rate of structure formation is faster, the smaller the particle size. A further consequence of particle agglomeration is that data superposition is not observed when data at a fixed particulate volume fraction are plotted in the manner of Fig. 10.8; data points lie above the corresponding hard sphere results. Furthermore, the suspension viscosity does not always decrease with increasing

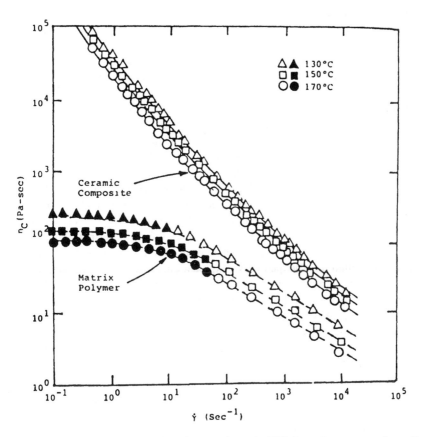

FIGURE 10.14 Shear viscosity of approximately 50% by volume ceramic particulates compounded in a thermoplastic polymer. (From Ref. 14.)

temperature; it can even increase, especially at low shear rates. This is because an increase in temperature leads not only to a decrease in the suspending medium viscosity but also to an increase in the rate of structure formation due to enhanced Brownian motion. Note that that the suspended particles can be properly dispersed and the system stabilized against flocculation by adsorbing or grafting polymer chains on the surface. If the suspending liquid is a good solvent for the polymer, polymer–solvent contacts are favored over polymer–polymer or particle–particle contacts, and the particles remain separated. This is known as *steric stabilization*; the use of a steric stabilizer can convert a gelled paste into a free-flowing liquid [49]. The addition of a poor solvent to a sterically stabilized suspension, though, can again cause flocculation.

Again, if the suspended particles carry like charge, then in a polar medium electrostatic repulsion results in a ordered lattice, which leads to a high zero-shear viscosity or even a yield stress [23]. Flow tends to disturb the lattice and, consequently, the viscosity decreases on increasing the shear rate. The addition of an electrolyte can shield the charges, reduce the viscosity level, and even lead to hard sphere behavior. Structures also build up when each particle carries charges of both signs, in which case a scaffold of house-of-cards structure can result. This happens, for example, with aqueous clay suspensions, which display mutual flocculation because of the presence of double layers of opposite sign on the edge and flat surfaces of the clay. Again one gets pronounced shear thinning and a reduction in the overall viscosity level by addition of charges or alteration of pH. This is shown in Fig. 10.15, where it is seen that the zero-shear viscosity of a 10 vol.% suspension of attapulgite clay in water depends quite sensitively on the pH level, which can be adjusted by adding various amounts of HCl or NaOH [50]. Viscosity differences are largest at low shear rates and are considerably reduced at the high-shear end due to differing amounts of shear thinning. It is interesting to plot the zero-shear viscosity as a function of pH; this is done in Fig. 10.16 [50]. There is complex behavior marked by the presence of a maximum and two minima. The qualitative explanation, however, is quite straightforward. The maximum at intermediate values of pH is due to the house-of-cards structure. Increasing pH causes a reversal of the positive charges on the edges, while decreasing pH causes a reversal of the negative charges on the faces. In both instances, the scaffolding structure is disrupted and the viscosity is lowered. At very high and very low pH, though, the ionic strength of the system increases, resulting in shrinkage of the electrical double layer around the particle surfaces. The reduced repulsion force then causes flocculation to recur, and this shows up as an increase in the viscosity.

In summary then, a flocculated suspension possesses a shear viscosity that is greater than the corresponding hard sphere value, and it may also exhibit a yield stress as well as thixotropic behavior. This is all the result of structure formation within the suspension. In this situation, viscosity levels can be lowered by adding charges or dispersant molecules to the system to prevent particulate

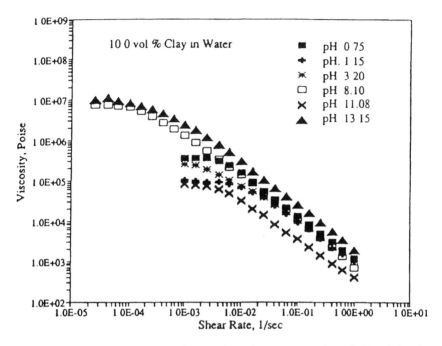

FIGURE 10.15 Shear viscosity of a 10% by volume suspension of attapulgite clay in water at various pH levels. (From Ref. 50.)

agglomeration. In the composites area, a method of achieving good dispersion is to modify the filler surface through the use of coupling agents. These are typically organometallic compounds based on titanium, silicon, or zirconium and are capable of reacting with both the polymer and the filler. This promotes polymer–filler contacts over filler–filler contacts. Titanates, for example, can be used to couple organic polymers with fillers like calcium carbonate, carbon black, sulfur, wood flour, pigments, aramid and carbon fibers, metals, and metal oxides. The amount of coupling agent employed is typically 0.5–2% by weight of the filler. We also remind the reader about the possibility of flow-induced agglomeration mentioned in Chap. 9. A simple analysis to evaluate this demixing resulting from interparticle collisions in laminar flow is described by Agarwal et al. [50a]. Reference 51 may be consulted for the theoretical aspects of coagulation and flocculation.

A. Yield Stress in Flocculated Suspensions

In a flocculated suspension, the agglomerates may be as small as doublets of primary particles, but most agglomerates of practical importance are much larger

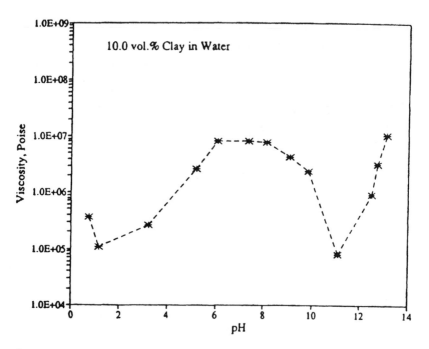

FIGURE 10.16 Data of Fig. 10.15 replotted as zero-shear viscosity versus pH. (From Ref. 50.)

than doublets. These agglomerates can be held together by various mechanisms, including mechanical interlocking of irregularly shaped particles, van der Waals forces, electrostatic forces, a variety of "glues," and water bridges. In a nonpolar medium, traces of water can collect at the interface between particles and hold them together by interfacial forces; these water bridges between particles are probably more important than is generally recognized. These agglomerates, regardless of the binding mechanism, can be broken down by flow, with an elongational flow field being more effective in breaking up weak agglomerates than a shear flow field. Recall that in a shear field, a particle experiences a tensile force that tends to pull the particle apart in a direction 45° to the plane of the moving plates causing the shear flow; the larger the particle, the greater is the tensile force acting on it, and the smaller is the shear stress required to break up the agglomerate. Agglomerate breakup is the cause of yielding, and it is found that, for a given suspension, the yield stress typically increases with decreasing size of the suspended particulates as well as with increasing solids loading. In a flocculated suspension, this yield stress must be exceeded before flow can take place. Below the yield stress, linear elastic behavior is observed.

Popular methods of measuring the yield stress of suspensions were discussed in the previous chapter. The results of these measurements have been represented by a variety of semi-empirical equations or equations based on experimental observations [52–54, for example]. In terms of theoretical approaches, Tanaka and White [55] used a cell model to compute the additional energy dissipated during shearing of spheres of radius R suspended in a Newtonian liquid in the presence of van der Waals and electrostatic forces. This led to the following expression for the yield stress due to the action of van der Waals forces alone:

$$\tau_1 = \frac{NA}{R^3} \left[\frac{(\phi/\phi_m)^{1/3}}{1 - (\phi/\phi_m^{1/3})} \right]^2 \tag{10.13}$$

where N is the total number of nearest neighbors of each sphere and A is Hamaker's constant, a measure of van der Waals forces. Clearly the yield stress increases with decreasing particle size and increasing solids volume fraction. The suspending liquid viscosity is not a relevant quantity.

When the suspending liquid is polar, electrostatic forces cannot be neglected, and Scales et al. [56] have proposed a model for polydisperse flocculated spheres that relates the yield stress to the surface chemistry of the suspended solids. Surface chemistry is measured through the zeta potential (or electrophoretic mobility), which is a function of the ratio of the surface charge density of the suspended colloid to the ionic strength of the suspension: the zeta potential is a unique function of the pH. The maximum yield stress at any given loading level is found to occur at a pH where the suspension is fully flocculated and at which point the zeta potential is zero: this is known as the *isoelectric point*. Typical yield stress behavior of aqueous alumina suspensions is depicted in Fig. 10.17 [56], and the curves of yield stress versus pH are essentially parabolic in nature: all the data collapse onto a single curve when the ordinate of each curve is divided by the corresponding maximum value of the yield stress. The model of Scales et al. calculates the mean force between particles lying in a specified plane and relates this to the yield stress: the expression for the electrical forces incorporates double-layer interactions. This model correctly predicts that the yield stress data normalized with respect to the maximum yield stress are independent of the particle size. The model also shows that the dependence of the normalized yield stress on the zeta potential is identical to that proposed by others in the past.

B. Data Representation

The most common equations employed to represent the shear data of yield stress fluids are the Bingham plastic [57] and Herschel–Bulkley [58] models. In one-dimensional form, these are, respectively:

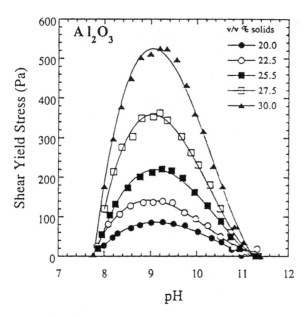

FIGURE 10.17 Yield stress versus volume fraction of a suspension of AKP-30 alumina particles in 0.01 M background electrolyte. (From Ref. 56. Reproduced with permission of the American Institute of Chemical Engineers.)

$$\tau = \tau_, + \eta_p\dot{\gamma} \qquad \tau \geq \tau_, \qquad\qquad (10.14)$$

$$\tau = \tau_y + K'\dot{\gamma}^n \qquad \tau \geq \tau_, \qquad\qquad (10.15)$$

It is seen that the Herschel–Bulkley model reduces to the Bingham model when the constant n equals unity. Also, the Bingham model is simply the Newtonian fluid with a yield stress term added to it, and for both the Bingham and Herschel–Bulkley fluids there is no flow at stress levels below the yield stress. Although the yield stress appearing in either equation can, in principle, be estimated using the theories presented in the previous section, in practice this is very difficult to do, for most filler particles are not monodisperse spheres, nor are values of the Hamaker constant and other surface characteristics readily available. As a consequence the yield stress has to be determined experimentally using any of the methods presented in Chap. 9. Once the yield stress is known, the other constants can be obtained by fitting Eq. (10.15) to viscometric data or to squeezing-flow data [59]. Note that Lipscomb and Denn have argued against the use of Eqs. (10.14) and (10.15) for the analysis of squeezing flow, because such an analysis unrealistically assumes the existence of an unsheared plug at the center of the gap between the two disks [60]. However, Gartling and Phan-Thien have demonstrated that this is not a serious concern, and the data analysis procedure outlined in Ref. 59 remains valid [61].

Another popular equation for concentrated plastic pastes and suspensions and also for blood is the Casson equation [62]:

$$\tau^{1/2} = \tau_{\iota}^{1/2} + K\dot{\gamma}^{1/2} \tag{10.16}$$

in which K is an empirical constant. If the suspending liquid is non-Newtonian in behavior, by itself, the constant K in Eq. (10.16) should be modified to [63]:

$$K = K\left(\frac{\eta_a}{\eta_0}\right)^{1/2} \tag{10.17}$$

where η_0 is the viscosity of the liquid at zero rate of shear and η_a is the apparent viscosity of the liquid at a rate of shear $\dot{\gamma}$.

Bird et al. have provided the tensorially correct three-dimensional forms of Eqs. (10.14)–(10.16), and have shown how these may be used to obtain analytical solutions to a large number of flow problems [64].

C. Elastic Effects

Most of the comments made concerning elastic effects in Sec. V carry over to flocculated suspensions as well. A consequence of the development of a yield stress is that the linear viscoelastic region shrinks considerably, and dynamic mechanical data become functions of the strain amplitude. This is shown in Fig. 10.18a, where the complex viscosity of a 70 vol% suspension of nearly colloidal silicon particles in a low-molecular-weight polyethylene is plotted as a function of frequency [65]. It is found that the data show a very pronounced strain dependence. However, all the curves can be made to superpose with one another and also with the steady-shear viscosity as a function of the shear rate if the abscissa is taken to be the product of the frequency and the strain amplitude. This is shown in Fig. 10.18b, which also clearly shows the occurrence of a yield stress—the lowest shear rate here is orders of magnitude smaller than the value at which a Newtonian plateau would generally have been observed. This data superposition is a very useful result, and it is analogous to the conventional Cox–Merz rule. However, unlike the empirical Cox–Merz rule, this modified Cox–Merz rule (now referred to as the Rutgers-Delaware correlation [65a]) has a theoretical basis. By a simple, but nontrivial, generalization of the Herschel–Bulkley equation (which incorporated a recoverable strain having a maximum value equal to the critical strain at which the material yields), Doraiswamy et al. were able to demonstrate the theoretical validity of this data superposition [65].

The presence of a yield stress in the shear data implies that extensional data should also exhibit a yield stress. This would also be predicted by the three-

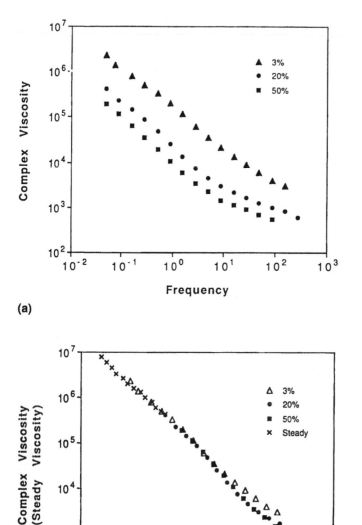

(a)

(b)

FIGURE 10.18 (a) Complex viscosity versus frequency data for a 70 vol% suspension of Si in polyethylene at 140°C for three different strain amplitudes. (b) Data in part (a) replotted versus the product of frequency and strain amplitude; also, superposition with shear viscosity data as a function of shear rate. (From Ref. 65.)

dimensional forms of Eqs. (10.14)–(10.16) and is, indeed, observed. Figure 10.19 shows the steady-extensional-viscosity data of Suetsugu and White on suspensions of calcium carbonate in molten polystyrene at a filler volume fraction of 0.3 [45]. Not surprisingly, the low-extension-rate viscosity increases rapidly with decreasing particle size, but this increase is significantly tempered upon coating the particles with stearic acid. Data tend to follow the Casson equation, and the ratio of the yield stress in extension to the yield stress in shear ranges from 1.4 to 1.9 [45].

Plucinski et al. measured the transient extensional viscosity of chocolate, a concentrated suspension of irregularly shaped, wide-size-distribution sugar and

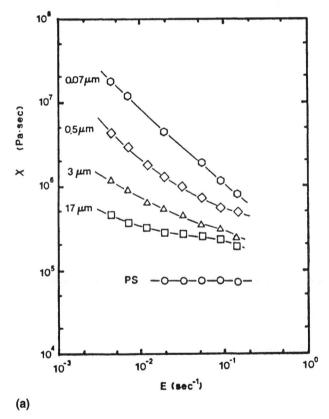

(a)

FIGURE 10.19 Steady-state elongational viscosity of a 30 vol% suspension of calcium carbonate in a polystyrene melt at 180°C as a function of stretch rate for different filler sizes: (a) uncoated particles and (b) particles coated with stearic acid. (From Ref. 45. Copyright © 1983 by John Wiley & Sons, Inc. Reprinted by permission of John Wiley & Sons, Inc.)

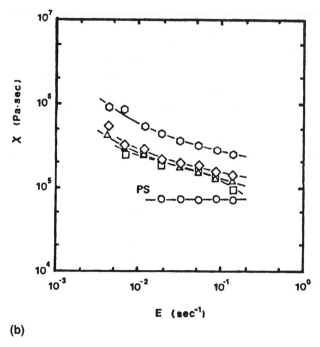

(b)

FIGURE 10.19 Continued.

cocoa solids in a Newtonian fat medium, by means of fiber spinning [66]. A commercial chocolate sample was used and experiments were done at 40°C; results are presented as open symbols in Fig. 10.20. Despite the fact that the stretch rate varies with position along the spinline and despite the fact that these are, a priori, not steady-state data, results from all the different runs superpose, and a unique curve is obtained. Figure 10.20 shows clear evidence of a yield stress at low stretch rates, extension thinning with increasing stretch rate, and an indication of a constant extensional viscosity at high stretch rates. It appears that while chocolate is elastic in the sense that it possesses a yield stress and has a measurable storage modulus, it is not elastic in the sense that it does not possess a memory for prior deformation. Indeed, filament-stretching experiments at constant stretch rate showed a very rapid attainment of the steady state, within about 1 sec of inception of stretching, and these results are shown as filled symbols. The two sets of data, though not overlapping in stretch rate, appear to be extensions of each other. It therefore appears that, in this case, the instantaneous apparent extensional viscosity is approximately the same as the true, steady-state extensional viscosity. If this turns out to be true for other suspensions or for a class

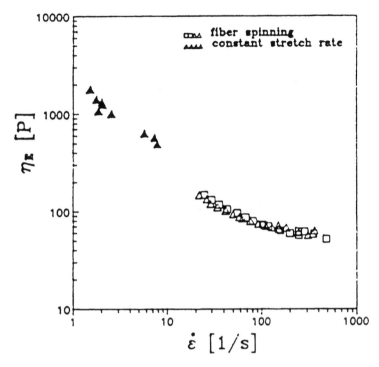

FIGURE 10.20 Extensional viscosity of molten chocolate at 40°C. (From Ref. 66.)

of suspensions, then the use of fiber spinning will afford a simple method of obtaining extensional viscosity data over a wide deformation rate range. These data could then be converted to shear data with the use of the three-dimensional forms of Eqs. (10.14)–(10.16). Such a procedure has the potential of yielding the entire flow curve from a single fiber-spinning run, and the results would not be corrupted with complications arising from wall slip, particle migration, or sample fracture.

D. Thixotropy

The structure formation that endows a colloidal suspension with a yield stress also leads to the phenomenon of thixotropy, or reversible rheological changes, if the structure can be broken down by flow but structural recovery is not instantaneous; very rapid recovery reveals itself through shear thinning alone. If a suspension is only weakly flocculated, the bonds holding the flocs together are weak, and these can be disrupted, in a progressive manner, by flow. On cessation of flow, Brownian motion helps to reestablish the flocculated rest state, although this process may take anywhere from a few minutes to a few hours to be complete. These structural rearrangements manifest themselves in a variety of ways during rheological testing. During shearing at a constant shear rate, for example, the shear stress continues to decrease with time and reaches a constant value only after an extended period of time. If shearing is stopped upon reaching equilibrium but then resumed, the curve of shear stress versus time lies below the earlier curve, unless a sufficient rest period is given between the two different runs. For the same reason, if the shear rate is ramped up and down, a thixotropic loop is obtained. This is shown in Fig. 10.21 for the shearing of a latex wall paint [67]; the area within the loop is taken to be a measure of the thixotropy of the sample.

Originally the term *thixotropy* denoted the reversible solid–liquid transition on agitating a gel to convert it to a sol. Today it is used to describe the continuous reduction in viscosity with time of shearing and the subsequent recovery on cessation of flow. Several excellent reviews are available on the topic, and the phenomenon is exhibited by many different classes of materials, including paints, detergents, and foodstuffs [68–70]. Fumed silica in paraffin oil is the prototypical thixotropic system, and use is made of this property in the formulation of nondrip paints, which contain a polyamide-modified thixotropic alkyd [71]. The amide

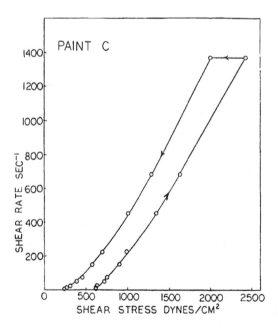

FIGURE 10.21 Thixotropic loop for a latex wall paint. (From Ref. 67.)

groups are bound within the alkyd backbone and form hydrogen bonds between themselves; thixotropy arises due to the shear-induced breakdown of these bonds and their slow reformation after the paint has been spread. This results in good leveling properties along with good sag resistance.

A large number of theories have been proposed to describe thixotopy mathematically [70]. In the simplest case, a structure parameter λ is used, and it has a value of unity for the completely built-up structure and a value of zero for the completely broken-down structure. These two limits also correspond to the upper and lower Newtonian viscosities, respectively. Under the influence of shearing, there is structure breakdown and recovery, and this process is represented by a first-order differential equation relating $d\lambda/dt$ to a term involving breakdown and a term involving buildup;these terms involve only the shear rate and the instantaneous value of λ. At a given shear rate, the equilibrium value of the structure parameter is determined by setting $d\lambda/dt$ equal to zero. The result is a number between zero and unity, and this corresponds to a value of the viscosity between the two Newtonian limits. Thixotropy arises from the time evolution of λ as it goes from one equilibrium state to another.

VII. ELECTRORHEOLOGY

The most fascinating development in rheology in recent years has been the widespread realization that the interaction of an electric field with a suspension of

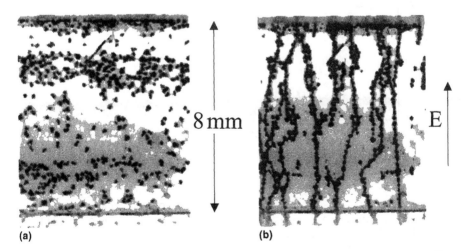

(a)   (b)

FIGURE 10.22 (a) Structure of a noncolloidal, 2 wt% suspension of alumina in silicone oil in the absence of an electric field. (b) Fibrous structure formed on switching on a 500-Hz electric field of 0.5 kV/mm rms field strength. (From Ref. 74, with permission from Elsevier Science.)

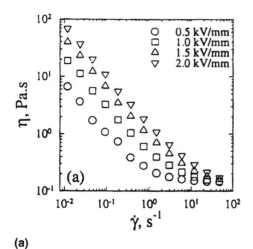

(a)

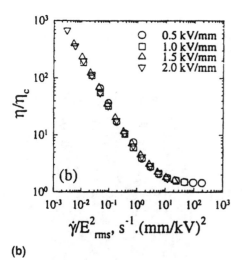

(b)

FIGURE 10.23 (a) Viscosity of a 20 wt% suspension of alumina in silicone oil as a function of shear rate and electric field strength. (b) Master curve of data in part (a). (From Ref. 74, with permission from Elsevier Science.)

hygroscopic particles, whether colloidal or noncolloidal, in a hydrophobic, electrically insulating liquid can result in a reversible sol-gel transition. Under appropriate conditions, the application of an electric field can cause a liquid-to-solid phase change to occur in microseconds and with a fraction of the power it takes to operate a light bulb [72]. Typical suspensions involve silica, alumina, or titania in silicone or hydrocarbon oils at loading levels up to 50% by volume. This phenomenon of instant solidification or liquefaction of a suspension by the application or removal of an electric field is known as *electrorheology*, and its potential in torque transmission, damping, and vibration control was first recognized by Winslow [73]. The potential of electrorheology has, however, not been realized, due to problems resulting from suspension instability, a limited operating range in terms of temperature, and time-dependent properties. Progress at overcoming these problems has been slow due to a lack of full understanding of the underlying mechanisms.

Several readable reviews on the subject of electrorheology are available [74 and references therein]. In essence, when an electric field is applied normal to the direction of flow, a disperse and free-flowing suspension is converted to a gelled structure consisting of fibrous columns perpendicular to the electrodes and spanning the entire gap between the electrodes. This is shown in Fig. 10.22; the columns thicken with increasing particle concentration, and they can be destroyed by increasing the applied pressure drop [74]. Shear viscosity data at a variety of field strengths for a 20 wt% suspension of porous spherical alumina particles in silicone oil are shown in Fig. 10.23a; these collapse onto a single curve (see Fig. 10.23b) when plotted as a function of the shear rate divided by the square of the electric field strength [74]. It is found that for a fixed particle concentration and fixed applied electric field strength, these and other electrorheological data can be described by the Bingham plastic equation, Eq. (10.14), in which the yield stress is zero in the absence of the electric field and increases essentially quadratically with increasing field strength.

The most popular explanation for the electric field–induced yield stress is electrostatic polarization. The applied electric field induces in the suspended particles a dipole that is aligned with the electric field. Attraction between neighboring dipole particles then produces the fibrous structure pictured in Fig. 10.22. Other possible mechanisms and mathematical models based on these mechanisms may be found in Ref. 74.

VIII. CONCLUDING REMARKS

Suspension rheology is clearly a complex subject, but it is also a very important one. This latter point can be gauged not only from the very large number of publications on the topic but also from the fact that the multitude of applications cut across disciplinary lines. Indeed, particulate technology, which includes suspension behavior, is considered a critically important technology by the largest chemical companies in the world from the viewpoint of success in global competition [75]. A recent industrial system of importance, for example, is the suspension of dendritic particles in a low-viscosity matrix; the space-filling nature of these additives induces yield stresses at concentrations of ~0.1% which makes it possible to use them as thickeners. Although much is now known about the behavior of model suspensions, these fluids are rarely encountered in industrial practice. What is worse is that our ability to theoretically predict the behavior of even model suspensions is quite limited. As a consequence, suspension rheology is likely to remain an area of active research for a long time to come.

REFERENCES

1. R.K. Gupta. Particulate suspensions. In: S.G. Advani, ed. Flow and Rheology in Polymer Composites Manufacturing. Elsevier, Amsterdam, 1994, pp. 9–51.
2. A. Einstein. Investigations on the Theory of the Brownian Movement. Dover, New York, 1956, pp. 36–54.
3. G.B. Jeffrey. The motion of ellipsoidal particles immersed in a viscous fluid. Proc. Roy. Soc. A102:161–179 (1922).
4. J. Happel, H. Brenner. Low Reynolds Number Hydrodynamics. Martinus Nijhoff, Dordrecht, 1983, pp. 431–473.
5. D.J. Jeffrey, A. Acrivos. The rheological properties of suspensions of rigid particles. AIChE J. 22:417–432 (1976).
6. A.B. Metzner. Rheology of suspensions in polymeric fluids. J. Rheol. 29:739–775 (1985).
7. L.A. Utracki. The rheology of two-phase flows. In: A.A. Collyer, D.W. Clegg, eds. Rheological Measurement. Elsevier, London, 1988, pp. 479–594.
8. P.M. Adler, A. Nadim, H. Brenner. Rheological models of suspensions. Adv. Chem. Eng. 15:1–72 (1990).
9. P.J. Carreau. Rheology of filled polymeric systems. In: R.P. Chhabra, D. DeKee, eds. Transport Processes in Bubbles, Drops, and Particles. Hemisphere, New York, 1992, pp. 165–190.
10. N.A. Frankel, A. Acrivos. On the viscosity of a concentrated suspension of solid spheres. Chem. Eng. Sci. 22:847–853 (1967).
11. J.S. Chong, E.B. Christiansen, A.D. Baer. Rheology of concentrated suspensions. J. Appl. Polym. Sci. 15:2007–2021 (1971).
11a. S.H. Maron, P.E. Pierce. Application of Ree-Eyring generalized flow theory to suspensions of spherical particles J. Colloid Sci. 11:80–95 (1956).
12. T. Kataoka, T. Kitano, M. Sasahara, K. Nishijima. Viscosity of particle filled polymer melts. Rheol. Acta 17:149–155 (1978).
13. T. Kitano, T. Kataoka, T. Shirota. An empirical equation of the relative viscosity of polymer melts filled with various inorganic fillers. Rheol. Acta 20:207–209 (1981).
14. A.J. Poslinski, M.E. Ryan, R.K. Gupta, S.G. Seshadri, F.J. Frechette. Rheological

behavior of filled polymeric systems. I. Yield stress and shear-thinning effects. J. Rheol. 32:703–735 (1988).

15. A.J. Poslinski, M.E. Ryan, R.K. Gupta, S.G. Seshadri, F.J. Frechette. Rheological behavior of filled polymeric systems. II. The effect of a bimodal size distribution of particulates. J. Rheol. 32:751–771 (1988).

16. R.C. Ball, P. Richmond. Dynamics of colloidal dispersions. Phys. Chem. Liq. 9:99–116 (1980).

17. I.M. Krieger. Rheology of monodisperse latices. Adv. Colloid Interface Sci. 3:111–136 (1972).

18. R.J. Farris. Prediction of the viscosity of multimodal suspensions from unimodal viscosity data. Trans. Soc. Rheol. 12:281–301 (1968).

19. R.K. Gupta, S.G. Seshadri. Maximum loading levels in filled liquid systems. J. Rheol. 30:503–508 (1986).

20. G.J. Jarzebski. On the effective viscosity of pseudoplastic suspensions. Rheol. Acta 20:280–287 (1981).

21. F.M. Chapman, T.S. Lee. Effect of talc filler on the melt rheology of polypropylene. SPE J. 26:37–40 (1970).

22. T.B. Lewis, L.E. Nielsen. Viscosity of dispersed and aggregated suspension of spheres. Trans. Soc. Rheol. 12:421–443 (1968).

23. M.E. Woods, I.M. Krieger. Rheological studies on dispersions of uniform colloidal spheres. I. Aqueous dispersions in steady shear flow. J. Colloid Interface Sci. 34:91–99 (1970).

24. Y.S. Papir, I.M. Krieger. Rheological studies on dispersions of uniform colloidal spheres. II. Dispersions in nonaqueous media. J. Colloid Interface Sci. 34:126–130 (1970).

25. G.K. Batchelor. The effect of Brownian motion on the bulk stress in a suspension of spherical particles. J. Fluid Mech. 83:97–117 (1977).

26. J.C. Van der Werff, C.G. de Kruif. Hard-sphere colloidal dispersions: the scaling of rheological properties with particle size, volume fraction, and shear rate. J. Rheol. 33:421–454 (1989).

27. C.R. Wildemuth, M.C. Williams. Viscosity of suspensions modeled with a shear-dependent maximum packing fraction. Rheol. Acta 23:627–635 (1984).

28. J.R. Van Wazer, J.W. Lyons, K.Y. Kim, R.E. Colwell. Viscosity and Flow Measurement. Interscience, New York, 1963.

29. L.A. Mondy, A.L. Graham, J.L. Jensen. Continuum approximations and particle interactions in concentrated suspensions. J. Rheol. 30:1031–1051 (1986).

30. R.P. Chhabra. Bubbles, Drops and Particles in Non-Newtonian Fluids, CRC Press, Boca Raton, FL, 1993, pp. 369–386.

31. R.L. Hoffman. Discontinuous and dilatant viscosity behavior in concentrated suspensions. I. Observation of a flow instability. Trans. Soc. Rheol. 16:155–173 (1972).

32. H.A. Barnes. Shear-thickening (''dilatancy'') in suspensions of nonaggregating solid particles dispersed in Newtonian liquids. J. Rheol. 33:329–366 (1989).

33. R.L. Hoffman. Discontinuous and dilatant viscosity behavior in concentrated suspensions. II. Theory and experimental tests. J. Colloid Interface Sci. 46:491–506 (1974).

34. B.J. Ackerson. Shear induced order and shear processing of model hard sphere suspensions. J. Rheol. 34:553–590 (1990).

35. P. D'Haene, G.G. Fuller, J. Mewis. Shear thickening effect in concentrated colloidal dispersions. Proc. XIth Int. Congr. on Rheology, Brussels, Belgium, 1992, pp. 595–597.

36. W.J. Frith, P. D'Haene, R. Buscall, J. Mewis. Shear thickening in model suspensions of sterically stabilized particles. J. Rheol. 40:531–548 (1996).

37. W.H. Boersma, J. Laven, H.N. Stein. Shear thickening (dilatancy) in concentrated dispersions. AIChE J. 36:321–332 (1990).

38. R.L. Hoffman. Explanations for the cause of shear thickening in concentrated colloidal suspensions. J. Rheol. 42:111–123 (1998).

39. C.D. Han. Rheology in Polymer Processing. Academic Press, New York, 1976.

40. S. Newman, Q.A. Trementozzi. Barus effect in filled polymer melts. J. Appl. Polym. Sci. 9:3071–3089 (1965).

41. W. Philippoff, F.H. Gaskins. The capillary experiment in rheology. Trans. Soc. Rheol. 2:263–284 (1956).

42. D.L. Faulkner, L.R. Schmidt. Glass bead–filled polypropylene. Part I: Rheological and mechanical properties. Polym. Eng. Sci. 17:657–665 (1977).

43. N. Minagawa, J.L. White. The influence of titanium dioxide on the rheological and extrusion properties of polymer melts. J. Appl. Polym. Sci. 20:501–523 (1976).

44. C.D. Han. Multiphase Flow in Polymer Processing. Academic Press, New York, 1981.

45. Y. Suetsugu, J.L. White. The influence of particle size and surface coating of calcium carbonate on the rheological properties of its suspensions in molten polystyrene. J. Appl. Polym. Sci. 28:1481–1501 (1983).

46. B.K. Aral, D.M. Kalyon. Viscoelastic material functions of noncolloidal suspensions with spherical particles. J. Rheol. 41:599–620 (1997).

47. J. Greener, J.R.G. Evans. Uniaxial elongational flow of particle-filled polymer melts. J. Rheol. 42:697–709 (1998).

48. W.B. Russel, D.A. Saville, W.R. Schowalter. Colloidal Dispersions. Cambridge University Press, Cambridge, 1989.

49. J. Rajaiah, E. Ruckenstein, G.F. Andrews, E.O. Forster, R.K. Gupta. Rheology of sterically stabilized ceramic suspensions. Ind. Eng. Chem. Res. 33:2336–2340 (1994).

50. S.H. Chang, M.E. Ryan, R.K. Gupta. The effect of pH, ionic strength and temperature on the rheology and stability of aqueous clay suspensions. Rheol. Acta 32:263–269 (1993).

50a. S. Agarwal, D. Doraiswamy, R.K. Gupta. Demixing effects in laminar shearing flows: a practically useful approach. Can. J. Chem. Eng. 76:511–515 (1998).

51. B. Dobias. Coagulation and flocculation. Marcel Dekker, New York, 1993.

52. D.G. Thomas. Transport characteristics of suspensions: III. Laminar flow properties of flocculated suspensions. AIChE J. 7:431–437 (1960).

53. A.S. Michaels, J.C. Bolger. The plastic behavior of flocculated kaolin suspensions. Ind. Eng. Chem. Fundam. 1:153–162 (1962).

54. E.C. Gay, P.A. Nelson, W.P. Armstrong. Flow properties of suspensions with high solids concentration. AIChE J. 15:815–822 (1969).

55. H. Tanaka, J.L. White. A cell model theory of the shear viscosity of a concentrated suspension of interacting spheres in a non-Newtonian Fluid. J. Non-Newt. Fluid Mech. 7:333–343 (1980).

56. P.J. Scales, S.B. Johnson, T.W. Healy, P.C. Kapur. Shear yield stress of partially flocculated colloidal suspensions. AIChE J. 44:538–544 (1998).

57. E.C. Bingham. Fluidity and Plasticity. McGraw-Hill, New York, 1922, pp. 215–218.

58. W.H. Herschel, R. Bulkley. Konsistenzmessungen von Gummi-Benzollosungen. Kolloid-Z. 39:291–300 (1926).

59. G.H. Covey, B.R. Stanmore. Use of the parallel-plate plastometer for the characterization of viscous fluids with a yield stress. J. Non-Newt. Fluid Mech. 8:249–260 (1981).

60. G.G. Lipscomb, M.M. Denn. Flow of Bingham fluids in complex geometries. J. Non-Newt. Fluid Mech. 14:337–346 (1984).

61. D.K. Gartling, N. Phan-Thien. A numerical simulation of a plastic fluid in a parallel plate plastometer. J. Non-Newt. Fluid Mech. 14:347–360 (1984).

62. N. Casson. In: C.C. Mill, ed. Rheology of Disperse Systems. Pergamon Press, London, 1959, pp. 84–104.

63. T. Matsumoto, A. Takashima, T. Masuda, S. Onogi. A modified Casson equation for dispersions. Trans. Soc. Rheol. 14:617–620 (1970).

64. R.B. Bird, G.C. Dai, B.J. Yarusso. The rheology and flow of viscoplastic materials. Rev. Chem. Eng. 1:1–70 (1982).

65. D. Doraiswamy, A.N. Mujumdar, I. Tsao, A.N. Beris, S.C. Danforth, A.B. Metzner. The Cox–Merz rule extended: a rheological model for concentrated suspensions and other materials with a yield stress. J. Rheol. 35:647–685 (1991).

65a. I.M. Krieger. Correspondence: comments on a manuscript by Doraiswamy et al. J. Rheol. 36:215–216 (1992).

66. J. Plucinski, R.K. Gupta, S. Chakrabarti. Extensional rheometry of chocolate. Proc. XIIth Int. Congr. on Rheology, Quebec City, Canada, 1996, pp. 775–776.

67. P.E. Pierce. Rheology of coatings. J. Paint Technol. 41:383–395 (1969).

68. J. Mewis. Thixotropy—a general review. J. Non-Newt. Fluid Mech. 6:1–20 (1979).

69. D.C.-H. Cheng. Thixotropy. Int. J. Cosmetic Sci. 9:151–191 (1987).

70. H.A. Barnes. Thixotropy—a review. J. Non-Newt. Fluid Mech. 70:1–33 (1997).

71. S. Rees. New developments in thixotropic coatings. Polym. Paint Colour J. 185 (4369):10–11 (1995).

72. F.E. Filisko. Electrorheological materials: smart materials of the future. Chem. Industry, 370–373, 18 May 1992.

73. W.M. Winslow. Induced fibration of suspensions. J. Appl. Phys. 20:1137–1140 (1949).

74. M. Parthasarathy, D.J. Klingenberg. Electrorheology: mechanisms and models. Mater. Sci. Eng. R17(2):57–103 (1996).

75. R.D. Nelson Jr., R. Davies, K. Jacob. Teach 'em particle technology. Chem. Eng. Education 29:12–16 (1995).

11

Short-Fiber Suspensions

I. INTRODUCTION

The addition of short glass, carbon, or aramid fibers to polymers such as nylons and polyesters can result in molded parts having increased toughness, temperature resistance, and dimensional stability. Stiffness and tensile strength also increase with increasing fiber content, which can be in excess of 20% by volume; due to cost considerations, glass is the most common reinforcement. The heat-deflection temperature (HDT) of an unreinforced polymer is typically about 20°C below the glass transition temperature; the use of glass reinforcement allows the HDT to easily exceed 100°C and to approach the polymer melting point, although there may be a reduction in impact strength. Glass-reinforced thermoplastics can be extruded, thermoformed, injection molded, or blow molded in the conventional way, and they are commonly utilized to make gears and other structural parts. Since they are used in engineering applications, these reinforced plastics are known as "engineering polymers." In short-fiber composites, the fiber length is typically 0.2 mm, while the *aspect ratio* (ratio of length to diameter) is about 15 [1]. Note that fibers having a length ranging from 13 to 25 mm are also used to make (thermosetting) sheet molding compounds and glass mats that are compression molded to produce automobile body panels [1]. Since the glass reinforcement is long and slender, it can be oriented by flow during processing; an extensional flow field is generally more effective compared to a shear field for this purpose. The fiber orientation gets frozen into the solid composite and makes the mechanical properties of the final part be anisotropic.

A discussion of the flow behavior of suspensions containing n fibers per unit volume, with each fiber having length L and diameter D, is logically divided into at least three concentration regimes. At one extreme, in the dilute region, individual fibers can rotate freely without encountering other fibers, and this makes the suspension viscosity cycle at the same frequency. This requires that n be less than $1/L^3$. At the other extreme, in the concentrated region, "logjams" can develop and the suspension behaves more like a solid than a liquid. This happens when the spacing between fibers becomes of the same order as the fiber diameter. For random orientation of the fibers, this requires that n approach $1/DL^2$; for fibers lying parallel to each other, n can be larger and equal $1/D^2L$ [2]. Since the fiber volume fraction, ϕ, is of the order of nD^2L, the semiconcentrated region for randomly oriented fiber suspensions is defined by:

$$\left(\frac{D}{L}\right)^2 < \phi < \left[\frac{D}{L}\right] \qquad (11.1)$$

and the region boundaries depend explicitly and strongly on the aspect ratio of the fibers. In our discussion here, we will neglect wall effects and assume that the fiber dimensions are such that Brownian motion can be neglected. Also we will assume that while the fiber orientation can change during flow, the fibers are always uniformly distributed within the suspension. In reality, we often observe fiber clustering, fiber migration, and the presence of air-filled voids [3,4]. Fiber migration is caused by the presence of normal stress gradients, while fiber clustering is exacerbated by increasing the loading level or by decreasing the deformation rate.

Data on fiber suspensions are much less extensive and much less definitive compared to data on particulate suspensions. This is due to the difficulty of obtaining reliable and repeatable results on well-characterized fiber suspensions. Although some of the problems can be traced to fiber dimensions being comparable to typical viscometer gaps, fiber flexibility, and the occurrence of mechanical degradation (breakage) of fibers during compounding and viscosity measurement [5], other problems arise due to the simple fact that the suspended fibers are oriented by flow. During capillary rheometry, for example, it is not surprising to find that fibers are randomly oriented at low shear rates but get highly aligned in the flow direction at high shear rates [6]. However, at a fixed shear rate, the extent of orientation increases as the capillary length decreases. This happens because fibers actually get aligned in the converging flow region at the die entrance, and this orientation is gradually lost during shear flow in the capillary; the loss of orientation with increasing capillary length is most apparent at low shear rates. Similarly but more strikingly, during diverging flow, as happens after entry into a mold, fiber alignment built up at the mold entrance is lost extremely rapidly, and it is replaced with fiber alignment in a direction perpendicular to the flow direction [6]. As might be expected, these changes in fiber orientation manifest themselves as changes in fluid rheology.

The preceding observations can be employed to qualitatively explain the observed viscosity behavior of short-fiber suspensions. It is found that the steady-shear viscosity of a polymeric liquid, as measured with a capillary viscometer, is increased upon increasing either the fiber length or the fiber concentration, with fiber concentration being the more important of the two variables. The increase can be appreciable but only at low shear rates; at high shear rates, the suspension and suspending liquid viscosities are virtually identical [7]. At low shear rates, since the fibers are randomly oriented, liquid is essentially forced to flow through a fiber mat, and this results in a high viscosity. At high shear rates, on the other hand, fiber alignment leads to an unsheared plug near the capillary axis and a significantly reduced resistance to flow. The presence of the unsheared plug also results in a blunt velocity profile, and this is observed as shear thinning even at shear rates where the suspending liquid has a constant viscosity. The effect of fiber addition on the activation energy for flow is found to be small, but the addition of fibers whose length is greater than twice the capillary diameter results in pressure fluctuations [7]. At high fiber loading levels, the suspension can become solidlike and exhibit a yield stress [8]. These and other observations made with fiber suspensions have been reviewed and explained in the literature [9–12]. In general, the rheological properties of fiber suspensions depend on variables like the fiber volume fraction, fiber aspect ratio and its distribution, deformation rate, temperature, suspending medium rheology, and the nature of the flow field.

II. DILUTE FIBER SUSPENSIONS

For the shearing flow of a dilute suspension of uniformly sized prolate spheroids in a Newtonian liquid, Jeffery solved the equations of motion and showed that, in the absence of any particle–particle interactions, the spheroids not only translated with the flow but also rotated about an axis that was perpendicular to both the direction of flow and the direction of the velocity gradient [13]. Thus, if shearing takes place in the x-y plane, the particles rotate about the z axis, but they do not always lie in the x-y plane [14,15]. The path of one end of a typical particle is shown in Fig. 11.1, in which the origin of the coordinate system is located at the center of the particle. For a large aspect ratio, the spheroids approximate rigid rods, and the time period of rotation T for flow at a constant shear rate is given by [12,13]:

$$T = \frac{2\pi L}{(D\dot{\gamma})} \tag{11.2}$$

where $\dot{\gamma}$ is the shear rate. Mason and coworkers used a Couette apparatus and verified the linear relationship between the time period of rotation and the recipro-

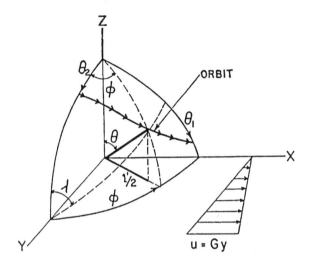

FIGURE 11.1 Motion of a rigid rod under the influence of shear flow. (From Ref. 15.)

cal of the shear rate as predicted by the Jeffery theory by making measurements on rigid cylinders suspended in a high-viscosity corn syrup [14,15]. However, due to differences between the geometry of cylinders and spheroids, the time period of rotation was only about two-thirds the value predicted by the use of Eq. (11.2).

Concerning the shear viscosity of a dilute suspension of short fibers, it is tempting to describe this using an equation similar to the Einstein equation, Eq. (10.1). Unfortunately, the viscosity depends on the microstructure, and, in the present case, the microstructure changes continuously in a periodic manner due to fiber rotation. Consequently, both the Einstein coefficient and the viscosity are time dependent, even when the shear rate is held constant. Indeed, as a result of fore–aft symmetry, both the shear viscosity and the normal stress differences should show undamped oscillations at a frequency equal to $2/T$ Hz. Verifying this result, though, is difficult due to the fact that viscosity enhancement upon addition of a small amount of short fibers is rather small, and the time-dependent viscosity measurement is difficult to carry out. Ivanov et al. studied the behavior of nearly uniform-sized rigid rods of aluminum-coated nylon suspended in a Newtonian castor oil [16]. They used three different fixture geometries, and their results for the suspension specific viscosity as a function of time for different concentrations are displayed in Fig. 11.2. As can be seen, the viscosity increases with concentration, and its behavior is oscillatory, as expected from theory. However, the oscillations are damped, and a steady state is ultimately reached; the

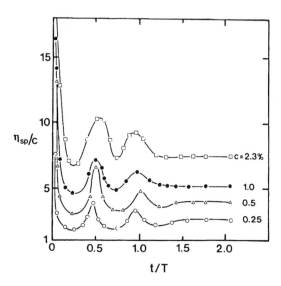

FIGURE 11.2 Observed variations in specific viscosity of suspensions of rods of various concentrations. The effective aspect ratio is 5.2 with a standard deviation of 0.9. The shear rate is 2.51 sec^{-1}. (From Ref. 16.)

suspension then behaves in a Newtonian manner. These authors showed that the damping was a result of a spread in the length of the suspended rods and possible particle–particle interactions [17]. Note that the presence of Brownian motion can also eliminate the time dependence in the rheological properties and make the Einstein coefficient be a determinate quantity [11,12].

Microstructural changes are not an issue in uniaxial elongational flow, since fibers align themselves very readily in the flow direction, and this alignment, which is independent of any initial condition, remains unchanged as long as the flow field is maintained. For this situation, Batchelor used slender-body theory to obtain an expression for the extensional viscosity of a fiber suspension in a Newtonian liquid as [18,19]:

$$\eta_E = \eta \left[3 + \frac{2\alpha\phi(L/D)^2}{3}\left(\frac{1 + 0.64\alpha}{1 - 1.5\alpha} + 1.659\alpha^2 \right) \right] \qquad (11.3)$$

In which α is $[\ln(2L/D)]^{-1}$ and η is the suspending liquid viscosity. Even though, with increasing aspect ratio, the second term on the right-hand side of Eq. (11.3) can become significant in comparison to the first term, the restriction to dilute suspensions precludes such a possibility; predicted stresses, therefore, are only modest perturbations of the stresses required to deform the suspending liquid by itself, and experimental validation of Eq. (11.3) is not easy [19]. Recall from Eq. (11.1) that if the fiber aspect ratio is 100, the suspension is not considered dilute if the volume fraction exceeds 0.0001!

When dilute fiber suspensions are formulated using non-Newtonian polymer solutions, the shear flow behavior is found to be similar, but not the same, as in the case of Newtonian suspending liquids. Karnis and Mason, for example, examined Poiseuille as well as Couette flow and found that fibers rotated in both aqueous polyacrylamide and polyisobutylene-in-decalin solutions, but tended to drift toward a preferred orientation [20]. The cause of this orbit drift is fluid elasticity. This was shown to be true theoretically by Leal by carrying out an analysis employing a second-order fluid model [21].

III. SEMICONCENTRATED FIBER SUSPENSIONS

As in the case of dilute fiber suspensions, it is easiest to consider the behavior of semiconcentrated suspensions in extensional flow, since one may assume perfect fiber alignment in the stretch direction at steady state. This was done by Batchelor, who used a cell model to determine the stress field around a fiber of interest when the average distance between fibers was much less than the fiber length but much greater than the fiber diameter [22]. The result for the suspension extensional viscosity is:

$$\eta_E = \eta \left[3 + \frac{4\phi(L/D)^2}{3 \ln (\pi/\phi)} \right] \tag{11.4}$$

which does not reduce to Eq. (11.3) with decreasing concentration. However, one may interpolate between Eq. (11.3) and (11.4) in an intermediate concentration range that is outside the region of validity of both these equations. More recently, Acrivos and Shaqfeh have proposed an effective medium theory that is valid over the entire concentration range [23]; in the concentration range of overlap with the cell model, the two theories predict essentially the same result for the extensional viscosity.

Equation (11.4) was successfully put to a test by Mewis and Metzner, who carried out fiber-spinning experiments using 0.1–1 vol % glass fibers suspended in a low-molecular-weight polybutene [19]. These authors found that the extensional viscosity of their fiber suspensions was independent of stretch rate, as predicted by Eq. (11.4), and all their results are shown in Fig. 11.3. Extensional viscosity values that were as many as 260 times larger than the suspending oil viscosity could be quantitatively explained by Eq. (11.4), which also correctly portrayed the separate influence of fiber concentration and fiber aspect ratio. Sridhar and Gupta measured tensile stresses down the spinline for the stretching of glass fibers suspended in a mixture of kerosene and polybutene, and found

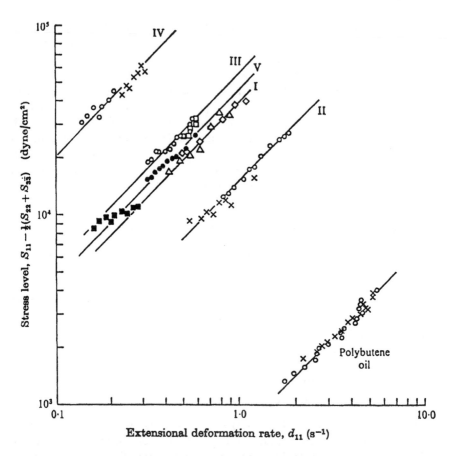

FIGURE 11.3 Stress–deformation rate relationships for fiber suspensions having different concentration levels and fiber aspect ratios. (From Ref. 19. Reprinted with permission of Cambridge University Press.)

that the Batchelor theory gave numerically correct results only near the spinneret; stresses increased by as much as an order of magnitude further down the spinline [24]. These authors hypothesized that this increase in stress was the result of decreasing interfiber distance, and they provided a simple correction to the Batchelor equation to account for this phenomenon. The influence of a distribution of fiber aspect ratios has been taken into account by Pittman and Bayram [25], while Goddard has considered the effect of non-Newtonian behavior of the suspending medium [26–28]. The use of a shear-thinning liquid as the suspending medium is predicted to result in a greatly diminished fiber contribution to the measured stress as compared to the Newtonian case [26–28].

Dinh and Armstrong used the work of Batchelor as the point of departure for developing a general stress constitutive equation for semiconcentrated fiber suspensions in a Newtonian liquid [2]. They used a cell model to focus attention on a single fiber that was acted upon by an anisotropic drag force resulting from a slipping motion relative to the bulk of the fluid. The resistance to flow of the fiber depended on whether it moved parallel or perpendicular to its axis. The drag coefficient for flow parallel to the fiber axis was the same as that in the Batchelor analysis, and this ensured that the extensional viscosity did not differ from the Batchelor result. The flow field was taken to be homogeneous so that the fiber orientation was a function of time but not of position within the suspension. For a suspension containing randomly oriented fibers, the shear viscosity is isotropic, and it was found to be given by [2]:

$$\eta_{sus} = \eta \left[1 + \frac{\pi n L^3}{90 \ln (2h/D)} \right] \tag{11.5}$$

in which h is the average lateral spacing between the fibers; for randomly oriented fibers h is $(nL^2)^{-1}$, while for aligned fibers it is $(nL)^{-1/2}$. Within the range of

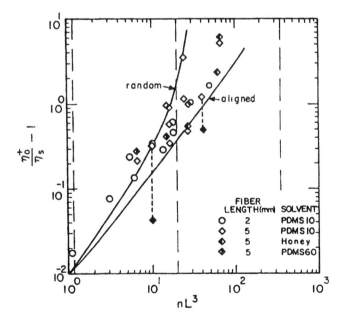

FIGURE 11.4 Dimensionless zero-shear viscosity of fiber suspensions as a function of dimensionless concentration. (From Ref. 30.)

validity of the theory, the choice of the expression for h is not particularly crucial. According to Ganani and Powell [29], the right-hand side of Eq. (11.5) also gives the low-frequency value of the dynamic viscosity when the random-orientation value of h is employed. Note that Eq. (11.5) represents the viscosity of the suspension as a function of volume fraction and fiber aspect ratio only upon inception of flow; flow tends to orient the fibers in the flow direction, and, at steady state, the suspension viscosity equals the suspending liquid viscosity. An important feature of the Dinh–Armstrong analysis is that suspension rheological properties are predicted to depend on the total strain rather than being functions of the strain rate and time separately; the strain rate determines only the rate of approach to the steady state. Both the shear viscosity and the two normal stress differences in shear are predicted to exhibit overshoots before attaining steady values; the extent of overshoot depends on the fiber aspect ratio and also on the fiber volume fraction [29]. The normal stress differences are zero to begin with, reach fairly large values relative to the shear stress, and again become zero at steady state.

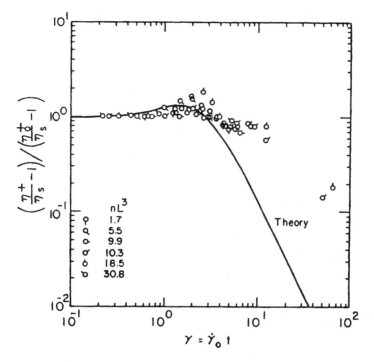

FIGURE 11.5 Dimensionless transient viscosity of fiber suspensions of Fig. 11.4 as a function of shear strain in the startup of shear flow at a constant shear rate. (From Ref. 30.)

Analytical expressions are not available for these material functions, and results have to be obtained using numerical techniques.

The main predictions of the Dinh–Armstrong theory were verified by Bibbo et al. [30]. Shown in Fig. 11.4 are zero-strain data on nylon as well as copper fibers suspended in Newtonian media. The solid lines represent the theory; curves marked aligned and random are obtained for the appropriate spacing between particles. Vertical dashed lines denote the limits of validity of the theory. Data for the longer fibers were corrected for wall effects, and the extent of correction is shown by the two solid points that are uncorrected data. As expected, the fiber contribution to the suspension viscosity is proportional to nL^3. Figure 11.5 shows how the viscosity of 2-mm-long nylon fibers suspended in silicone oil changes with the total strain. Clearly, the data depend uniquely on strain and exhibit a slight maximum. However, there is a mismatch between theory and experiment at large strain values. In accord with theory, these authors also found that the steady-state-shear viscosity was essentially the same as the suspending liquid viscosity.

In contrast to the behavior of particulate suspensions, when semiconcentrated fiber suspensions are formulated using non-Newtonian liquids, the curve of the zero-shear-rate relative viscosity as a function of the fiber volume fraction is found to lie above the corresponding curve obtained with Newtonian liquids [29]. This behavior also carries over to the dynamic viscosity measured at low frequencies. Additionally, the storage modulus is found to increase quadratically with frequency at low frequencies [29].

IV. CONCENTRATED FIBER SUSPENSIONS

Fiber suspensions encountered in the processing of composite materials are almost always in the concentrated region, and the suspending liquid is invariably a polymer melt. Much of the available data from before 1984 have been reviewed by Ganani and Powell [10]. A comprehensive experimental study of the rheological properties of such composite systems has been published by Chan et al. [31]. Glass fibers having a diameter of 12.7 μm were suspended in melts of high-density polyethylene and polystyrene at a fiber loading of 20 and 40 wt % and subjected to testing in both shear and extension; fiber length was 2 mm at the lower loading level and 1 mm at the higher loading level. The steady-shear viscosity as a function of shear rate is displayed in Fig. 11.6 for the suspensions in polystyrene. Also shown are data on the neat resin. It is seen that the presence of fibers increases the viscosity at each shear rate, but the increase is relatively higher at low shear rates; all three curves seem to come together at the high shear rate end. Further, increasing the loading level increases the viscosity and moves the point of onset of shear thinning to lower shear rates. Note, though, that viscos-

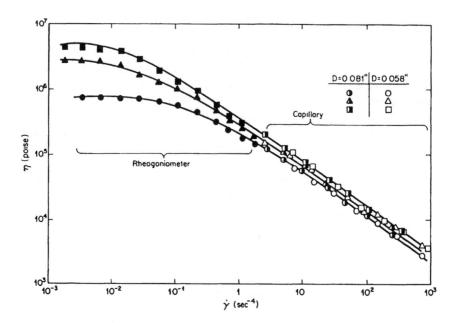

FIGURE 11.6 Shear viscosity as a function of shear rate for polystyrene melts with 0 (circles), 20 (triangles), and 40 (squares) wt % glass fibers at 180°C. (From Ref. 31.)

ity enhancement on adding a certain volume fraction of fibers is usually less than the enhancement caused by an equivalent amount of particulate fillers [32]. Additionally, as mentioned in Sec. I, increasing the fiber aspect ratio at a fixed loading level also increases the suspension viscosity. These general conclusions are independent of the type of fibers used, whether glass, cellulose, or aramid [33].

First normal stress difference in shear data corresponding to the viscosities shown in Fig. 11.6 are presented in Fig. 11.7. The addition of fibers leads to a considerable increase in the first normal stress difference over the entire shear rate range of measurement; increasing the fiber aspect ratio leads to still higher normal stress values [33]. This increase is significantly larger than the enhancement in the viscosity or, equivalently, the shear stress. As a result, when N_1 data are replotted as a function of the shear stress, lines for the filled fluid lie above the line for the neat polymer [33]. This is in contrast to the behavior of particulate suspensions seen in Chap. 10, Sec. V, where the relative position of the different lines was found to depend on whether data were plotted in terms of the shear

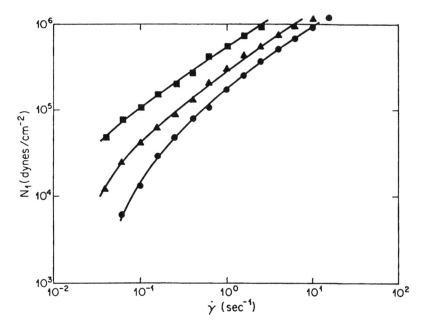

FIGURE 11.7 First normal stress difference as a function of shear rate for the suspensions of Fig. 11.6. (From Ref. 31.)

rate or the shear stress. As in the case of pure melts (see Chap. 5, Sec. II), the N_1 versus shear stress plot is independent of the temperature of measurement and is again described by Eq. (5.3) (but see Ref. 34):

$$N_1 = A\tau^a \tag{11.6}$$

where the constant A depends on volume fraction, aspect ratio, and fiber modulus, but a is independent of these quantities [33]. For other fiber suspensions, though, a may depend on the fiber volume fraction [35]. White and coworkers [31] have speculated that the large increase in N_1 upon the addition of short fibers to a molten polymer is a consequence of particle–particle interactions and not the elastic nature of the suspending liquid. In support of this assertion, they cite the fact that fiber suspensions exhibit a strong Weissenberg effect (rod climbing) even when the suspending liquid is inelastic [19]. Note, however, that fluid elasticity, as measured by the occurrence of extrudate swell, is found to be severely diminished by the presence of the fibers [7]. Also, entrance pressure drops in Poiseuille flow appear not to be influenced by the presence of fibers [36].

The steady extensional viscosity of fiber suspensions of Figs. 11.6 and 11.7 is shown in Fig. 11.8. This material function is very high at low stretch rates, but then it stretch thins considerably, always remaining above the extensional viscosity of the unfilled melt. The Batchelor theory is not expected to apply to these very concentrated suspensions, and, indeed, it predicts stress values that are very different and much higher than those that are observed [31]. Kamal et al. have also measured the steady extensional viscosity of glass-fiber-filled polymer melts [37]. Their results are similar to those of Chan et al. except that they observed yield stresses at high concentrations; the yield stress was evident in extensional data but not in the corresponding shear data. When a yield stress is observed in the shear flow of fiber suspensions, its value typically follows a power law in the fiber mass fraction [8,12].

Although there are published data on the dynamic behavior of fiber suspensions at high loading levels [38], such data are not extensive. This is probably because of the fact that the linear viscoelastic region shrinks to very low strain values upon the addition of a significant volume fraction of fibers.

Although attempts have been made to propose constitutive theories that

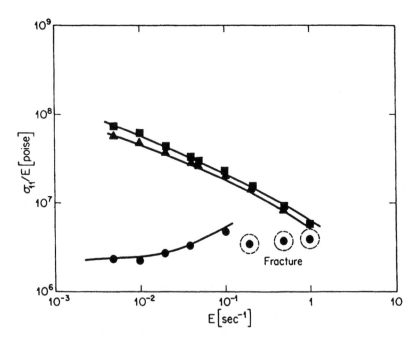

Figure 11.8 Steady-state extensional viscosity as a function of stretch rate for the suspensions of Fig. 11.6. (From Ref. 31.)

would explain all or most of the experimental observations just described (see, for example, Refs. 39 and 40), there is, at present, no universally accepted constitutive equation for fiber suspensions. Some material functions, such as the viscosity, can, however, be related empirically to shear rate, fiber volume fraction, and fiber aspect ratio in limited ranges of these variables. The best known such expression is for the relative viscosity as a function of the fiber volume fraction. It is based on the work of Maron and Pierce [41], and, as adapted to fiber suspensions by Kitano et al. [42], it reads [see also Eq. (10.4)]:

$$\eta_R = [1 - (\phi/A)]^{-2} \tag{11.7}$$

in which the relative viscosity is determined as the ratio of the viscosity of the filled system to that of the medium at the same value of the shear stress, and A is a single constant that is obtained by fitting Eq. (11.7) to experimental data. The success of Eq. (11.7) is demonstrated in Fig. 11.9, which shows data on a wide variety of polymer/fiber systems where the fiber aspect ratio varies from 6 to 27. The numbers against each curve are the best-fit values of A. As shown in Fig. 11.10, A depends uniquely on the fiber aspect ratio. In his review, however, Metzner cautions that Fig. 11.10 should not be extrapolated in either direction [32].

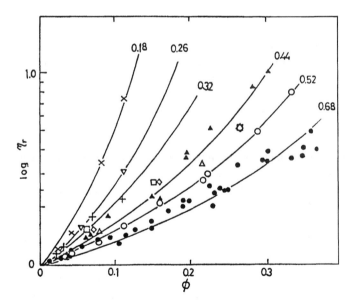

FIGURE 11.9 Relative viscosity evaluated at the same shear stress versus volume fraction for spherical and anisotropic solids suspended in molten polymers. (From Ref. 42. Used by permission of Steinkopff Publishers, Darmstadt, FRG.)

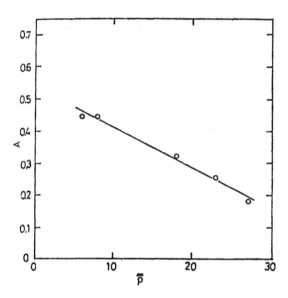

FIGURE 11.10 Dependence of the parameter A in Eq. (11.7) on the fiber aspect ratio. (From Ref. 42. Used by permission of Steinkopff Publishers, Darmstadt, FRG.)

In closing this section, we mention that at extremely high concentrations a fiber suspension must show fiber alignment even in the rest state. In other words, with increasing fiber concentration, a situation must arise that is similar to the formation of polymeric liquid crystals; this was considered earlier, in Chap. 8. The similarity between the two situations suggests the application of liquid crystal theory to concentrated fiber suspensions. This was done by Becraft and Metzner |4| by modifying the work of Doi |43|. The result is a constitutive equation involving two constants that have to be determined by comparison with experimental data. Becraft and Metzner showed that the application of this theory to glass-filled polypropylene allowed them to theoretically represent the entire flow curve of the suspension.

V. CONCLUDING REMARKS

In this chapter, we have, for the most part, been concerned with the question of how fiber orientation affects the observed rheological behavior of short-fiber suspensions. Equally important, from a practical standpoint, is the question of how fluid rheology affects fiber orientation during the filling of a mold in, say, injection molding. Here, one often observes a skin-core morphology in the molded part; the skin contains fibers aligned in the direction of flow, while fibers in the core are aligned transverse to the flow direction [1,6]. One of the more popular models in the literature for predicting this flow-induced orientation is the one due to Folgar and Tucker [44]. This is a phenomenological model that takes into account fiber–fiber interactions, but it also includes elements of the Jeffery theory of dilute fiber suspensions. In shear flow, this model predicts less than perfect fiber orientation in the flow direction, and it also predicts that fiber orientation is not reversible when the flow is reversed. Recent progress in this area has been reviewed by Tucker and Advani, who have also considered fiber orientation development during composites processing by extrusion, injection molding, and compression molding [1].

REFERENCES

1. C.L. Tucker, S.G. Advani. Processing of short-fiber systems. In: Flow and Rheology in Polymer Composites Manufacturing (S.G. Advani, ed.). Elsevier, Amsterdam, 1994, pp. 147–202.
2. S.M. Dinh, R.C. Armstrong. A rheological equation of state for semiconcentrated fiber suspensions. J. Rheol. 28:207–227 (1984).
3. S. Wu. Order–disorder transitions in the extrusion of fiber-filled poly(ethylene terephthalate) and blends. Polym. Eng. Sci. 19:638–650 (1979).
4. M.L. Becraft, A.B. Metzner. The rheology, fiber orientation, and processing behavior of fiber-filled fluids. J. Rheol. 36:143–174 (1992).
5. A. Vaxman, M. Narkis, A. Siegmann, S. Kenig. Short-fiber-reinforced thermoplastics. Part III: Effect of fiber length on rheological properties and fiber orientation. Polym. Compos. 10: 454–462 (1989).
6. R.J. Crowson, M.J. Folkes, P.F. Bright. Rheology of short glass fiber-reinforced thermoplastics and its application to injection molding. I. Fiber motion and viscosity measurement. Polym. Eng. Sci. 20:925–933 (1980).
7. R.J. Crowson, M.J. Folkes. Rheology of short glass fiber-reinforced thermoplastics and its application to injection molding. II. The effect of material parameters. Polym. Eng. Sci. 20:934–940 (1980).
8. C.P.J. Bennington, R.J. Kerekes, J.R. Grace. The yield stress of fiber suspensions. Can. J. Chem. Eng. 68:748–757 (1990).
9. R.O. Maschmeyer, C.T. Hill. The rheology of concentrated suspensions of fibers. I. Review of the literature. Adv. Chem. Ser. 134:95–105 (1974).
10. E. Ganani, R.L. Powell. Suspensions of rodlike particles: literature review and data correlations. J. Compos. Mater. 19:194–215 (1985).
11. R.L. Powell. Rheology of suspensions of rodlike particles. J. Stat. Phys. 62:1073–1094 (1991).
12. W.J. Milliken, R.L. Powell. Short-fiber suspensions. In: Flow and Rheology in Polymer Composites Manufacturing (S.G. Advani, ed.). Elsevier, Amsterdam, 1994, pp. 53–83.
13. G.B. Jeffery. The motion of ellipsoidal particles immersed in a viscous fluid. Proc. Roy. Soc. A102:161–179 (1922).

14. B.J. Trevelyan, S.G. Mason. Particle motions in sheared suspensions. I. Rotations. J. Colloid Sci. 6:354–367 (1951).

15. W. Bartok, S.G. Mason. Particle motions in sheared suspensions. V. Rigid rods and collision doublets of spheres. J. Colloid Sci. 12:243–262 (1957).

16. Y. Ivanov, T.G.M. van de Ven, S.G. Mason. Damped ocillations in the viscosity of suspensions of rigid rods. I. Monomodal suspensions. J. Rheol. 26:213–230 (1982).

17. Y. Ivanov, T.G.M. van de Ven. Damped oscillations in the viscosity of suspensions of rigid rods. II. Bimodal and polydisperse suspensions. J. Rheol. 26:231–244 (1982).

18. G.K. Batchelor. Slender-body theory for particles of arbitrary cross-section in Stokes flow. J. Fluid Mech. 44:419–440 (1970).

19. J. Mewis, A.B. Metzner. The rheological properties of suspensions of fibers in Newtonian fluids subjected to extensional deformations. J. Fluid Mech. 62:593–600 (1974).

20. A. Karnis, S.G. Mason. Particle motion in sheared suspensions. XIX. Viscoelastic media. Trans. Soc. Rheol. 10:571–592 (1966).

21. L.G. Leal. The slow motion of slender rod-like particles in a second-order fluid. J. Fluid Mech. 69:305–337 (1975).

22. G.K. Batchelor. The stress generated in a non-dilute suspension of elongated particles by pure straining motion. J. Fluid Mech. 46:813–829 (1971).

23. A. Acrivos, E.S.G. Shaqfeh. The effective thermal conductivity and elongational viscosity of a nondilute suspension of aligned slender rods. Phys. Fluids 31:1841–1844 (1988).

24. T. Sridhar, R.K. Gupta. Application of the Batchelor theory to fiber spinning of suspensions. Proc. 4th National Conf. on Rheol., Adelaide, Australia, 1986, pp. 185–190.

25. J.F.T. Pittman, J. Bayram. Extensional flow of polydisperse fiber suspensions in free-falling liquid jets. Int. J. Multiphase Flow 16:545–559 (1990).

26. J.D. Goddard. Tensile stress contribution of flow-oriented slender particles in non-Newtonian fluids. J. Non-Newtonian Fluid Mech. 1:1–17 (1976).

27. J.D. Goddard. The stress field of slender particles oriented by a non-Newtonian extensional flow. J. Fluid Mech. 78:177–206 (1976).

28. J.D. Goddard. Tensile behavior of power-law fluids containing oriented slender fibers. J. Rheol. 22:615–622 (1978).

29. E. Ganani, R.L. Powell. Rheological properties of rodlike particles in a Newtonian and a non-Newtonian fluid. J. Rheol. 30:995–1013 (1986).

30. M.A. Bibbo, S.M. Dinh, R.C. Armstrong. Shear flow properties of semiconcentrated fiber suspensions. J. Rheol. 29:905–929 (1985).

31. Y. Chan, J.L. White, Y. Oyanagi. A fundamental study of the rheological properties of glass-fiber-reinforced polyethylene and polystyrene melts. J. Rheol. 22:507–524 (1978).

32. A.B. Metzner. Rheology of suspensions in polymeric fluids. J. Rheol. 29:739–775 (1985).

33. L. Czarnecki, J.L. White. Shear flow rheological properties, fiber damage, and masti-

cation characteristics of aramid-, glass-, and cellulose-fiber-reinforced polystyrene melts. J. Appl. Polym. Sci. 25:1217–1244 (1980).

34. A.T. Mutel, M.R. Kamal. The effect of glass fibers on the rheological behavior of polypropylene melts between rotating parallel plates. Polym. Compos. 5:29–35 (1984).

35. T. Kitano, T. Kataoka, Y. Nagatsuka. Shear flow rheological properties of vinylon- and glass-fiber reinforced polyethylene melts. Rheol. Acta 23:20–30 (1984).

36. A.T. Mutel, M.R. Kamal. Rheological behavior and fiber orientation in slit flow of fiber reinforced thermoplastics. Polym. Compos. 12:137–145 (1991).

37. M.R. Kamal, A.T. Mutel, L.A. Utracki. Elongational behavior of short glass fiber reinforced polypropylene melts. Polym. Compos. 5:289–298 (1984).

38. T. Kitano, T. Kataoka, Y. Nagatsuka. Dynamic flow properties of vinylon fiber and glass fiber reinforced polyethylene melts. Rheol. Acta 23:408–416 (1984).

39. G. Ausias, J.F. Agassant, N. Vincent, P.G. Lafleur, P.A. Lavoie, P.J. Carreau. Rheology of short glass fiber reinforced polypropylene. J. Rheol. 36:525–542 (1992).

40. A. Ramazani, A. Ait-Kadi, M. Grmela. Experimental study and model predictions of rheological behavior of short fiber composites. Proc. SPE ANTEC 45:2714–2718 (1999).

41. S.H. Maron, P.E. Pierce. Application of Ree–Eyring generalized flow theory to suspensions of spherical particles. J. Colloid Sci. 11:80–95 (1956).

42. T. Kitano, T. Kataoka, T. Shirota. An empirical equation of the relative viscosity of polymer melts filled with various inorganic fillers. Rheol. Acta 20:207–209 (1981).

43. M. Doi. Molecular dynamics and rheological properties of concentrated solutions of rodlike polymers in isotropic and liquid crystalline phases. J. Polym. Sci.: Polym. Phys. Ed. 19:229–243 (1981).

44. F.P. Folgar, C.L. Tucker. Orientation behavior of fibers in concentrated suspensions. J. Reinf. Plast. Compos. 3:98–119 (1984).

12

Emulsions

I. INTRODUCTION

Emulsions are small drops of one liquid dispersed in another liquid. Everyday examples include paints, cosmetics, and foods; mayonnaise and salad dressing, for instance, are concentrated oil-in-water emulsions. In the area of plastics technology, we find that polymers are routinely blended with each other with a view toward producing materials with improved properties. A specific example of the success of this approach is the improvement in impact strength of plastics such as nylons and polyesters by the addition of thermodynamically immiscible, rubber-type polymers [1, for example]. High-impact polystyrenes and ABS polymers generally contain microgel rubber particles in which the rubber particles are also graft polymers, and the morphology of the dispersed particles may be quite complex.

The ability of one polymer to toughen another polymer depends, in part, on how fine a dispersion can be generated. In particular, it is necessary that the minor phase be dispersed in the major phase in the form of micron- or submicron-sized spherical droplets; this is typically achieved in corotating twin-screw extruders. A novel application of emulsions that is related directly to their rheological properties is the pipeline transport of heavy crude oils having a viscosity as high as 500 Pa-sec [2]. Ordinarily, to economically pump such viscous liquids, we would have to either heat or dilute the crude oils, but these crudes can be pumped easily in the form of concentrated (but low-viscosity) oil-in-water emulsions having an oil volume fraction in excess of 0.7. When the ratio of the dispersed-phase viscosity to the continuous-phase viscosity is this high, the emulsion behaves essentially as a suspension, and its viscosity can be predicted using the equations presented in Chap. 10 [3]. The emulsion viscosity is typically low because of the very large value of the maximum packing fraction that results from a wide distribution of droplet sizes. Note that such emulsions are often produced naturally during enhanced oil recovery using water flooding.

In a laboratory, an emulsion can be formed by mechanically agitating, in laminar or turbulent flow, a liquid mixture using devices such as mixers, colloid mills, and homogenizers [4]. Alternatively, one may employ ultrasonic techniques that promote mixing through the mechanism of cavitation. Initially, large droplets are formed, but these are subsequently subdivided into smaller ones. The resistance to drop breakup comes from the Laplace pressure that exists across a curved interface. If the drop radius is R and the interfacial tension is σ, the pressure inside the drop exceeds the pressure outside the drop by $2\sigma/R$, and the pressure gradient needed to disrupt the drop is $2\sigma/R^2$. This pressure gradient can be reduced by reducing the interfacial tension with the help of an emulsifier, which is a surface-active agent that typically accumulates at the interface between the two liquids. Increasing the emulsifier concentration progressively reduces the interfacial tension but only up to a limiting value of the interfacial tension. The most effective emulsifier may be selected based on the concept of hydrophilic–lipophilic balance, a scale that ranges from 1 to 40 [4]. An emulsifier not only promotes the breakup of droplets; it also forms an elastic film around the dispersed phase and provides stability against flocculation (sticking) and coalescence. It should be realized, though, that all emulsions are thermodynamically unstable and, given enough time, will break or phase separate.

As in the case of suspensions, the viscosity of an emulsion depends strongly on the size distribution of droplets. The size distribution can be measured in a variety of ways. One may simply spread the emulsion on a slide and observe the droplet sizes with a microscope or use a more automated device that might work on the principle of laser light scattering or the measurement of electrical conductivity. The radial distribution function that characterizes the droplet packing structure can be obtained using the Kossel diffraction technique [5]. In the case of immiscible polymer blends, it is common practice to freeze the structure, fracture the sample, and examine the morphology with the help of a scanning electron microscope.

The measurement of rheological properties of emulsions is beset with the same problems encountered with suspensions: slip at the viscometer walls, emulsion instability, and droplet migration. Creaming and sedimentation resulting from density differences, though, can be reduced by increasing the external-phase viscosity. If polymers are used for this purpose, it has to be ensured that the polymer molecules do not bridge two droplets, or a network will be formed, and this will result in the appearance of a yield stress. Although any of the viscometers introduced in Chap. 2 may be used to measure emulsion viscosity, it appears that the use of a coaxial cylinder viscometer may give more reliable data compared to a cone-and-plate viscometer when the densities of the two phases are not matched [6]. In any case, it is incumbent upon the rheologist to check for internal consistency of data and to verify that the same results are obtained regardless of the instrument used, the fixture geometry employed, or the fixture size chosen.

As might be expected, the flow properties of an emulsion depend on a very large number of variables. These include shear rate, temperature, the volume fraction of the dispersed phase, the droplet size and size distribution, the kind and amount of emulsifier, and the viscosity and density of the two phases. Results are qualitatively similar to those expected with suspensions; representative data at different concentration levels are shown in Fig. 12.1 [7]. These and similar data can be fitted to the Newtonian, power-law, Carreau, or Bingham plastic equations, as appropriate. A major quantitative difference between the viscosity of suspensions and emulsions, however, results from the deformable nature of liquid drops: droplet deformation and liquid circulation within the droplets lead to a reduction in viscosity relative to otherwise identical but rigid suspensions. Liquid drops, though, become more rigid as their size decreases, and these show higher viscosity values compared to larger drops at the same volume fraction of

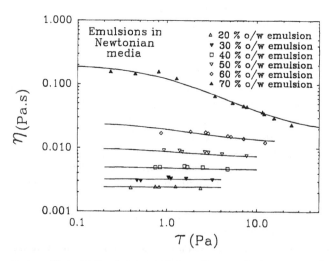

FIGURE 12.1 Plots of apparent viscosity versus shear stress for emulsions of kerosene oil in an aqueous surfactant (Triton x-100). (From Ref. 7. Copyright 1992 Overseas Publishers Association N.V. Reprinted by permission of Gordon and Breach Publishers.)

the dispersed phase [8]. A narrowing of the size distribution also results in an increase in viscosity.

II. VISCOSITY OF A DILUTE EMULSION OF SPHERICAL DROPLETS

The presence of a sphere in a shear field disturbs the streamlines so that additional energy is dissipated by the system. This additional dissipation of energy manifests itself as an increase in viscosity. As seen in Chap. 10, the relative viscosity of a dilute suspension of rigid spheres having a volume fraction ϕ of solids is

$$\eta_R = 1 + k_E\phi \tag{12.1}$$

The Einstein coefficient, k_E, is 2.5 for rigid spheres if there is no slippage of the liquid at the surface of the sphere. If there is perfect slippage of liquid at the interface, k_E becomes unity [9]. In emulsions there is circulation of fluid within the spheres in addition to displacement of the streamlines of the drop. The circulation within the drops allows relative motion to take place in the neighborhood of the interface in a manner similar to what takes place during slippage at the interface. Thus, intuitively, one would expect the Einstein coefficient to be about 2.5 when the viscosity of the fluid in the drops is much greater than the viscosity of the continuous liquid, and k_E should be 1.0 when the viscosity of the continuous liquid is much greater than that of the drops. Taylor derived the following equation for the Einstein coefficient of emulsions at low shear rates [10]:

$$k_E = 2.5\left(\frac{\eta_2 + (2/5)\eta_1}{\eta_2 + \eta_1}\right) \tag{12.2}$$

in which η_2 is the viscosity of the dispersed liquid and η_1 is the viscosity of the matrix. This equation is plotted in Fig. 12.2; it has been experimentally verified by Nawab and Mason [11]. Equation (12.2) reduces to the Einstein equation as the dispersed-phase viscosity becomes very large; at the other extreme of negligible droplet phase viscosity, such as might occur with gas bubbles, it still predicts an increase in viscosity above the value of the continuous phase. It should be pointed out that the drops rotate in a simple shear field at the same time that there is circulation within the drop itself [4]; the angular rotation rate is one-half the shear rate.

Most emulsions contain surfactants that can affect the rheology by means of at least three mechanisms: (1) The surfactant may form a "skin" around the drops that reduces the amount of circulation within the drops; (2) the surfactant forms a coating on the drops that makes the drops appear larger than they would be if there were no surfactant; (3) some of the surfactant may dissolve in the

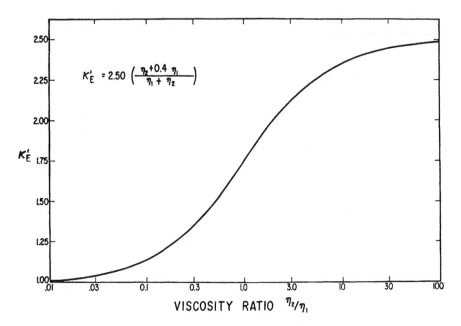

$$K'_E = 2.50 \left(\frac{\eta_2 + 0.4\, \eta_1}{\eta_1 + \eta_2} \right)$$

VISCOSITY RATIO η_2/η_1

FIGURE 12.2 Einstein coefficient as a function of the ratio of the viscosity of the dispersed phase to that of the continuous phase.

suspending fluid and change its viscosity. In general, all of these factors tend to increase the viscosity of the system.

III. DROP DEFORMATION AND BREAKUP

It is interfacial tension that keeps the suspended liquid in the shape of spheres when there is no flow. However, since liquid drops are easily deformed, the drops may become distorted from the spherical shape in a shear field. Thus Eq. (12.1) and (12.2) are generally applicable only at low shear rates. Taylor [12] and Mason and coworkers [13,14] have studied the deformation of liquid drops in flow fields. If the rate of shear is so small as to distort the drops only slightly, the deformation is given by [4,12]:

$$\frac{L - W}{L + W} = \frac{D\eta_1 \dot{\gamma}}{2\sigma} \left(\frac{\eta_1 + 1.1875\eta_2}{\eta_1 + \eta_2} \right) \qquad (12.3)$$

The dimensions of the resulting ellipsoid are L and W for the length of the major and minor axes, respectively. The diameter of the undistorted spheres is D; σ is the interfacial tension. Note that the quantity in parenthesis in Eq. (12.3) varies only between unity and 1.187 over the entire range of p, the ratio of the dispersed-phase viscosity to the suspending-liquid viscosity, and it is clear that large drops are more easily deformed than small ones. The stretching action of the flow field is counteracted by the interfacial tension, which tries to restore the ellipsoid to a sphere. In polymer melts, the elastic modulus may be the major factor in place of interfacial tension in resisting deformation.

At higher rates of shear, the deformed drops may either break up into smaller drops or stretch out into long filaments. Typical types of behavior for Newtonian fluids in a shear field are shown in Fig. 12.3 [13]. At low viscosity ratios, a drop initially aligns itself with the flow field in the form of an ellipsoid that develops pointed tails on increasing the shear rate; these tails then break off to form satellite drops. At an intermediate viscosity ratio, dumbbell-shaped drops can be observed. These break apart with increasing shear rate, giving two large drops and some small drops. At large viscosity ratios, there is no drop breakup

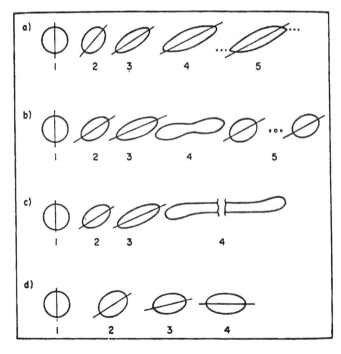

FIGURE 12.3 Shape of drops in shear fields as the intensity of the rate of shear increases from left to right. The ratio of the dispersed-phase viscosity to the continuous-phase viscosity is (a) 2×10^{-4}, (b) 1.0, (c) 0.7, (d) 6.0 (From Ref. 13.)

regardless of the shear rate imposed; there is only drop elongation. Somewhat similar behavior is observed in elongational flow fields. As the ratio p increases, there is less breakup and more filament formation. Whether breakup occurs or not depends on a balance between the shear stress tending to deform the drop and the interfacial stress tending to restore the drop to a spherical shape. The ratio of these two stresses, $\eta_1 \dot{\gamma} R/\sigma$, is known as the *capillary number* (Ca). At a given value of p and for small values of the capillary number, an ellipsoidal drop is stable; but as the capillary number is increased beyond a critical value, no stable drop shape exists. The drop stretches and ultimately breaks into fragments. Results for the flow-induced breakup of a single drop of a Newtonian liquid suspended in another Newtonian liquid have been obtained in an experimental manner by Grace [15]; these are displayed in Fig. 12.4 [16] in terms of the critical capillary number versus the viscosity ratio p. It is seen that in shear flow, it is especially easy to cause drop breakup when p ranges from 0.1 to 1 but is impossible when p exceeds approximately 4. As opposed to this, elongational flow is always more effective than shear flow, and drop breakup is possible for all values of p.

Most of these results can be predicted theoretically, and progress in this area has been reviewed by Rallison [17]. Note, though, that quasi-equilibrium conditions have been assumed in all the work that has been discussed so far.

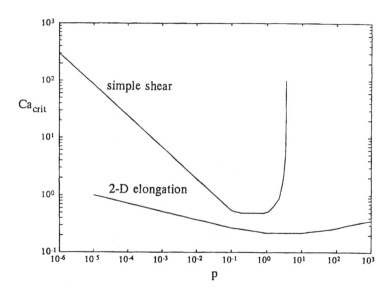

FIGURE 12.4 Critical capillary number for drop breakup versus viscosity ratio p. (From Ref. 16.)

Results obtained under transient conditions can be different from those depicted in Fig. 12.4 [16]. A very readable account of the actual mechanisms of drop breakup may be found in the Ph.D. dissertation of Janssen [18]; only a knowledge of the breakup mechanism allows one to calculate the time needed for drop breakup and the resulting droplet size. Many emulsions contain a surface-active agent, and its presence can alter the process of drop deformation and breakup. The influence of added surfactants has been examined by Stone and Leal [19] and reviewed by Stone [20]. Other authors have considered the behavior of Newtonian drops in a viscoelastic matrix [21], that of viscoelastic drops in Newtonian liquids [22], and polymeric liquids dispersed in another polymeric liquid [23–25]. In general, non-Newtonian polymer melts have a greater tendency to form filaments than do Newtonian liquids. This is due to the elastic nature of polymer melts.

In summary, changes in deformation rate lead to changes in the extent of drop deformation and eventually result in drop breakup. An understanding of these phenomena is clearly important for the study of laminar mixing. From our point of view, a change in the dispersed-phase morphology must necessarily be reflected in changes in rheology of the dilute emulsion. In particular, the time-dependent nature of drop distortion and also alignment with the flow direction have to be observable as (1) transients in the measurement of rheological properties at a fixed shear rate and (2) a shear-thinning viscosity. Additionally, since changes in interfacial area imply changes in the stored energy of the emulsion owing to a nonzero value of the interfacial tension, shear-induced drop deformation endows an emulsion of two Newtonian liquids with fluid elasticity [26].

IV. ELASTICITY OF DILUTE EMULSIONS

Schowalter et al. [27] considered the flow of a Newtonian liquid around a deformable drop of another Newtonian liquid and solved the creeping-flow equations for the velocity and stress profiles both inside and outside the drop. This was done in terms of a (small) perturbation parameter that was proportional to the capillary number and represented the departure of the drop from a spherical shape. By matching the inner and outer solutions, these authors determined that in steady laminar shearing flow, the viscosity of an emulsion of monodisperse drops was still given by Eqs. (12.1) and (12.2), but the presence of interfacial tension resulted in the appearance of normal stress differences. At low values of the viscosity ratio p, these normal stress differences are [28]:

$$N_1 = \left(\frac{32}{5}\right) \text{Ca} \eta_1 \phi \dot{\gamma} \tag{12.4}$$

$$N_2 = -\left(\frac{20}{7}\right) \text{Ca} \eta_1 \phi \dot{\gamma} \tag{12.5}$$

At large values of p, the corresponding results are:

$$N_1 = \left(\frac{361}{40}\right)Ca\eta_1\phi\dot{\gamma} \tag{12.6}$$

$$N_2 = -\left(\frac{551}{280}\right)Ca\eta_1\phi\dot{\gamma} \tag{12.7}$$

Thus it is seen that, as in the case of polymeric fluids, the first normal stress difference is positive while the second normal stress difference is negative. Also, the first normal stress difference is always greater in magnitude than the second normal stress difference. In addition, Eqs. (12.4)–(12.7) are valid only for small capillary numbers, since they are the result of a perturbation analysis.

Elasticity in a dilute emulsion also shows up during dynamic mechanical analysis. For small-amplitude sinusoidal deformation, the analysis of Schowalter et al. predicts that the storage and loss moduli are, respectively [29]:

$$G'(\omega) = \frac{\eta_1 R\phi}{80\sigma}\left(\frac{19p + 16}{p + 1}\right)^2\omega^2 \tag{12.8}$$

$$G''(\omega) = \eta_1\left[1 + \left(\frac{5p + 2}{2p + 2}\right)\phi\right]\omega \tag{12.9}$$

and the variation of both moduli with frequency, at low frequencies, is similar to that of viscoelastic liquids. Furthermore, the shear viscosity calculated from Eq. 12.9, assuming the validity of linear viscoelasticity, is identical to that given by Eqs. (12.1) and (12.2).

An earlier theory of dilute emulsions is due to Oldroyd [30–31]. Here again, the shear viscosity does not differ from the Taylor result, but the expressions for G' and G'' are slightly different from those given in Eqs. (12.8) and (12.9). This theory can account for the presence of an interfacial film around the dispersed phase; the film resists being deformed due to either internal friction or elasticity. The theory has been used by Graebling and Muller [32] to model the small-amplitude dynamic behavior of an emulsion of two viscoelastic liquids. Data were found to compare well with the model predictions.

V. DROPLET COALESCENCE DURING SHEAR FLOW

When a truly dilute emulsion of two Newtonian liquids is subjected to shear flow, there are no droplet–droplet interactions, and the observed equilibrium drop size can be predicted with the help of Fig. 12.4 [33]. For nondilute emulsions, though, the equilibrium drop size is generally larger than that calculated based on the

Taylor theory; mismatch by an order of magnitude is not uncommon, and the difference increases with increasing dispersed-phase concentration. This large deviation from the single-drop analysis is the result of droplet coalescence: when drops collide with each other, either due to Brownian motion or as a result of flow, they can coalesce, and this results in a coarsening of the morphology even when the dispersed-phase concentration is as low as 0.5% [34,35]. This is shown in Fig. 12.5 as the ratio of the instantaneous drop diameter to the initial drop diameter for the laminar shearing flow at 11 sec^{-1} of a 1 vol% emulsion of castor oil in silicone oil having an initial average drop size of 8.1 μm; the viscosity ratio is about 15, so drop breakup is prohibited. Since the average dispersed-phase size and its distribution affect the rheology of concentrated suspensions and emulsions, these quantities have to be either predicted or measured if we wish to relate rheology to blend morphology. In this connection, it is worth noting that efforts have been made to predict emulsion morphology from the measured rheology [36].

The theory of droplet coalescence in shear flow has been reviewed by Chesters [37]. In general, for two drops to coalesce into one, they must collide during

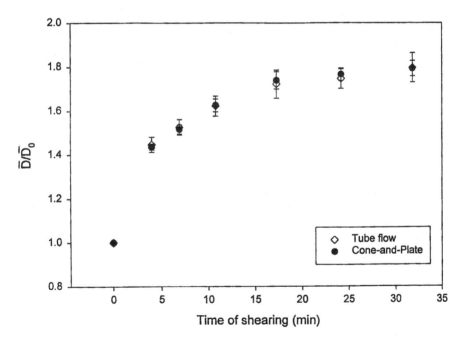

FIGURE 12.5   Drop size evolution during shearing of an emulsion of castor oil in silicone oil. (From Ref. 35. Used by permission of Steinkopff Publishers, Darmstadt, FRG.)

the time period of the experiment, and the film of liquid trapped between them must drain away. Provided that the time scale of the experiment is not the limiting factor, the time during which two drops interact is approximately the reciprocal of the shear rate. Film drainage must occur within this time interval, or else the two drops go their separate ways. The film drainage time can be estimated by assuming the existence of a squeezing flow in the film under the influence of a constant force that is taken to be the Stokes drag force; the initial film thickness is taken to be the separation at which the approach velocity equals $\dot{\gamma}R$; the final film thickness for rupture is governed by the van der Waals force [18]. By equating the drainage time (which depends on the drop radius R) to the interaction time, an expression is obtained for the critical drop size below which coalescence takes place and above which it does not. Although the specific form of this expression depends on interface mobility, whether fully mobile, partially mobile, or immobile, in each case a double logarithmic plot of the equilibrium droplet size versus the shear rate is predicted to be a straight line with a negative slope. This is, indeed, observed to be the case [35].

Since coalescence does not occur above a certain droplet size and breakup does not happen below a critical droplet size, morphological hysteresis can arise depending on the relative values of these two limiting sizes [38]. In other words, the average droplet size need not always be a single-valued function of the shear rate. When single valuedness does occur, microstructure evolution resulting from changes in the shear rate can be calculated using differential population balances [39].

VI. RHEOLOGY OF CONCENTRATED EMULSIONS

As we have seen, dilute emulsions exhibit a Newtonian viscosity that increases linearly with dispersed-phase concentration. With increasing concentration, though, this increase becomes nonlinear, even though the viscosity remains Newtonian. At still higher concentrations, approaching 40 or 50% by volume, the viscosity becomes shear thinning. The nonlinear viscosity increase and shear thinning are both the result of droplet aggregation that traps some of the continuous-phase liquid, resulting in a larger effective dispersed-phase volume fraction; this liquid is released at high shear rates, giving a lower viscosity [4]. When the dispersed-phase volume fraction reaches about 0.74, the emulsion becomes solidlike and exhibits a yield stress. There can also be phase inversion. Since liquid drops can deform, one cannot really talk about a maximum packing fraction, and it is possible to formulate emulsions containing more than 90% by volume of the dispersed phase.

Many of the equations presented in Chap. 10 can be employed to represent the shear viscosity of an emulsion as a function of concentration. An especially popular equation is due to Mooney [40]:

$$\eta = \eta_1 \exp\left(\frac{k_E \phi}{1 - \phi/\phi_m}\right) \tag{12.10}$$

in which k_E and ϕ_m are empirical constants obtained by comparison with data. For a suspension of monodisperse spheres, of course, if k_E is taken to be 2.5, ϕ_m ranges from 0.52 to 0.74. Regarding the shear rate dependence of the viscosity, this can be expressed over the entire range of shear rates (but at a fixed concentration) as [41]:

$$\eta = \frac{\eta_0 - \eta_\infty}{(\dot{\gamma} + 1)^n} + \eta_\infty \tag{12.11}$$

where n is a constant.

Pal and Rhodes have proposed a semiempirical equation to account for both the concentration and the shear rate dependence of the viscosity of emulsions having a dispersed-phase volume fraction less than 0.74, i.e., in the absence of a yield stress. This equation is [42]:

$$\frac{\eta}{\eta_1} = \left[1 + \frac{\phi/(\phi)_{\eta_{r^-} 100}}{1.187 - \phi/(\phi)_{\eta_{r^-} 100}}\right]^{2.492} \tag{12.12}$$

where $(\phi)_{\eta_{r^-} 100}$ is the dispersed-phase concentration at which the relative viscosity becomes 100; this unusual parameter is used to get around the difficulty of defining a maximum packing fraction for emulsions. If the emulsion viscosity at a given concentration depends on shear rate, this parameter must also be obtained at the same shear rate. The success of Eq. 12.12 in representing experimental data is demonstrated in Fig. 12.6, in which 16 sets of data are plotted. The procedure for obtaining the non-Newtonian viscosity at any concentration and shear rate with the help of Fig. 12.6 is described in the original paper [42].

In terms of theoretical expressions for the rheological properties of concentrated emulsions, Choi and Schowalter [3] used a cell model to account for changes in droplet shape due to both flow and hydrodynamic interactions between drops of one Newtonian liquid dispersed inside another Newtonian liquid. The calculated viscosity was found to be Newtonian and to increase quadratically with increasing concentration. Their viscosity expression is similar to that obtained in a different manner by Oldroyd [30], and it is given by:

$$\frac{\eta}{\eta_1} = 1 + \phi \frac{5p + 2}{2(p + 1)}\left[1 + \phi \frac{5(5p + 2)}{4(p + 1)}\right] \tag{12.13}$$

which properly reduces to Eqs. (12.1) and (12.2) as concentration approaches zero. Expressions for the first and second normal stress differences are somewhat complicated, but these show that $N_1/N_2 < 0$.

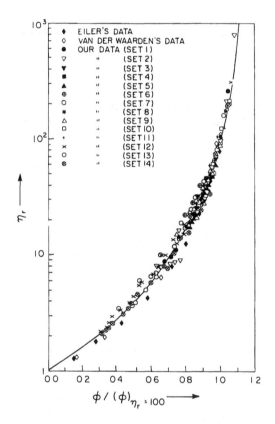

FIGURE 12.6 Relative viscosity of emulsions as a function of the dimensionless concentration. (From Ref. 42.)

As in the case of dilute emulsions, the work of Choi and Schowalter predicts a storage modulus in oscillatory flow. Expressions for the storage and loss moduli are [29]:

$$G'(\omega) = \frac{\eta\omega^2(h_1 - h_2)}{1 + \omega^2 h_1^2} \qquad (12.14)$$

$$G''(\omega) = \frac{\eta(\omega^3 h_1 h_2 + \omega)}{1 + \omega^2 h_1^2} \qquad (12.15)$$

in which h_1 and h_2 are constants that depend upon the dispersed- and continuous-phase viscosities, dispersed-phase size and concentration, and the interfacial tension. The major difference between Eq. 12.8 for dilute emulsions and Eq. 12.14

for concentrated emulsions is that Eq. 12.14 predicts a plateau in G' with increasing frequency; in this sense, it is similar to the elastic behavior of the single-mode Maxwell model used for polymer melts. Note that since the origin of elasticity in an emulsion of two Newtonian liquids is interfacial tension, Eq. 12.14 cannot portray the high-frequency behavior of an emulsion of viscoelastic liquids; here the elastic response at high frequencies is related to the viscoelastic behavior of the two phases themselves. This is shown in Fig. 12.7 for a 20 vol% emulsion of Newtonian polyoxyethylene dispersed in viscoelastic polydimethylsiloxane [43]. Clearly, the model of Choi and Schowalter can explain the increase of emulsion elasticity at low frequencies, but it is totally inadequate at high frequencies.

To incorporate the elastic properties of blend components into an emulsion model, Palierne [44] modified the model of Oldroyd [30] by including dipole-type particle interactions and also a distribution of droplet sizes. As Carreau et al. show, a simplified version of the model has no adjustable parameters, and explicit expressions can be derived for storage and loss moduli of immiscible polymer blends [45]. Model inputs are the moduli of both phases, the particle

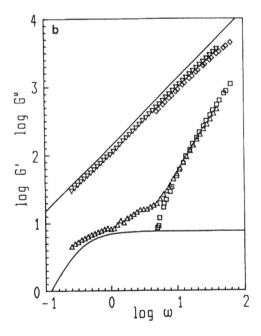

FIGURE 12.7 Dynamic moduli measured in units of Pa versus frequency in rad/sec for a 20 vol% emulsion of polyoxyethylene in PDMS. Solid lines represent Eqs. (12.14) and (12.15). The square symbols are PDMS data. (From Ref. 43.)

size and size distribution, the volume fraction of droplets, and the interfacial tension. Model predictions are found to agree with linear viscoelastic data on a number of polymer–polymer systems over the entire frequency range [36,45]. In accord with experimental data, the Palierne model also predicts that the zero-shear-rate viscosity of an emulsion can be significantly larger than the zero-shear viscosity values of both blend components [45]. Similarly, the first normal stress difference in shear is found to exceed the value predicted from the normal stresses of the two components using a linear mixing rule [36]. Attempts have been made to explain these observations in steady shearing flow by deriving general constitutive equations that consider the time evolution of the area and orientation of the interface in a flow field [46,47]. For a 50 vol% mixture of two immiscible fluids having the same viscosity and density, Doi and Ohta determined theoretically that the shear viscosity should be independent of shear rate, but the first normal

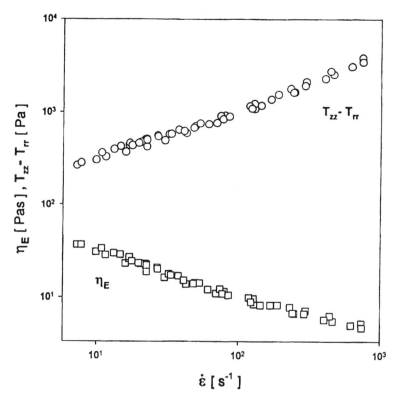

FIGURE 12.8 Extensional stress and extensional viscosity as a function of stretch rate for mayonnaise. (From Ref. 48. Used by permission of Steinkopff Publishers, Darmstadt, FRG.)

stress difference in shear should be proportional to the magnitude of the shear rate [46]; the latter prediction has turned out to be true [36].

At very high dispersed-phase concentrations, emulsions possess a yield stress and act like soft solids. A consequence of the solidlike character of an emulsion is the possibility of loss of adhesion at the viscometer wall during rheological measurement. If the material continues to be deformed in the bulk, data have to be corrected for slip effects. If, however, there is total loss of adhesion and there is no bulk deformation, a viscosity cannot be computed from the measured forces. This latter situation was encountered by Plucinski et al. for the flow of full-fat mayonnaise in a number of geometries [48]. The problem was so severe that rotational instruments could not be used even at extremely low shear rates. Extensional data, however, could be obtained with ease using a fiber-spinning device, and these data are displayed in Fig. 12.8 [48] As in the case of the chocolate data shown in Fig. 10.20, there is data superposition for different runs showing the absence of memory effects. Also, the extensional viscosity is stretch thinning, and there is an indication of the presence of a yield stress. At present, there is no theory we can use to predict the extensional stresses shown in Fig. 12.8. Note, however, that expressions for the yield stress of concentrated emulsions are identical to those for "dry" foams [49]; these are considered in chapter 13.

VII. CONCLUDING REMARKS

As this chapter demonstrates, a great deal of progress has been made in understanding and predicting the rheological properties of emulsions, but much still remains to be done if we want to describe the flow behavior of mixtures of polymers, block polymers, and graft polymers. There are several complicating factors: (1) The dispersed phase is not always spherical in shape, and its morphology may change with concentration and shear rate. For example, block polymers may have cylindrical domains, or both phases may be continuous. Figure 12.9 illustrates two kinds of domains found in block polymers and shows how the flow of block polymers differs from that of homopolymers. Spherical and cylindrical morphology exist when one of the components is less than about 30% of the total volume. Lamellar morphology is found when both components are more equal in volume. There can also be an inversion of phases. In a flowing system, the low-viscosity component typically tends to become the continuous phase and to encapsulate the high-viscosity component, reducing the blend viscosity in the process. (2) The interfacial effects characteristic of emulsions may be completely overwhelmed by elastic effects in polymer melts. (3) Block and graft polymers have parts of each molecule in each of the two liquid phases. When large deformations take place, strands of one kind of polymer are pulled through the other material. These filaments may remain as filaments, they may break up into small drops, or they may connect to each other to give an interconnected network.

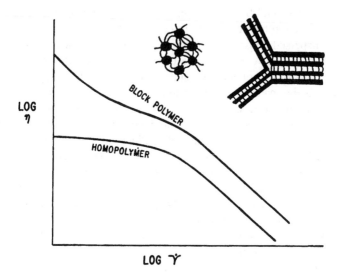

LOG η

BLOCK POLYMER

HOMOPOLYMER

LOG γ

FIGURE 12.9 Comparison of the viscosity of a block polymer with a typical homo-polymer that has the same molecular weight as the total molecular weight of the block polymer. Inserts show two typical morphologies of block polymers with cylindrical (or spherical) domains or lamellar domains.

Under some conditions, ribbons or sheets of one polymer can be formed inside of the other polymer. These phenomena can greatly enhance the viscosity of such materials. In addition, other complex rheological behavior may be observed. At constant temperature and constant shear rate, the viscosity of polyblends as a function of composition may go through a minimum or even both a minimum and a maximum [50]. The minimum occurs in a concentration range where phase inversion is expected to occur and where both polymers are more or less continuous phases. The first normal stress difference goes through a maximum at about the same concentration as where the viscosity goes through a minimum. It is extremely difficult to predict such behavior.

REFERENCES

1. L. A. Utracki. Polymer Alloys and Blends. Hanser, Munich, 1989.
2. A. Pilehvari, B. Saadevandi, M. Halvaci, P. E. Clark. Oil/water emulsions for pipe-line transport of viscous crude oils. Soc. Pet. Eng. paper #18218 presented at SPE 63rd annual technical conference and exhibition, Houston, TX, Oct. 2-5 (1988).
3. S. J. Choi, W. R. Schowalter. Rheological properties of nondilute suspensions of deformable particles. Phys. Fluids 18:420–427 (1975).

4. P. Sherman, ed. Emulsion Science. Academic Press, London, 1968.
5. D. T. Wasan, K. Koczo, A. D. Nikolov. Interfacial characterization of food systems. In: A. G. Gaonkar, ed. Characterization of Food: Emerging Methods. Elsevier Science, Amsterdam, 1995, pp. 1–22.
6. R. Pal. Anomalous effects in the flow behavior of oil-in-water emulsions. Chem. Eng. J. 63:195–199 (1996).
7. R. Pal. Rheological properties of emulsions of oil in aqueous non-Newtonian polymeric media. Chem. Eng. Commun. 111:45–60 (1992).
8. R. Pal. Effect of droplet size on the rheology of emulsions. AIChE J. 42:3181–3190 (1996).
9. H. L. Frisch, R. Simha. The viscosity of colloidal suspensions and macromolecular solutions. In: F. R. Eirich, ed. Rheology. Vol. 1. Academic Press, New York, 1956, pp. 525–613.
10. G. I. Taylor. The viscosity of a fluid containing small drops of another fluid. Proc. Roy. Soc. A138:41–48 (1932).
11. M. A. Nawab, S. G. Mason. The viscosity of dilute emulsions. Trans. Faraday Soc. 54:1712–1723 (1958).
12. G. I. Taylor. The formation of emulsions in definable fields of flow. Proc. Roy. Soc. A146:501–523 (1934).
13. F. D. Rumscheidt, S. G. Mason. Particle motions in sheared suspensions. XII. Deformation and burst of fluid drops in shear and hyperbolic flow. J. Colloid Sci. 16: 238–261 (1961).
14. H. L. Goldsmith, S. G. Mason. The microrheology of dispersions. In: F.R. Eirich, ed. Rheology. Vol. 4. Academic Press, New York, 1967, pp. 85–250.
15. H. P. Grace. Dispersion phenomena in high viscosity immiscible fluid systems and application of static mixers as dispersion devices in such systems. Chem. Eng. Commun. 14:225–277 (1982).
16. J. M. H. Janssen, H. E. H. Meijer. Droplet breakup mechanisms:stepwise equilibrium versus transient dispersion. J. Rheol. 37:597–608 (1993).
17. J. M. Rallison. The deformation of small viscous drops and bubbles in shear flows. Annu. Rev. Fluid Mech. 16:45–66 (1984).
18. J. M. H. Janssen. Dynamics of liquid–liquid mixing. Ph.D. dissertation, University of Technology, Eindhoven, The Netherlands, 1993.
19. H. A. Stone, L. G. Leal. The effects of surfactants on drop deformation and breakup. J. Fluid Mech. 220:161–186 (1990).
20. H. A. Stone. Dynamics of drop deformation and breakup in viscous fluids. Annu. Rev. Fluid Mech. 26:65–102 (1994).
21. R. W. Flumerfelt. Drop breakup in simple shear fields of viscoelastic fluids. Ind. Eng. Chem. Fundam. 11:312–318 (1972).
22. P. P. Varanasi, M. E. Ryan, P. Stroeve. Experimental study on the breakup of model viscoelastic drops in uniform shear flow. Ind. Eng. Chem. Res. 33:1858–1866 (1994).
23. H. Vanoene. Modes of dispersion of viscoelastic fluids in flow. J. Colloid Interface Sci. 40:448–467 (1972).
24. J. J. Elmendorp, R. J. Maalcke. A study on polymer blending microrheology: Part 1. Polym. Eng. Sci. 25:1041–1047 (1985).

25. J. Levitt, C. W. Macosko. Influence of normal stress difference on polymer drop deformation. Polym. Eng. Sci. 36:1647–1655 (1996).

26. D. C. Peters. Dynamics of emulsification. In: N. Harnby, M. F. Edwards, A. W. Nienow, eds. Mixing in the Process Industries. 2nd ed. Butterworth Heinemann, Oxford, 1992, pp. 294–321.

27. W. R. Schowalter, C. E. Chaffey, H. Brenner. Rheological behavior of a dilute emulsion. J. Colloid Interface Sci. 26:152–160 (1968).

28. W. R. Schowalter. Mechanics of Non-Newtonian Fluids. Pergamon Press, Oxford, 1978, pp. 264–289.

29. P. Scholz, D. Froelich, R. Muller. Viscoelastic properties and morphology of two-phase polypropylene/polyamide 6 blends in the melt. Interpretation of results with an emulsion model. J. Rheol. 33:481–499 (1989).

30. J. G. Oldroyd. The elastic and viscous properties of emulsions and suspensions. Proc. Roy. Soc. A218:122–132 (1953).

31. J. G. Oldroyd. The effect of interfacial stabilizing films on the elastic and viscous properties of emulsions. Proc. Roy. Soc. A232:567–577 (1955).

32. D. Graebling, R. Muller. Rheological behavior of polydimethylsiloxane/polyoxyethylene blends in the melt. Emulsion model of two viscoelastic liquids. J. Rheol. 34:193–205 (1990).

33. L. A. Utracki, Z. H. Shi. Development of polymer blend morphology during compounding in a twin-screw extruder. Part I: Droplet dispersion and coalescence—a review. Polym. Eng. Sci. 32:1824–1833 (1992).

34. J. J. Elmendorp, A. K. Van der Vegt. A study on polymer blending microrheology: Part IV. The influence of coalescence on blend morphology origination. Polym. Eng. Sci. 26:1332–1338 (1986).

35. A. Al-Mulla, R. K. Gupta. Droplet coalescence in the shear flow of model emulsions. Rheol. Acta 39:20–25 (2000).

36. I. Vinckier, P. Moldenaers, J. Mewis. Relationship between rheology and morphology of model blends in steady shear flow. J. Rheol. 40:613–631 (1996).

37. A. K. Chesters. The modelling of coalescence processes in fluid–liquid dispersions: a review of current understanding. Trans. IChemE 69A:259–270 (1991).

38. M. Minale, J. Mewis, P. Moldenaers. Study of the morphological hysteresis in immiscible polymer blends. AIChE J. 44:943–950 (1998).

39. S. A. Patlazhan, J. T. Lindt. Kinetics of structure development in liquid–liquid dispersions under simple shear flow. J. Rheol. 40:1095–1113 (1996).

40. M. Mooney. The viscosity of a concentated suspension of spherical particles. J. Coll. Sci. 6:162–170 (1951).

41. L. Djakovic, P. Dokic, P. Radivojevic, V. Kler. Investigation of the dependence of rheological characteristics on the parameters of particle size distribution at O/W emulsions. Colloid Polym. Sci. 254:907–917 (1976).

42. R. Pal, E. Rhodes. Viscosity/concentration relationships for emulsions. J. Rheol. 33:1021–1045 (1989).

43. D. Graebling, D. Froelich, R. Muller. Viscoelastic properties of polydimethylsiloxane–polyoxyethylene blends in the melt. Emulsion model. J. Rheol. 33:1283–1291 (1989).

44. J. F. Palierne. Linear rheology of viscoelastic emulsions with interfacial tension. Rheol. Acta 29:204–214 (1990).
45. P. J. Carreau, D. C. R. De Kee, R. P. Chhabra. Rheology of Polymeric Systems. Hanser, Munich, 1997, pp. 338–344.
46. M. Doi, T. Ohta. Dynamics and rheology of complex interfaces. I. J. Chem. Phys. 95:1242–1248 (1991).
47. H. M. Lee, O. O. Park. Rheology and dynamics of immiscible polymer blends. J. Rheol. 38:1405–1425 (1994).
48. J. Plucinski, R. K. Gupta, S. Chakrabarti. Wall slip of mayonnaises in viscometers. Rheol. Acta 37:256–269 (1998).
49. H. M. Princen. Rheology of foams and highly concentrated emulsions. II. Experimental study of the yield stress and wall effects for concentrated oil-in-water emulsions. J. Colloid Interface Sci. 105:150–171 (1985).
50. C. D. Han, Y. W. Kim. Dispersed two-phase flow of viscoelastic polymeric melts in a circular tube. Trans. Soc. Rheol. 19:245–269 (1975).

13

Gas-Containing Melts and Foams

I. INTRODUCTION

A foam is a two-phase system made up of gas bubbles trapped in a liquid; the desired quantity of bubbles of the requisite shape and size may be generated by any one of several methods. Often the gas volume fraction is very large, and this converts the gas bubbles into cells of a regular or irregular shape. Such cellular plastic foams can be formulated using either thermoplastic or thermosetting polymers. In the latter case, the processes of foaming and curing usually occur simultaneously. The final solid product can be rigid, semirigid, or flexible. Indeed, we are all familiar with semirigid polystyrene foam used in coffee cups and flexible urethane foam employed in seat cushions and carpet backing. The cells in the foam may be open (interconnected), permitting the transport of fluids, or closed and therefore impermeable. Open cell foams are good acoustical insulators, and open cell foams can also be used as filter media. Closed cell foams, on the other hand, are good thermal insulators, especially when the average bubble size is of the order of the mean free path of the gas. An example of closed cell, but rigid, foam is microcellular structural foam used to make automobile bumpers. Here, the foam contains a very large number of tiny bubbles (having a size less than or equal to 10 μm) of a gas such as carbon dioxide or nitrogen, and this makes the solid foam lightweight and resilient, and energy absorbing. The gas is typically injected in the supercritical state into the melt zone of the extruder barrel (see, for example, Goel and Beckman [1]). The resulting single-phase solution has a significantly lowered viscosity, and it can be processed at a lower temperature. An important application area for structural foams is equipment housing and thin-walled parts, and injection molding is the preferred manufacturing process, although structural foam can be thermoformed, blow molded, and also extruded into rods, tubes, and sheets. Common polymers, both filled and unfilled, used in structural foam are nylons, polyolefins, polyurethane, polystyrene, ABS, and PPO, and the main attraction of structural foam is the high strength-to-weight ratio [2]. However, transparent parts cannot be produced, since the cells act as opacifiers.

Instead of using a supercritical fluid for foaming a polymer, one may employ a chemical blowing agent such as a bicarbonate or a hydrazide [3]. This is preblended with the polymer during injection molding of structural foam. The blowing agent decomposes on heating in the extruder, generating a gas that dissloves in the melt. When the gas-containing melt is injected into the mold, the system pressure is reduced, nucleating gas bubbles that grow by diffusion of gas from within the melt to the bubbles. The bubbles also coalesce as they grow and ultimately yield the desired cellular structure. Bubble growth rates depend on the solubility and diffusivity of the gas in the polymer [4]. The cell size and size distribution are also found to be functions of the blowing agent concentration, the pressure, the injection speed, the melt temperature (as this controls melt viscosity and elasticity), any temperature and pressure gradients within the mold, and the presence of bounding surfaces [4,5]. Under quiescent, isothermal conditions, and in the absence of interactions between neighboring bubbles, the bubble radius is found to increase with time in accordance with the power law [5]. However, deviations that also lead to a distribution in bubble sizes are found to occur when the bubbles are closely spaced [6]. Other complicating effects are those resulting from fluid elasticity [7], varying temperature [8], and simultaneous chemical reaction [9]. A model of bubble growth during microcellular foam generation has been proposed [10].

In the fluid state, foams find use in fire fighting and in oil field applications involving drilling, completion, fracturing, and cleanup of wells [11–13]. Additionally, foams are employed as mobility control agents for oil displacement from porous media as part of enhanced oil recovery operations [14]. The flow of foam is interesting and challenging in that it has characteristics both of a liquid and a solid—the presence of yield stress, a finite shear modulus, slip at solid surfaces, and a shear-thinning viscosity. Foams are also compressible fluids, and their properties can be time dependent because of their metastable nature. Further complexities arise when the length scale of the dispersed phase becomes comparable to channel dimensions, as in flow through porous media; in such a case, it may be meaningless to talk about a shear viscosity because the continuum hypothesis breaks down [11].

II. VISCOSITY OF GAS-CONTAINING POLYMERS

In order to model and optimize the operation of an extruder used to process polymer melts prior to foam generation, it is necessary to know the viscosity versus shear rate relationship for polymeric fluids containing dissolved gases. The primary effect of the presence of the gas is plasticization of the polymer: there is an increase in the free volume, and this causes a reduction in the polymer glass transition temperature. This effect can be quite significant; the addition of 12% by weight of carbon dioxide to polymethyl methacrylate (PPMA), for example, results in a lowering of the PMMA glass transition temperature from 105°C to room temperature [1]. An important consequence of the lowering in T_g is viscosity reduction as predicted by Eq. (7) of Chapter 3; this has been observed for a number of polymer melts [15–17].

Gerhardt et al. [18] used a model system to study the influence of gas content on liquid viscosity. They measured the viscosity as a function of wt% carbon dioxide of a polydimethyl siloxane (PDMS) having the consistency of molten polymers at processing temperatures. Measurements were made at 50 and 80°C for shear rates ranging from 40 to 2300 s^{-1}, and for gas contents ranging from 0 to 21 wt%. A capillary viscometer, modified to operate as a sealed, pressurized unit, was employed, and representative results are displayed in Figure 13.1. It is seen that all curves have a similar shape, and the presence of the

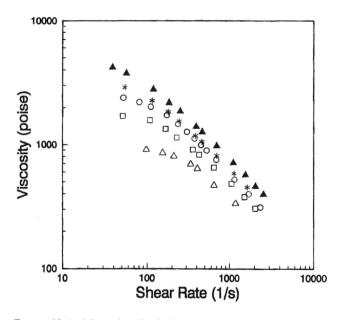

FIGURE 13.1   Viscosity of polydimethylsiloxane (PDMS) containing carbon dioxide as a function of shear rate at 50°C. The uppermost curve is pure PDMS. The subsequent curves represent behavior of PDMS containing 4.84, 9.03, 14.4, and 20.7 wt% CO_2. (From Ref. 18. Copyright © 1997 by John Wiley & Sons, Inc. Reprinted by permission of John Wiley & Sons, Inc.)

gas reduces the viscosity of PDMS quite substantially, especially at high gas concentrations and low shear rates.

Just as flow curves at different temperatures can be superposed by shifting on to a reference curve, the curves in Figure 13.1 can be reduced to a master curve by means of a combined horizontal and vertical shift. This is shown in Figure 13.2 in which the reference curve is taken to be the one corresponding to pure PDMS. This shift factor a_c is the same for both ordinate and abscissa, and it is given by

$$a_c = \frac{\eta_0}{\eta_{p,0}} \tag{13.1}$$

where η_0 is the low shear rate solution viscosity and $\eta_{p,0}$ is the zero shear rate PDMS viscosity. Clearly, the shift factor is a number less than unity, and the procedure is similar to that used to obtain Figure 3.11. The dependence of the shift factor on gas concentration is shown in Figure 13.3 and is found to be independent of the temperature of measurement. A knowledge of the data in Figure 13.3 allows us to construct the flow curve for any polymer–gas composition if we know the

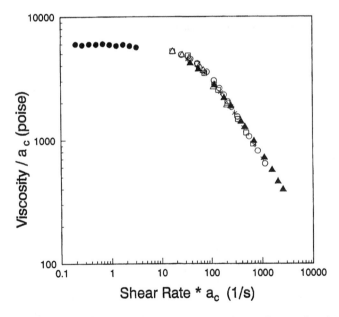

FIGURE 13.2 Viscosity master curve corresponding to the data shown in Figure 13.1. Solid circles represent measurements of the limiting low shear rate viscosity of pure PDMS. Other symbols have the same meaning as in Figure 13.1. (From Ref. 18. Copyright © 1997 by John Wiley & Sons, Inc. Reprinted by permission of John Wiley & Sons, Inc.)

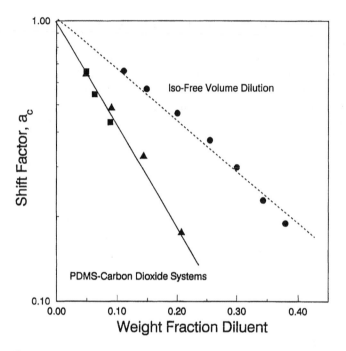

FIGURE 13.3 Variation of the shift factor with gas concentration (solid line) for the data of Figures 13.1 and 13.2. The two different symbols represent two different measurement temperatures (From Ref. 18. Copyright © 1997 by John Wiley & Sons, Inc. Reprinted by permission of John Wiley & Sons, Inc.)

viscosity curve of the pure melt at the same temperature and pressure. The composition shift factors can also be computed using free volume theories [19].

III. YIELDING OF FOAM

In considering the flow behavior of suspensions and emulsions, we had found it convenient to begin with the limiting situation of vanishingly small dispersed-phase concentrations. In foams, on the other hand, it is the limiting situation of "dry" foams or of the dispersed-phase volume fraction tending to unity that is of greater practical and theoretical interest. Here it is found that a knowledge of the structure of foam is essential to understanding its rheological behavior. However, the true three-dimensional structure is rather complex, and progress has been made by developing micromechanical theories based on two-dimensional models such as those proposed by Princen [20,21]. In two dimensions, foam is taken to be composed of uniformly sized cells or prisms of hexagonal cross section, as shown in Figure 13.4: these cells repeat themselves in the x-y plane. As explained by Kraynik [12], the hexagonal coordination minimizes the surface free energy. If we define a Cartesian coordinate system (x, y), cell orientation is given by the angle θ. Other dimensions of importance are cell size or length of each side of the hexagon a, the film thickness h, and the radius of curvature r at the Plateau border or the film junction region. Because of the presence of surface tension, the pressure in the Plateau borders is less than that in the thin films. This pressure difference would ordinarily force all the liquid to drain into the Plateau borders, but an equilibrium is maintained due to the presence of a repulsive or "disjoining" pressure that arises in the thin films because of molecular, ionic-electrostatic, and static interaction effects. This equilibrium endows the foam with a solidlike structure that manifests itself in the form of a measureable yield stress.

The process of foam deformation under the influence of a shear stress can be visualized to go through the series of steps shown in Figure 13.5, where the cell orientation angle is taken to be zero [21]. The cells deform affinely, and the foam behaves in an elastic manner with a stress that is a unique function of the applied stain. As the strain increases, however, two Plateau borders coalesce as in Figure 13.5c. This is an unstable structure which quickly rearranges itself into a structure that looks identical to the starting structure (Fig. 13.5a). The cells that were initially in contact, however, find themselves separated by one cell at the end of the deformation process, and it apprears that the foam has yielded.

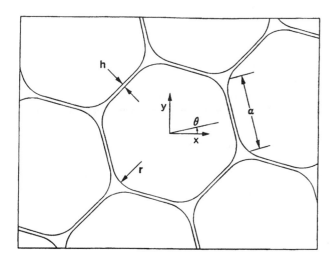

FIGURE 13.4   Equilibrium structure of an idealized two-dimensional foam. (From Ref. 12.)

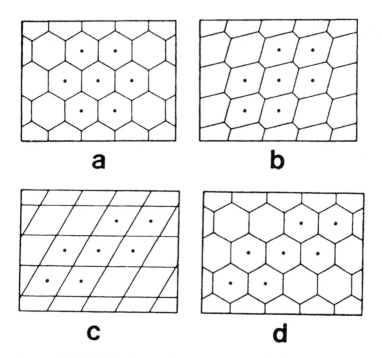

a   b

c   d

FIGURE 13.5 Model showing the strain dependence of cell morphology for simple shear resulting in the separation of cells that were orginally in contact and completing the deformation cycle. (From Ref. 22.)

The yield stress is the stress in the material at the time of film rearrangement, and this stress is unrelated to and not bounded by the stress values within the two fluids making up the foam. The deformation, though, is clearly strain periodic, and further straining results in a repetition of the events depicted in Figure 13.5.

According to Princen [21], the shear stress in the foam is the projection onto the shear plane of the tension in an initially vertical film. This, together with a knowledge of the magnitude of the angle between the film and the horizontal at the time of structure rearrangement, give the yield stress as

$$\tau_y = \frac{\alpha\sigma}{R} \tag{13.2}$$

in which α has the value 0.525, σ is the interfacial tension, and R is the radius of the largest circle that can be drawn inside one of the hexagonal cells making up the foam in the rest state; from geometry, R equals $0.9094a$. This approach to determining the yield stress remains valid even as the foam quality or the gas volume fraction is reduced to 0.9069, the value corresponding to the maximum packing fraction of cylindrical bubbles. The value of α, however, decreases significantly with decreasing foam quality, becoming approximately 0.1 when the foam quality reaches 0.9069. Nonetheless, the yield stress predicted by Eq. (13.2) can be significant and measurable, especially for small bubbles.

Khan and Armstrong rederived Eq. (13.2) by equating the work done in shearing a unit cell of foam to the change in free energy of the microstructure [23]. They, however, found that the numerical value of α depended on the cell orientation angle. These authors also examined the influence of bubble size distribution and liquid viscosity on the shear stress versus shear strain behavior of dry foams [24]. They found that polydispersity did not influence either the yield stress or the critical strain at the point of yielding. Further, the influence of viscous forces in the liquid films was likely to be negligible for aqueous foams but might have to be accounted for in polymeric foams involving extremely viscous liquids. The viscous contribution to the total stress could affect the cell shape and alter the yield stress and strain. These complications, however, are of somewhat academic interest because real foams are not two-dimensional. For three-dimensional foams, and these are difficult to analyze theoretically, all that one can hope for is that the form of Eq. (13.2) remains valid; the numerical value of the coefficient would then have to be obtained by experiment. Khan et al [25] have carried out experiments on a polymer-surfactant-based aqueous foam having a mean bubble diameter of 65 μm. A parallel plate viscometer was used, and slip at the wall was eliminated by affixing sandpaper to the fixture surfaces. It was found that the yield stress was quite large, being approximately 18 Pa, and, as expected, it decreased with increasing liquid content in the foam. Not surprisingly, the two-dimensional model did not fit the data quantitatively; the theory overpredicted the yield stress.

A major consequence of the presence of a large yield stress in foam is the possibility of plug flow and the absence of bulk deformation during viscosity measurement by capillary rheometry. If the applied wall shear stress is lower than the yield stress, foam will not deform. However, if there is a thin liquid layer beteen the viscometer wall and the body of the foam, apparent slip can take place due to shearing of this liquid layer. In such a case, the foam can move at a constant velocity as a solid plug; this has indeed been observed [11,12], and no conclusions can be drawn about foam viscosity from a knowledge of the measured pressure drop.

IV. FOAM VISCOSITY

The experimental determination of the shear viscosity of foam is fraught with all of the problems discussd in Chapters 9–12, which are characteristic of two-phase systems. In addition, there is the very real possibility of bubble collapse during the process of measurement. A rotational viscometer that was filled with fluid at the beginning of the experiment ends up being only partially full at a later time. This phenomenon, coupled with the occurrence of a wall slip, can make data reproducibility difficult, and different instruments can give different results. Similarly, fluid compressibility effects can play havoc during tube flow through long tubes; unless total pressure drops are small, bubbles can expand and the foam quality can change along the length of the tube. Indeed, according to Heller and Kuntamukkula [11] most published data of foam viscosity in the technical literature are in error for on reason or another.

Shown in Figure 13.6 on logarithmic coordinates are the shear viscosity versus shear rate data, obtained using a parallel plate viscometer by Khan et al. [25] on the same foam whose yield stress measurements were discussed in the previous section. It is seen that over the shear rate of measurement, foam is shear thinning with a viscosity that is extremely large—much larger than the constant 5 mPa-s viscosity of the liquid making up the foam. Increasing foam quality results in a higher viscosity, and at each gas volume fraction, data lie on a straight line of slope − 1. The last observation reveals that foam behaves as a material whose shear stress is constant, independent of the shear rate. This constant shear stress is the yield stress, and foam deformation involves only elastic effects; viscous effects, if any, are small. The conclusion is strengthened by the observation that the storage modulus of the foam is independent of frequency and much larger

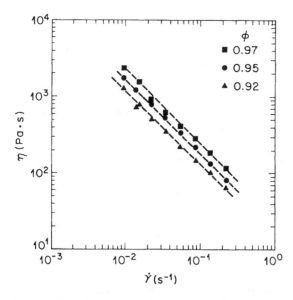

Figure 13.6 Foam viscosity as a function of shear rate for three different gas volume fractions. (From Ref. 25.)

than the loss modulus. It is the large value of the yield stress that is responsible for the large and shear-thinning viscosity. Since the yield stress increases with foam quality, so does the shear viscosity.

Consistent with the preceding observations, theories of foam flow [12,22,23] predict that the shear viscosity of foam is given by the well-known Bingham plastic equation, according to which

$$\eta = \frac{\tau_y}{\dot{\gamma}} + \eta_p \qquad (13.3)$$

where the plastic viscosity η_p arises due to dissipation in the liquid film. Clearly, foam viscosity is made up of the sum of an elastic and a viscous contribution. The dissipation term becomes important for wet foams or for foams made with extremely viscous liquids. For foams formulated with Newtonian liquids, the magnitude of the plastic viscosity is directly proportional to the liquid viscosity and the liquid volume fraction. Recent data of Saint-Jalmes and Durian show that elastic effects disappear entirely as the gas volume fraction is reduced to

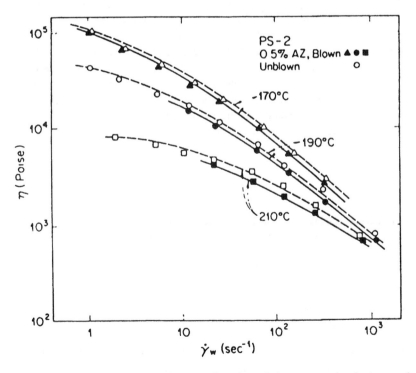

FIGURE 13.7 Apparent viscosity as a function of shear rate of polystyrene foam at three different temperatures, with and without a blowing agent. (From Ref. 27. Copyright © 1979 by John Wiley & Sons, Inc. Reprinted by permission of John Wiley & Sons, Inc.)

about 0.63. Similarly, for polymeric foams, the elastic term in Eq. (13.3) is negligible, and the foam viscosity is comparable to the polymer viscosity. This is demonstrated in Figure 13.7, which shows the flow of curves of polystyrene foamed with an azodicarbonamide blowing agent and obtained using a capillary viscometer [27]. As expected, polymer melt viscosity decreases with increasing temperature and increasing shear rate. The addition of the blowing agent also reduces the viscosity, but this reduction has not been explained on a theoretical basis. It is probably related to the gas solubility effects discussed in Section II.

V. CONCLUDING REMARKS

The short length of this chapter must have indicated to the reader the paucity of reliable information, both theoretical and experimental, on the flow behavior of foams. As we have seen, foam viscosity can be predicted under some circumstances, and limited data exist with which to compare these predictions. Expressions have also been derived for the first normal stress difference in shear and for stresses in extensional flow [23]; measuring these material functions, however, is a very daunting task that is yet to be accomplished. This is the prevailing situation for the flow of foam when the size of the bubbles is much smaller than the length scale characterizing the flow geometry. As opposed to this, there are no theories that one can apply to foam flow when the bubble size is comparable to the channel dimensions, a situation encountered routinely in flow through a porous medium. Many data sets exist, especially in the petroleum literature, but these are all thought to be corrupted by wall slip or foam stability effects [11]. Clearly, much work remains to be done.

REFERENCES

1. S.K. Goel, E.J. Beckman. Generation of microcellular polymeric foams using super-critical carbon dioxide. I: Effect of pressure and temperature on nucleation. Polym. Eng. Sci. 34:1137–1147 (1994).
2. N. Peach. Plastic foams: options, methods, and materials. Plast. Eng. 40(8):19–24 (1984).
3. J.L. Throne, H.G. Griskey. Structural thermoplastic foam: low energy processed material. Polym. Eng. Sci 15:747–756 (1975).
4. C.D. Han. Multiphase Flow in Polymer Processing, Academic Press, New York, 1981, pp. 257–340.
5. C.A. Villamizar, C.D. Han. Studies on structural foam processing. II. Bubble dynamics in foam injection molding. Polym. Eng. Sci. 18:699–710 (1978).
6. M. Amon, C.D. Denson. A study of the dynamics of foam growth: analysis of the growth of closely spaced spherical bubbles. Polym. Eng. Sci. 24:1026–1034 (1984).
7. J.R. Street. The rheology of phase growth in elastic liquids. Trans. Soc. Rheol. 12:103–131 (1968).

8. A. Arefmanesh, S.G. Advani. Nonisothermal bubble growth in polymeric foams, Polym. Eng. Sci. 35:252–260 (1995).
9. J.R. Youn, H. Park. Bubble growth in reaction injection molded parts foamed by ultrasonic excitation, Polym. Eng. Sci. 39:457–468 (1999).
10. S.K. Goel, E.J. Beckman. Nucleation and growth in microcellular materials: supercritical CO2 as foaming agent, AIChE J. 41:357–367 (1995).
11. J.P. Heller, M.S. Kuntamukkula. Critical review of the foam rheology literature. Ind. Eng. Chem. Res. 26:318–325 (1987).
12. A.M. Kraynik. Foam flows, Ann. Rev. Fluid Mech. 20:325–357 (1988).
13. V.G. Constien. Fracturing fluid and proppant characterization, Reservoir Stimualtion (M.J. Economides and K.G. Nolte, eds.), 2nd ed., Prentice Hall, Englewood Cliffs, NJ, 1989, Chap. 5.
14. F.I. Stalkup, Jr. Miscibile Displacement, Society of Petroleum Engineers, New York, 1983.
15. L.L. Blyer, Jr., T.K. Kwei. Flow behavior of polyethylene melts containing dissolved gases, J. Polym. Sci.: Part C 35:165–176 (1971).
16. C.D. Han, C.A. Villamizar. Studies on structural foam processing. I. The rheology of foam extrusion, Polym. Eng. Sci. 18:687–698 (1978).
17. C.D. Han, C.-Y. Ma. Rheological properties of mixtures of molten polymer and fluorocarbon blowing agent. I. Mixtures of low-density polyethylene and fluorocarbon blowing agent, J. Appl. Polym. Sci. 28:831–850 (1983).
18. L.J. Gerhardt, C.W. Manke, E. Gulari. Rheology of polydimethylsiloxane swollen with supercritical carbon dioxide, J. Polym. Sci. B: Polym. Phys. 35:523–534 (1997).
19. L.J. Gerhardt, A. Garg, C.W. Manke, E. Gulari. Concentration-dependent viscoelastic scaling models for polydimethylsiloxane melts with dissolved carbon dioxide, J. Polym. Sci. B: Polym. Phys. 36:1911–1918 (1998).
20. H.M. Princen. Highly concentrated emulsions. I. Cylindrical systems, J. Colloid Interface Sci. 71:55–66 (1979).
21. H.M. Princen. Rheology of foams and highly concentrated emulsions. I. Elastic properties and yield stress of a cylindrical model system, J. Colloid Interface Sci. 91:160–175 (1983).
22. A.M. Kraynik, M.G. Hansen. Foam rheology: A model of viscous phenomena, J. Rheol. 31:175–205 (1987).
23. S.A. Khan, R.C. Armstrong Rheology of foams: I. Theory for dry foams, J. Non-Newtonian Fluid Mech. 22:1–22 (1986).
24. S.A. Khan, R.C. Armstrong. Rheology of foams: II. Effects of polydispersity and liquid viscosity for foams having gas fraction approaching unity, J. Non-Newtonian Fluid Mech. 25:61–92 (1987).
25. S.A. Khan, C.A. Schnepper, R.C. Armstrong. Foam rheology: III. Measurement of shear flow properties, J. Rheol. 32:69–92 (1988).
26. A. Saint-Jalmes, D.J. Durian. Vanishing elasticity for wet foams: Equivalance with emulsions and role of polydispersity, J. Rheol. 43:1411–1422 (1999).
27. Y. Oyanagi, J.L. White. Basic study of extrusion of polyethylene and polystyrene foams, J. Appl. Polym. Sci. 23:1013–1026 (1979).

14

Rheology of Powders and Granular Materials

I. INTRODUCTION

The flow of powders and granular materials, often termed *particulate solids* and having a solids volume fraction that typically ranges from 0.3 to 0.6, is encountered in many aspects of everyday life. It is also important in several practical applications involving powder metallurgy, soil mechanics, ceramic science, and polymer engineering. Plastic pellets and powders, for example, are stored in bins and hoppers, and these containers have to be designed for emptying in a free-flowing manner. In recent years, powder coatings have become important and have replaced liquid surface coatings in many applications. The flow characteristics of such powders are essential to the successful use of these coatings. Another place where the flow of polymer powders is encountered is in rotational molding. In addition, granular polymers and polymer pellets are melted in extruders and are also compounded, either with rubber to form toughened polymers or with glass fibers to give reinforced polymers. The rheological properties of these granular materials before they become melted down to a liquid are important to the proper performance of extruders and injection molding machines. In particular, an appreciable fraction of the energy used to operate these machines can go into the polymer while it is in the granular state.

The science of powder rheology, however, is not as well developed as the rheology of liquidlike materials. The practical importance of the subject, though, is such that there has been a concerted effort of late aimed at narrowing this gap [1–3]. The flow of powders and pellets is very different from the flow of liquids. Indeed, the flow characteristics of powders often are completely different from

what would be expected on the basis of experience with liquids. The discharge
rate of granular materials through an orifice at the bottom of a filled container,
for example, is independent of the height of the vessel. More surprisingly, fine
particles may not flow at all, even through large openings due to the tendency
of a powder to form an arch [4]. Available expressions for discharge rates of
granular materials from hoppers, along with the associated theory, have been
reviewed by Nedderman et al. [5]. The time taken for a given quantity of powder
to flow through an orifice or a hollow cone with a hole at the apex is often used
for the purpose of powder characterization [6].

II. INSTRUMENTS

Although a few of the instruments used to study granular materials will be familiar to the rheologist who works with liquids, most of the techniques used with
these materials are very different from those used with liquids. Some instruments
are similar to coaxial cylinder viscometers [7]. However, smooth rotor surfaces
cannot be used, because powders cannot wet the surface of the instrument to
achieve adhesion between the powder particles and the parts of the instrument.
Thus rotor surfaces must be roughened or contain grooves or teeth in order to
transmit forces into the mass of powder. Such a rotational instrument is illustrated
in Fig. 14.1 in which the torque required to stir the powder is measured either

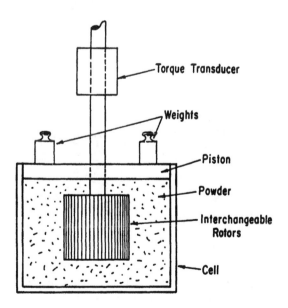

FIGURE 14.1 Schematic diagram of a rotational rheometer for powders.

as a function of speed of rotation or as a function of the total number of turns of the rotor. Although the viscosity of a liquid is only slightly dependent upon hydrostatic pressure, the torque required to stir a powder is extremely dependent upon the pressure exerted on the surface of the powder. For this reason, powder rheometers are generally equipped to apply a normal force to the surface of the powder. However, this force will vary throughout the bulk of the powder, since hydrostatic pressure cannot be transmitted uniformly through a powder as in a liquid. Rankine [8] derived an equation for relating the horizontal pressure p_H resulting from the vertical pressure p_V applied to the top surface of a powder.

$$p_H = p_V \frac{1 - \sin \alpha}{1 + \sin \alpha} \qquad (14.1)$$

The angle α is the coefficient of internal friction and can be estimated from a graph in which the yield locus of the powder is drawn tangent to the Mohr circle [9]. It can also be measured in a compaction experiment [9,10] and is sometimes approximated by the angle of repose. For loose solids, the angle of repose typically assumes values between 30° and 40° [4].

A second type of powder rheometer is a shear cell [6,7,11]. One version of such an instrument is shown schematically in Fig. 14.2. The top part of a split shear cell containing the powder is pushed at a uniform speed; the force required to produce this motion is measured by a dynamometer or force transducer. The split cell shears the powder along only a single plane. The shear force is strongly dependent upon the normal load applied to the top surface of the powder. The

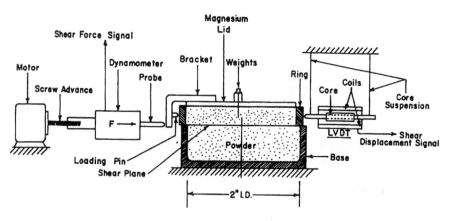

FIGURE 14.2 A modified Jenike shear cell apparatus. Shear cell motion is measured by the linear variable-differential transducer (LVDT). (From Ref. 7.)

shear force is measured either as a function of the displacement of the top of the shear cell or as its equivalent, the time from the start of the experiment. Another version of the shear cell is shown in Fig. 14.3 [12]. It is an annular shear cell, and it consists of two concentric horizontal circular disk assemblies mounted on a common shaft. The bottom disk is the movable member, and it has an annular trough to contain the sample. The upper disk is stationary, and it has a lipped annular protrusion that fits in the trough in the bottom disk. It is also capable of vertical motion under the influence of both internal stresses and externally applied loads. The annular shear cell is designed to operate at high shear rates and at known values of the mean solids concentration.

A third type of measurement made on powders is the *angle of repose* [6]. The angle of repose can be measured by many techniques. The simplest method is to form a pile of the powder on a flat surface and measure the angle formed by the sides of the pile and the horizontal surface. Intuitively, one suspects that powders that flow readily should form a smaller angle of repose than powders that do not flow so readily. Yet other kinds of measurements, such as the force required to withdraw a plate from a cylindrical bed of dry powder, have been described by Lee [13].

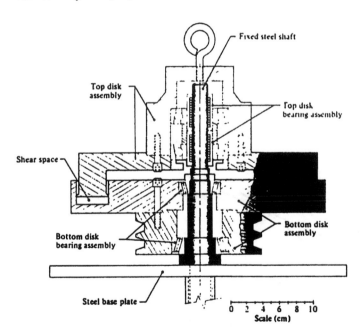

FIGURE 14.3 Cross-sectional view of an annular shear cell. (From Ref. 12. Reprinted with the permission of the Cambridge University Press.)

III. FLOW BEHAVIOR OF PARTICULATE SOLIDS

The flow behavior of a powder or granular material is governed by particle–particle friction, collisions amongst particles, and interparticle forces. It is, however, not influenced by the nature of the gas that fills the interstitial space between the solid particles [14]. This is in contrast to the behavior of (dilute) solid-in-liquid suspensions, where the particulate solid serves to modify the rheology of the matrix fluid. Depending on the shear rate, two distinct regimes of flow behavior are observed, both at high solid volume fractions. At low shear rates, the stress in the material is determined by interparticle friction. This situation is exemplified by the discharge of particulates from hoppers or bins. At the other extreme of high shear rates, momentum transfer occurs by a translational/collisional mechanism, and the material superficially resembles a dense gas. This so-called grain-inertia regime is important in flow down an inclined plane, as, for example, in rock falls and avalanches.

Powders are further classified as being either cohesive or noncohesive. The difference between the two classes is schematically illustrated in Fig. 14.4. For noncohesive powders, the steady-state torque in the case of rotational rheometers or the force in the case of a shear cell extrapolates to zero at zero normal load on the powder. Cohesive powders have a finite yield stress and do not extrapolate

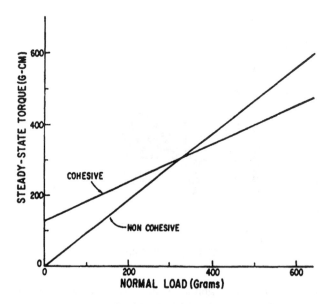

FIGURE 14.4 Typical data obtained with a rotational rheometer on cohesive and noncohesive powders. The speed of rotation is constant.

to zero steady-state torque or displacement force as the normal load approaches zero. However, in both cases, the torque or displacement force is nearly a linear function of the normal load. Thus, data from shear cells also give graphs similar to Fig. 14.4 except the steady-state force to move the shear cell instead of the torque is plotted against the normal load on the powder.

Cohesive powders do not flow readily under the action of small forces. They tend to form a uniform coherent pile with large angles of repose. In contrast, a noncohesive powder, such as a mass of small ball bearings, readily flows under the action of small forces, such as gravity, and it is more difficult to form a distinct pile of the powder.

The cause of coherence in cohesive powders can be due to several factors [15]. These include: (1) rough particle surfaces and interlocking of irregularly shaped particles; (2) sticky coatings; (3) interparticle attractions, such as magnetism in iron particles, electrostatic charges on nonconducting particles, or van der Waals' forces between neutral atoms and molecules. Since the magnitude of these forces depends, in general, on the average interparticle distance, the extent of coherence is likely to be a function of the bulk density of the material; (4) the presence of a liquid or "glue" at points of particle–particle contacts. Due to the presence of atmospheric humidity, moisture may be present as adsorbed vapor at low relative humidities and as liquid bridges at higher humidities. Liquid may also be generated by the melting of the granular material.

Curves such as those shown in Fig. 14.4 are examples of the Coulomb equation:

$$F_s = CA + \alpha F_n \qquad (14.2)$$

or

$$\tau = C + \alpha \sigma_n \qquad (14.3)$$

F_s is the total shear force in a shear cell, or it is the torque in a rotational rheometer. The constants C and α are the coherence of the powder and the coefficient of internal friction, respectively. The normal force or load applied to the surface of the powder is F_n. The constant A is the area of the shear cell in shear cell instruments; in rotational instruments, A is a constant dependent upon the dimensions of the rotor. The shear stress and normal stress applied to the powder are τ and σ_n, respectively. The coefficient of internal friction is sometimes related to the angle of repose, but, in general, the two quantities are not identical. In Fig. 14.4, the intercept on the torque axis is proportional to the coherence, while the slope is proportional to the coefficient of internal friction. For perfectly free-flowing powders, that is, for noncoherent powders, the coherence C is zero. Then

$$\tau = \alpha \sigma_n \qquad (14.4)$$

Equation (14.4) is identical to Amonton's law, which defines the coefficient of friction between two solid surfaces in sliding contact, and it clearly shows the solidlike character of the flow behavior of granular materials—the shear stress is proportional to the normal stress rather than to the rate of deformation. However, Eq. (14.4) holds as an equality only during flow. In the absence of flow, particulate solids do sustain a shear stress, but the magnitude of this stress is indeterminate; all that one can say is that the shear stress is less than or equal to the value given by Eq. (14.4).

Many cases are known, as is shown in Fig. 14.4, where a cohesive powder has a smaller coefficient of internal friction than does a noncohesive powder. In such cases, the cohesive powder becomes easier to stir than the noncohesive powder at high normal loads. This unexpected ease of stirring cohesive powders results from their inability to pack as densely as noncohesive powders. It has been found that the apparent density of a granular material depends on the same quantities that determine the friction conditions between powder particles [16]; low bulk densities are the result of high interparticle friction.

Not all powders obey the Coulomb equation, but they then follow a more general equation [17]:

$$\left(\frac{\tau}{C}\right)^n = \frac{\sigma_n}{\sigma_B} + 1 \tag{14.5}$$

The tensile strength of the powder is σ_B; it is the intercept on the horizontal axis of Fig. 14.4 if the curve for the cohesive powder is extended in the negative normal load direction. The constant n generally varies from 1.0 to 2.0; when $n = 1$, the Coulomb equation is recovered.

Strongly cohesive powders cannot flow through orifices or from bins and hoppers. Noncohesive powders and weakly cohesive powders can flow through funnels or other types of orifices. As mentioned in Sec. I, short-stemmed funnels are often used to simulate crudely the flow behavior of a powder and to estimate the ease with which the powder will flow from hoppers. It should be emphasized that the rate of flow of a powder from a funnel is essentially independent of the height of the powder in the funnel [7]. This is in contrast to a Newtonian, liquid where the rate of flow is directly proportional to the hydrostatic pressure of the liquid at the orifice.

The flow of a powder through an orifice depends upon the relative diameters of the orifice and the particles making up the powder. Figure 14.5 is typical of the flow behavior of powders through an orifice. There is a maximum rate of flow at some particle size. Below a critical particle size, the powder tends to form arches over the orifice that stop the flow. As the particle size approaches the diameter of the orifice, flow again becomes blocked. For a number of powders,

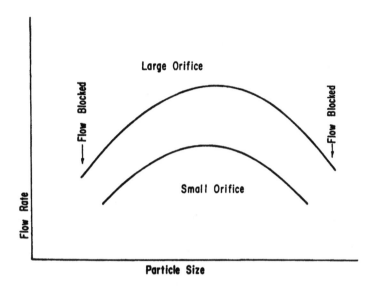

FIGURE 14.5   Typical results for the flow of powders through an orifice or a funnel as a function of particle size for orifices of two different diameters.

the following equation holds approximately for the flow of powders through a circular orifice [18]:

$$D = (0.52 D_p + 1.97) \left[\frac{4W}{(60\pi\rho\sqrt{g})} \right]^{0.4} + 0.838D_p^{0.7} \qquad (14.6)$$

The orifice diameter and the average particle diameter are D and D_p, respectively. The flow rate is W (g/min), ρ is the particle density, and g is the acceleration of gravity. The exponent 0.4 varies somewhat for different powders, but the exponent generally is limited to between 0.3 and 0.5.

In closing this section, we note that the discussion thus far has been limited to situations where neither the application of a normal load to a granular material nor subsequent flow results in a change in the bulk density or in a fracture of the individual particles. A large change in the bulk density can arise, for example, during the manufacture of composites by powder compaction [19]; in this situation, both the shear and bulk viscosities are found to exhibit a power law dependence on the solids volume fraction [20,21].

A. Slow Frictional Flow

At low shear rates, the flow behavior of a powder is dominated by interparticle friction effects, and in a rotational rheometer this means that the steady-state torque is independent of the speed of rotation; this is analogous to solid–solid friction, where the force required to overcome the friction of a solid block pulled across a flat surface is nearly independent of the speed of sliding. The general behavior of a powder in a rotational viscometer is illustrated in Fig. 14.6; it is seen that the steady-state torque remains nearly constant as the shear rate is increased; in some cases, the torque may even decrease, and this would be reminiscent of the difference between static friction and dynamic friction. One of the factors that determines the magnitude of the torque is the particle shape. Spherical particles tend to give a smaller torque than rough, granular particles. Another characteristic of powders during shearing action is the presence of a well-defined slip plane or shear plane in the powder. In other words, all the deformation takes place within this narrow slip plane, and a true rate of shear cannot be calculated. Thus, a viscosity cannot be computed, since the deformation does not vary uniformly across the annular gap of the rheometer as it does in the case of liquids. Again, this behavior is more characteristic of solid–solid friction than of liquid viscosity.

When a rotational rheometer containing powder is started, the steady-state torque is not achieved immediately, as is illustrated in Fig. 14.7. Several rotations of the rotor are required before the torque becomes constant. As might be expected, the steady-state torque increases significantly and linearly with the normal

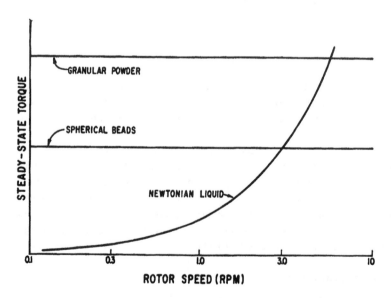

FIGURE 14.6 Typical data obtained with a rotational rheometer for a granular powder, beads and a liquid. The normal load on the powders is a constant.

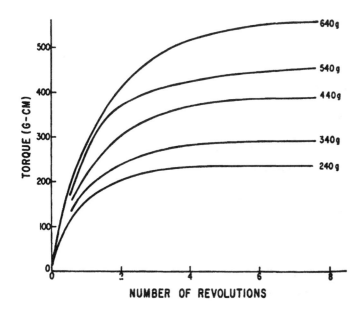

FIGURE 14.7 Rotational rheometer data with glass beads used as a model powder. The numbers refer to the normal load applied to the beads.

load applied to the surface of the powder. The initial density of packing of the granular material also has a strong influence on the startup torque of a rheometer. as shown in Fig. 14.8. Loosely packed powders tend to have a monotonically increasing torque as the number of revolutions increases. Consolidated powders, produced by tapping or vibrating the powders to increase the density of packing, produce a maximum in the torque which can be considerably greater than the steady-state torque.

Shear cells give data similar to those shown in Figs. 14.7 and 14.8. which are obtained with rotational rheometers. With shear cells, the displacement of the shear cell takes the place of the number of rotations and the force required to move the shear cell takes the place of the torque in the graphs. Thus, shear cells produce data from which shearing force is plotted against the displacement of the cell for various normal loads on the cell.

In summary, even though granular materials tend to flow like liquids, they display many features that are characteristic of solids. At rest, they can withstand a shear stress, exhibit a nonisotropic pressure distribution, and show yielding. Following yield, the shear stress is independent of the rate of deformation. These features are responsible for several observations that go counter to thinking based on expectations of liquidlike behavior. During plunger-driven tube flow, in partic-

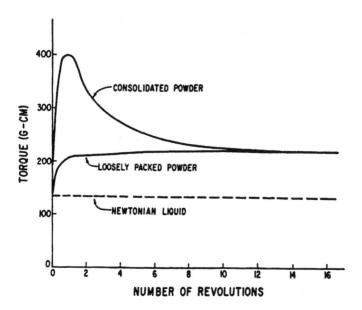

FIGURE 14.8 Rotational rheometer data on a loosely packed powder and on a consolidated powder as a function of the number of revolutions of the rotor. The normal load on the powders is constant.

ular, the shear stress at the wall is proportional to the pressure at the wall rather than to the pressure gradient; this leads to an exponential decrease of the axial stress with axial distance instead of the more familiar linear decrease [9].

B. Rapid Granular Flow

When a granular material is sheared at increasingly high shear rates, the individual particles making up the granular material begin to move independently. Under the influence of the velocity gradient, the faster-moving particles continually collide with the slower-moving particles in the adjacent layer and transfer momentum to them. As explained by Bagnold [22], the number of collisions as well as the amount of momentum transferred per collision are each proportional to the relative velocity between the layers. Consequently, the stresses generated ought to be proportional to the square of the shear rate $\dot{\gamma}$. Additionally, the stresses were predicted to be proportional to the mass density ρ_p of the particles and the square of the particle diameter D_p. Thus, the extra-stress tensor has the form (see also [23]):

$$\tau_{ij} = \rho_p D_p^2 f(\phi) \, \dot{\gamma}^2 \qquad\qquad (14.7)$$

in which $f(\phi)$ is a function of the solids volume fraction ϕ. Note that this functional form can be obtained with the help of dimensional analysis (see Ref. 1, for example).

The preceding predictions have, to a large extent, been borne out by experimental work. Displayed in Figs. 14.9a and 14.9b are representative steady-shearing data of Savage and Sayed obtained on polystyrene spheres using the shear cell shown in Fig. 14.3 [12]. In these experiments, the normal stress and the shear stress were both measured simultaneously as a function of the imposed shear rate at different constant values of the solids volume fraction. In accord with Eq. (14.7), it is seen that both stress values increase essentially quadratically with the shear rate; typically, an increase in shear rate was accompanied by an increase in the shear stress as well as an increase in the volume of the powder. The powder volume was brought back to its original level by the imposition of weights to the top of the cell—hence the increase in the normal stress. In a different run, the solid volume fraction was changed by beginning with a different set of weights. Care had to be taken to ensure that there was uniform shearing across the entire gap. This was accomplished by limiting the gap thickness to between 5 and 10 particle diameters. At low shear rates, the curves deviate from a slope of 2, and tend to flatten out. This, no doubt, is due to a transition to slow frictional flow that was considered in the previous section. Even though Fig. 14.9 explores only the effect of shear rate and solid concentration, other variables are also known to influence the measured stresses. These variables include the coefficient of restitution, the surface friction coefficient, the size distribution of the solid particles, and their angularity [1].

The data of Fig. 14.9 are replotted in Fig. 14.10 as the ratio of the shear to the normal stress (at the same value of the concentration) as a function of the dimensionless shear rate, and it is found that the ratio increases slightly with shear rate and decreases slightly with solids concentration. Also shown in Fig. 14.10 is the quasi-static value of the stress ratio determined in a separate shear test—a light load was put on top of the upper disk assembly and the torque needed to move the lower disk assembly slowly by hand was measured. This latter value has a similar magnitude, and this says that Eq. 14.4 is valid even in the grain-inertia regime.

In the last 20 years, significant progress has been made in formulating microstructural theories of rapid granular flow, and several reviews are available in the literature [1,2,14,24]. The essence of any such theory is to consider the motion of a granular material to be similar to that of a dense gas. Thus although the particulates all move in a direction that is governed by the imposed velocity gradient, flow-induced collisions give rise to random, fluctuating velocity components. The mean square value of these fluctuations $\langle c^2 \rangle$, in analogy with the kinetic theory of gases, is denoted the particle temperature T_p, defined as $T_p = \langle c^2 \rangle / 3$, and the flux of energy is taken to be proportional to the gradient of this temperature.

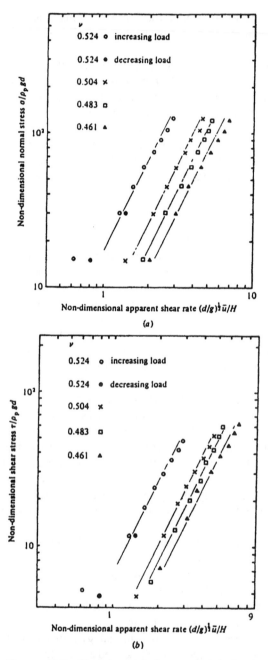

FIGURE 14.9 Experimental shear cell data for 1-mm-mean-diameter polystyrene spheres: (a) normal stress, (b) shear stress. (From Ref. 12. Reprinted with the permission of the Cambridge University Press.)

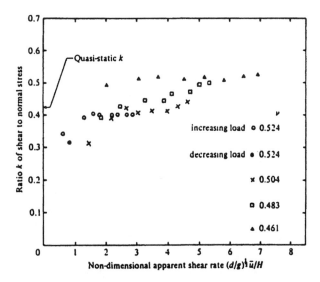

FIGURE 14.10   Ratio of shear to normal stress for the data of Fig. 14.9. (From Ref. 12. Reprinted with the permission of the Cambridge University Press.)

T_p enters the stress constitutive equation, and its numerical value, for any flow situation, is determined by solving a "pseudo-thermal energy equation" simultaneously with the mass and momentum balance equations. Unlike gas molecules, collisions between particles are inelastic; consequently, this energy balance equation contains a dissipation term that accounts for the irreversible conversion of the kinetic energy of the particles into true thermal energy. During steady laminar shearing flow, at a fixed solids concentration, this theory properly predicts that both the shear and normal stresses increase quadratically with the shear rate.

Furthermore, the ratio of these two stresses is also predicted to be a constant, independent of the imposed shear rate [14,24]. This theory has also been applied to the two-dimensional gravity flow of granular materials down an inclined plane. Savage has summarized the available data for concentration and velocity profiles [1] as well as the theoretical predictions [24]; agreement between theory and practice appears to be qualitative rather than quantitative. Recent progress has been made with the help of computer simulations [2,25], which, for example, have been able to explain the decrease, with increasing solids concentration, of the ratio of the shear to normal stresses in shear flow that was seen in the data of Fig. 14.10. Computer simulations also allow us to explore the influence of variables such a nonspherical shape and a distribution of sizes.

IV. CONCLUDING REMARKS

The flow of powders and granular materials is usually divided into slow frictional flow and rapid granular flow. In slow flow, particles slide on top of each other and interact by surface friction. The result is solidlike behavior that has traditionally been modeled using concepts of soil mechanics. In rapid flow, particles collide with each other over short time scales and dissipate energy by surface inelasticity. The result is behavior that superficially resembles the motion of a dense gas. As might be expected, the form of the equations that describe stresses generated in these two different flow regimes is quite dissimilar, and so are the results. In steady laminar shearing flow, in particular, the shear stress is found to be independent of the shear rate at low shear rates but to increase with the second power of the shear rate at high shear rates. Attempts have, therefore, been made in the literature to reconcile these two different viewpoints. On the one hand, Tardos [3] has attempted to recast the equations governing slow flow into a form similar to the Navier–Stokes equations; on the other hand, Savage [24] and others have tried to combine the two approaches. Results have been tested against experimental data. These, however, are not very plentiful. Also, they are often not reproducible, because subtle differences in packing can produce large differences in flow behavior, especially at high solids concentration [12].

REFERENCES

1. S.B. Savage. The mechanics of rapid granular flows. Adv. Appl. Mech. 24:289–366 (1984).
2. C.S. Campbell. Rapid granular flows. Annu. Rev. Fluid Mech. 22:57–92 (1990).
3. G.I. Tardos. A fluid mechanistic approach to slow, frictional flow of powders. Powder Technol. 92:61–74 (1997).
4. A.W. Jenike. Storage and flow of solids. Bull. No. 123, Utah Engineering Experiment Station, University of Utah, Salt Lake City, 1964.
5. R.M. Nedderman, U. Tuzun, S.B. Savage, G.T. Houlsby. The flow of Granular materials—I. Discharge rates from hoppers. Chem. Eng. Sci. 37:1597–1609 (1982).
6. R.L Brown, J.C. Richards. Principles of Powder Mechanics. Pergamon Press, Oxford, 1970.
7. M. Takano, L.E. Nielsen, R.W. Buchanan. Rheology of some powders and granular materials. Organic Coatings and Plastics Chem. Div., ACS Preprints 33(2):447–454 (1973).
8. W.J.M. Rankine. Phil. Trans. Roy. Soc. London 146:9 (1856).
9. Z. Tadmor, C.G. Gogos. Principles of Polymer Processing. Wiley, New York, 1979, pp. 246–256.
10. R.A. Thompson. Mechanics of powder pressing: I. Model for powder densification. Ceram. Bull. 60:237–243 (1981).
11. A.W. Jenike. Gravity Flow of Bulk Solids, Bull. No. 108, Utah Engineering Experiment Station, University of Utah, Salt Lake City, 1961.

12. S.B. Savage, M. Sayed. Stresses developed by dry cohesionless granular materials sheared in an annular shear cell. J. Fluid Mech. 142:391–430 (1984).
13. B.-L. Lee. Low pressure rheology of granular powders using a drawing plate technique. Polym. Eng. Sci. 28:469–476 (1988).
14. R. Jackson. Some features of the flow of granular materials and aerated granular materials. J. Rheol. 30:907–930 (1986).
15. N. Harnby, M.F. Edwards, A.W. Nienow. Mixing in the Process Industries. 2nd ed. Butterworth-Heinemann, Oxford, 1992, pp. 79–98.
16. H.H. Hausner. Friction conditions in a mass of metal powder. Int. J. Powder Metallurgy 3(4):7–13 (1967).
17. M.D. Ashton, D.C.-H. Cheng, R. Farley, F.H.H. Valentin. Some investigations into the strength and flow properties of powders. Rheol. Acta 4: 206–218 (1965).
18. T.M. Jones, N. Pilpel. The flow properties of granular magnesia. J. Pharm Pharmac. 18:81–93 (1966).
19. L.S. Stel'makh, A.M. Stolin, B.M. Khusid. Extrusion rheodynamics for a viscous compressibe material. Inz. Fiz. Zh. 61(2):268–276 (1991).
20. L.M. Buchatskii, A.M. Stolin. Determining the rheological properties of compressible powder materials in the high-temperature range. Inz. Fiz. Zh. 57(4):645–653 (1989).
21. J.A. Puszynski, S. Miao, B. Stefansson, S. Jagarlamundi. In-situ densification of combustion synthesized coatings. AIChE J. 43:2751–2759 (1997).
22. R.A. Bagnold. Experiments on a gravity-free dispersion of large solid spheres in a Newtonian fluid under shear. Proc. Roy. Soc. Lond. A225:49–63 (1954).
23. P.K. Haff. A physical picture of kinetic granular fluids. J. Rheol. 30:931–948 (1986).
24. S.B. Savage. Granular flows at high shear rates. In: R.E. Meyer, ed. Theory of Dispersed Multiphase Flow. Academic Press, New York, 1983, pp. 339–358.
25. C.S. Campbell. The stress tensor for simple shear flows of a granular material. J. Fluid Mech. 203:449–473 (1989).

15

Chemorheology and Gelation

I. INTRODUCTION

Chemorheology is the study of rheological changes in a system due to the occurrence of chemical reactions. Such reactions are an integral part of thermoset polymer processing. The reactants are mixed together, and curing is typically initiated by raising the temperature of the system or by irradiating it. The process itself may be run batchwise or continuously, and the reaction may involve either step-growth or chain-growth polymerization. As the reaction proceeds, the molecular weight of the reactants increases, and, if one of the starting materials is a prepolymer, branches begin to form. These branches are crosslinking points, and as the branches become longer with time they ultimately interconnect to form a three-dimensional network called a *gel*.

The point of gelation can be predicted if the reaction chemistry is known [1], and it can be detected with the help of rheological techniques. Note that chemical reactions continue beyond the point of gelation and cease permanently only when the limiting reactant or reactive groups are completely consumed. Increasing diffusional resistances, though, can lead to glass formation and a temporary extinction of the chemical reaction, whether before or after gelation, if the polymer glass transition temperature increases to equal the curing temperature [2]; reaction resumes on raising the processing temperature. The practical consequence of gelation is that the material can neither flow nor dissolve in a solvent. It gets set. As a consequence, it is not very meaningful to talk about polymer viscosity beyond the point of gelation; it is the modulus of the solidlike material that is relevant.

A technologically important application in which thermosetting polymers are employed is as matrices in fiber-reinforced polymeric composites or fiber-reinforced plastics (FRPs) (Refs. 3,4 for example). Fiber-reinforced plastics consist of as much as 70% by volume high-strength and high-modulus fibers embedded in a polymeric matrix that is usually a thermoset. The fibers, and these are generally glass, graphite, or aramid, are the primary load-bearing elements, while the matrix serves to protect the fibers and acts as a load transfer medium. The most important selling points in favor of FRPs are their very high modulus-to-weight ratio and their very high strength-to-weight ratio. These properties have led to FRPs being used widely as both structural and nonstructural members in the aerospace and automotive industries. It should be noted though that for aerospace applications, production volumes are low, and labor-intensive methods such as hand layup and spray-up are usually employed. Conversely, in the more price-sensitive automotive applications, the production volumes are large and automated techniques such as compression molding, resin transfer molding, reaction injection molding, pultrusion and filament winding are employed [3,4].

Figure 15.1 shows, in schematic form, the process of pultrusion that is commonly used to manufacture products such as luggage racks, tail gates, hand rails, and bridge decks [5]. Here, fiber reinforcements in the form of rovings and/ or fabric are continuously pulled through a liquid resin bath and a heated die in sequence. The die serves to squeeze out excess resin, shape the composite, and cure the polymer. The solidified FRP profile is cut to size by a saw. The goal of any pultrusion processor is to increase pulling speeds and to produce parts having a complex geometry. If not done correctly, however, this can lead to poor quality products owing to nonuniform curing of polymer and non-uniform wetting of fibers by the resin; the presence of voids and dry spots can lower strength and modulus values quite significantly. Consequently, polymer processing models (and not just for pultrusion) that incorporate the correct curing kinetics and appro-

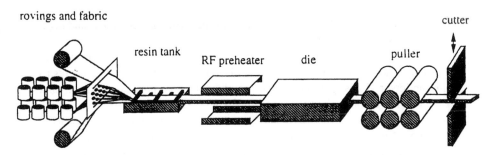

FIGURE 15.1 Schematic of the pultrusion process. (From Ref. 5. Reproduced with permission of the American Institute of Chemical Engineers.)

priate viscosity models have been formulated in recent years (Refs. 5–7, for example).

Although a variety of thermosetting resins are utilized in the polymer-composites industry, the most common ones are unsaturated polyesters, vinyl esters, epoxies, phenolics, and polyurethanes. Unsaturated polyesters are less expensive compared to vinyl esters and epoxies and account for almost 75% of total thermoset usage in FRPs [3]. Typically one reacts an unsaturated dicarboxylic acid (such as maleic anhydride) with a dihydric alcohol (such as ethylene glycol) to form a (linear) prepolymer containing double bonds in the prepolymer backbone. The prepolymer is then dissolved in an almost equal amount by weight of a reactive monomer such as styrene to give a low-viscosity liquid. It is this liquid that would be contained in the resin bath shown in Figure 15.1. Crosslinking is carried out with the help of an organic peroxide such as benzoyl peroxide that decomposes on heating to give free radicals. These free radicals attack the unsaturation sites (double bonds) in the prepolymer and also react with the styrene, resulting in the growth of polystyrene branches which ultimately unite all the prepolymer molecules into a gel.

The time to gelation is known as the working life or pot life, and, in the case of polyesters, it is fairly short, being of the order of minutes, due to the chain growth nature of the reaction. The gel time can be varied by changing the temperature or the initiator concentration or by the addition of an accelerator. Before gelation, there is little viscosity increase; at gelation, only about 5% of the original unsaturation may have reacted [3]. By contrast, epoxies gel at much higher conversion levels. Here, the prepolymer has the three-member epoxide group containing one oxygen and two carbon atoms at each of the two ends. Crosslinking is now done using a polyamine ''hardener'' in which the amine groups react with the epoxide groups by a step growth mechanism. Consequently, the pot life can be in excess of 1 hour, but the viscosity of the reaction mass increases continuously with conversion. The reaction itself can be carried out at room temperature or at elevated temperatures. Epoxies are often preferred to polyesters because they have superior mechanical, corrosion-resistance, and high-temperature properties, but they are more difficult to handle due to higher viscosities. Vinyl esters, which are obtained by reacting epoxies with an unsaturated acid such as acrylic acid or methacrylic acid, combine the benefits of epoxy resins with the lower viscosity and faster curing of unsaturated polyesters. Since vinyl esters contain a double bond (instead of the epoxide group) at each end of the molecule, they can be cured in the same way as polyesters after being dissolved in styrene. Note that mineral fillers such as calcium carbonate and glass microspheres are sometimes added to the reaction mass in order to reduce shrinkage and also to lower the cost of the FRP.

Another industrially important situation involving chemical reactions between polymer molecules arises during hydraulic fracturing of oil wells [8,9].

Oil well productivity can be enhanced by injecting into the formation, under pressure, a polymer solution containing particulates. This creates extended fractures that have to be kept open once the injection pressure subsides. This is achieved using proppants, typically sand, that are transported to the fracture in the form of a slurry that is kept in suspension in an aqueous polymer solution; the polymer used is most commonly the polysaccharide hydroxypropylguar. Since the temperature inside an oil well can be substantially higher than at the surface, the polymer solution viscosity can decrease and the proppant, which can constitute up to 20 lb per gallon of the fracturing fluid [9], can settle out of the suspension. To prevent this from happening, the guar molecules can be crosslinked with the help of metal ions such as borate, titanium, and zirconium. Now, however, the viscosity at the wellhead can be so high that pumping costs can become excessive. The compromise is to use a combination of a slow and a fast crosslinking agent. The fast crosslinker ensures adequate viscosity at the wellhead, while the slow crosslinker builds up viscosity by the time the fluid reaches the wellbore.

Yet other applications of crosslinked polymers are in the rubber industry, in the packaging of circuit boards [10,11] and in low-solvent coatings [12]. In each case, we wish to know how the rheological properties, and particularly the viscosity, change with time. This is usually done by relating the viscosity to the degree of cure and temperature, and the degree of cure to time; here the degree of cure can be considered to be the analog of molecular weight for thermoplastics. For some polymer systems, such as epoxies and phenolics, it may also be necessary to consider shear-thinning effects; in still other cases, the influence of fillers may need to be accounted for in an explicit manner [11]. The degree of cure can be followed by measuring thermal effects in a differential scanning calorimeter (DSC) or by following changes of mass (for condensation polymers) in a thermogravimetric analyzer (TGA) [13]. Alternately, dynamic mechanical analysis (DMA) can be employed to follow changes in the glass transition temperature T_g, a quantity related uniquely to the extent of cure. An advantage of using rheological techniques is that the gel point can be precisely determined; DSC and TGA are inherently incapable of doing this.

II. THERMAL CHARACTERIZATION OF CURING REACTIONS

During the course of a curing reaction, heat is generally liberated. Provided that the rate of energy generation is neither too high nor too low, the total energy evolved per unit mole of reactive groups or double bonds $\Delta H_{R_{xn}}$ and the instantaneous rate of evolution dH/dt can be determined using a DSC under isothermal conditions. If we define the extent of reaction α as the fraction of reactive groups or double bonds that have reacted, then

$$\frac{d\alpha}{dt} = \frac{dH/dt}{\Delta H_{Rxn}} \tag{15.1}$$

in which ΔH_{Rxn} may depend on temperature [13]. Here it is assumed that the heat generated by the curing reaction is proportional to the extent of reaction; clearly, $\alpha(t)$ is $H(t)/\Delta H_{Rxn}$.

In a typical isothermal DSC measurement, the rate of heat evolution dQ/dt is measured directly as a function of time at the chosen temperature of cure. The sample size is small, usually a few milligrams, in order to ensure constant temperature conditions, especially at high temperatures, where heat evolution can be quite rapid. Isothermal DSC data as a function of time are shown in Fig. 15.2 for the curing of an unsaturated polyester resin at four different temperatures [14]. It is seen that the total amount of heat evolved Q_T, which is the area under the cure-versus-time curve in Fig. 15.2, increases as the cure temperature increases. This is because the resin does not cure completely at lower temperatures. As a consequence, additional energy Q_R is released when a sample cured at a given temperature is heated to higher temperatures to complete the curing pro-

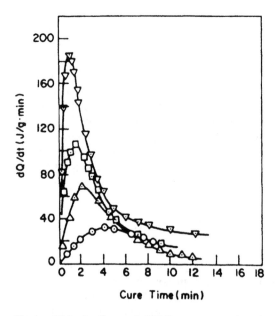

FIGURE 15.2   Isothermal dQ/dt versus cure time for a general-purpose unsaturated polyester resin. Curing temperatures are ○, 40; △, 45; □, 55; and ▽, 60°C. (From Ref. 14.)

cess. Figure 15.3 shows that the sum of Q_T and Q_R is constant at Q_{UT}, independent of the temperature chosen for the isothermal cure. The extent of cure α at any temperature, therefore, is $Q(t)/Q_{UT}$. Calculated values of α using the data of Figs. 15.2 and 15.3 are plotted in terms of time in Fig. 15.4 and as $d\alpha/dt$ versus α in Fig. 15.5. The results of Fig. 15.5 can be portrayed by the following equation for the kinetics of the overall curing process [14–16]:

$$\frac{d\alpha}{dt} = (k_1 + k_2\alpha^m)(1 - \alpha)^n \tag{15.2}$$

in which the k's are rate constants that depend on temperature in an exponential manner and m and n are constants that are independent of temperature; for a second-order reaction, $m + n = 2$. By separating the variables in Eq. (15.2) and integrating, we can explicitly relate α to time at a given temperature; the resulting expression is the analytical representation of the data shown in Fig. 15.4. A major utility of isothermal curing kinetics data is in being able to predict the extent of cure as a function of time under conditions of varying temperature [14,17]. In the simplest situation, since the temperature dependence of k_1 and k_2 is known, the instantaneous rate of curing at any temperature and extent of cure is again

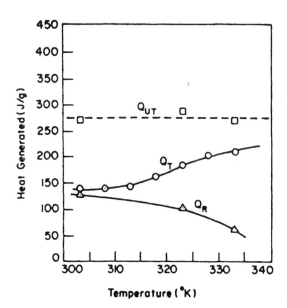

FIGURE 15.3 Heat generated versus temperature for the resin of Fig. 15.2. (From Ref. 14.)

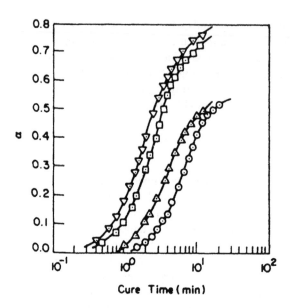

FIGURE 15.4 Isothermal degree of cure versus time. Symbols have the same meaning as in Fig. 15.2. (From Ref. 14.)

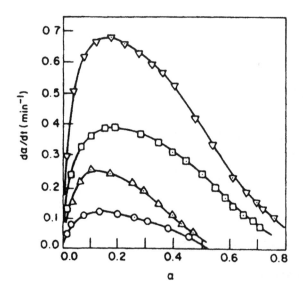

FIGURE 15.5 Data of Fig. 15.2–15.4 replotted as $d\alpha/dt$ versus α. Symbols have the same meaning as in Fig. 15.2. (From Ref. 14.)

given by Eq. (15.2). Note that most thermosetting cure reactions can be represented by this equation [13].

In all of the foregoing, it has been assumed that the rate of the curing reaction is chemically controlled. This is not the case in each instance. The polymer glass transition temperature always increases with increasing conversion, and if T_g begins to approach the cure temperature, the reaction rate can become diffusion controlled. The polymer vitrifies, resulting in a very considerable slowing down in the progress of the reaction, and this is the most likely reason why the limiting conversion values in Fig. 15.4 increase with increasing temperature. If curing is done in a DSC, the reaction can be quenched at any time simply by lowering the sample temperature and the T_g determined by a regular DSC scan. Hale et al. [18], among others, have shown that the polymer glass transition temperature is a unique function of conversion and is independent of the temperature at which curing was carried out. T_g is also a very sensitive measure of the degree of cure, especialy in the final stages of curing, when the reaction slows down and the rate of heat evolution diminishes.

In closing this section, we mention that isothermal DSC measurements become inaccurate for fast-curing reactions, especially when the sample has to be heated rapidly to an elevated temperature. This is because it takes some time for the instrument baseline to stabilize, and part of the exotherm can be lost during this time interval. A solution is to use nonisothermal DSC measurements, which have the added benefit that they require less time as compared to the isothermal measurements. The theory for obtaining isothermal results from nonisothermal measurements may be found elsewhere [13,19], and Ref. 13 may be consulted for the finer points of thermal characterization of thermosetting polymers.

III. MECHANISTIC MODELS OF CURING

Although Eq. (15.2) is a simple equation and is easy to use, it is an empirical equation. Its biggest drawback is that the constants appearing in it have to be obtained by experiment each time the resin formulation is changed. This can be avoided with the help of mechanistic models. For chain-growth polymers, the process involves analyzing crosslinking based on initiation, inhibition, propagation, and termination steps, with each step having its own rate constant. The result is a set of coupled equations involving concentrations of the different reactants and whose solution gives the conversion and temperature as a function of time [5,20–23, for example]; conversion can also be related to an appropriately defined average molecular weight for branched polymers. With this information, either the viscosity can be related to molecular weight using expressions of the kind introduced in Chapters 3 and 4, or it can be (more easily) determined at a given conversion and temperature, using models of the kind discussed next.

IV.   VISCOSITY AS A FUNCTION OF CONVERSION

Shown in Fig. 15.6 are isothermal data for the change in viscosity with time
resulting from the curing a high-temperature Novalac-epoxy resin at three different temperatures [24]. As expected, the higher the curing temperature, the lower
is the initial viscosity. However, due to the enhanced rate of reaction (with an
attendant viscosity rise) with increasing temperature, the different curves cross
each other fairly quickly. The initial portion of each curve is linear, but the onset
of gelation causes the viscosity to increase without bound. Also, the gel time
decreases with increasing temperature. A further observation is that all the iso-

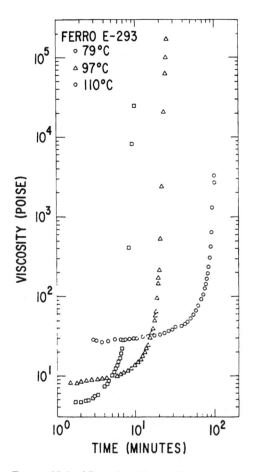

FIGURE 15.6   Viscosity–time profiles for the isothermal curing of an epoxy resin.
(From Ref. 24.)

thermal data can be made to collapse onto a single master curve by means of both a horizontal and a vertical shift. This is shown in Fig. 15.7. A knowledge of the shift factors allows us to determine the viscosity at any other temperature and time.

In terms of mathematically representing the data shown in Fig. 15.6, the following simple equation is found to correlate all the isothermal viscosity results [24,25]:

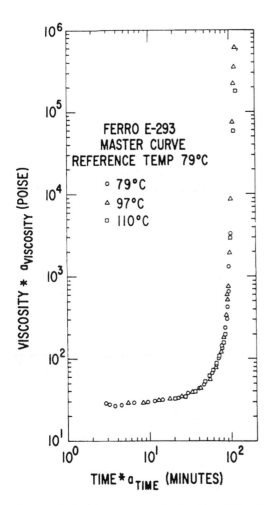

FIGURE 15.7   Master curve obtained by shifting the data in Figure 15.6. (From Ref. 24.)

$$\ln \eta(t) = \ln \eta_0 . kt \tag{15.3}$$

in which η_0 and k are constants whose values depend on the temperature of measurement and t is time. Far away from the polymer glass transition temperature, these two constants depend on temperature in an Arrhenius manner [25]:

$$\eta_0 = \eta_\infty \exp \frac{\Delta E_\eta}{RT} \tag{15.4}$$

$$k = k_\infty \exp \frac{\Delta E_k}{RT} \tag{15.5}$$

and finally we have for isothermal conditions:

$$\ln \eta(t) = \ln \eta_\infty + \frac{\Delta E_\eta}{RT} + tk_\infty \exp \frac{\Delta E_k}{RT} \tag{15.6}$$

Which is a four-parameter model. To determine these four constants, we first plot $\ln \eta$ versus time at each temperature. The slope and intercept of these plots yield k and η_0 respectively. The natural logarithm of each of these quantities is then plotted as a function of $1/T$; the required constants are obtained from the slope and intercept of these two staight lines according to Eq. (15.4) and (15.5).

If the temperature varies with time, the viscosity at any time and temperature can be calculated using a modified version of Eq. (15.6) [25]:

$$\ln \eta(T,t) = \ln \eta_\infty + \frac{\Delta E_\eta}{RT} + \int_0^t k_\infty \exp \frac{\Delta E_k}{RT} dt \tag{15.7}$$

provided that the reaction does not become diffusion controlled at any time. Under isothermal conditions, the onset of the phenomenon is revealed by the plot of $\ln \eta$ versus time becoming nonlinear. When this happens, viscosity should be related to temperature using the WLF equation [Eq. (3.7)] instead of the Arrhenius equation [26]. This is discussed later in this section. Another likely point of concern is the shear rate dependence of the viscosity. In the pregel liquid, though, this is generally not important, since Newtonian behavior is usually observed. [27,28].

Although Eq. (15.7) is a useful result, we normally take a Eulerian approach in the simulation of polymer processing operations of the kind shown in Fig. 15.1. Here, we do not follow the polymerization process in time, but instead focus attention on a fixed position in space. In order to compute quantities such as the pulling force in the process of pultrusion, we need to know the spatial variation of shear stress or the viscosity. This is best done by relating viscosity to conversion or the extent of cure.

As shown in Fig. 15.6, we can measure viscosity as a function of time during the course of a curing reaction at a selected temperature. Since we can also determine the extent of cure at each time instant with the help of a DSC, we can plot viscosity versus conversion under isothermal conditions. This is done in Fig. 15.8 for an epoxy resin [29]; the result is the appearance of straight lines on a semilogarithmic plot. Consequently, we can say that

$$\ln \eta = \ln A + K\alpha \tag{15.8}$$

where K is a constant and $\ln A$ is a parameter that depends on temperature. A can generally be written in Arrhenius form as:

$$A = \eta_\infty \exp \frac{\Delta E}{RT} \tag{15.9}$$

so

$$\eta(T,\alpha) = \eta_\infty \exp \left(\frac{\Delta E}{RT} + K\alpha \right) \tag{15.10}$$

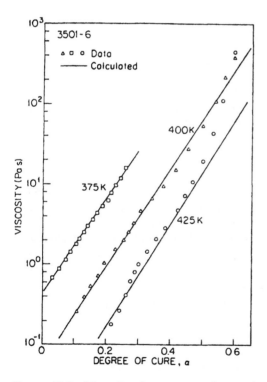

FIGURE 15.8 Viscosity of an epoxy resin as a function of the degree of cure. The solid lines represent Eq. (15.10). (Reprinted from Ref. 29, with permission from Technomic Publishing Co., Inc., copyright 1982.)

and it is seen from Fig. 15.8 that Eq. (15.10) does a good job of fitting the data. Also as shown in Ref. 29, viscosity values measured under nonisothermal conditions agree with the predictions of Eq. (15.10) employing instantaneous values of α and T.

During the curing of polyesters and vinyl esters and also in the later stages of curing of epoxy resins, the curing rate is diffusion controlled. Under these circumstances, the cure temperature is close to T_g, and it is appropriate to represent polymer viscosity with the WLF equation (see Chapter 3) [30]:

$$\ln \frac{\eta(T)}{\eta(T_g)} = \frac{-C_1(T - T_g)}{C_2 + (T - T_g)} \tag{15.11}$$

where, as mentioned in Sec. II, T_g is a unique function of the degree of cure α. For the curing of polyesters, Lee and Han [31] measured T_g as a function of the degree of cure, allowed C_2 to have the universal value 51.6, but made C_1 and $\eta(T_g)$ depend on α. They rewrote Eq. (15.11) as

$$\ln \eta(T) = \ln \eta(T_g) - C_1 + \frac{51.6C_1}{51.6 + T - T_g} \tag{15.12}$$

and plotted $\ln \eta(T)$ versus the reciprocal of $(51.6 + T - T_g)$ for fixed values of α. They obtained straight lines from which C_1 and $\eta(T_g)$ could be obtained and related to α. When all the results are introduced back into Eq. (15.11), polymer viscosity is related uniquely to temperature and degree of cure. This is the analog of Eq. (15.10) for curing reactions that are diffusion controlled.

V. POLYMER GELATION

A logical consequence of the progress of curing reactions is the formation of a gel, i.e., the conversion of the entire reaction mass into a single molecule. An indicator of the onset of gelation is an unbounded increase in the shear viscosity of the polymer. This is due to a transition of the material from a liquid to a solid. Since flow stops upon gel formation, it is essential that all processing be complete before this increase in viscosity arises. A liquid-to-solid transition also implies that complete stress relaxation is impossible, and the polymer develops a nonzero equilibrium modulus. This modulus increases monotonically with time as the curing reaction proceeds beyond gelation, and its numerical value is given by the theory of rubber elasticity [23]. Both these phenomena are shown schematically in Fig. 15.9 [32].

For a well-characterized system, the point of gelation can be predicted theoretically as a function of crosslink functionality and stoichiometric ratio [1]. Gelation occurs at a fixed conversion that is independent of the reaction temperature; an estimate of this conversion can be obtained using branching theories or perco-

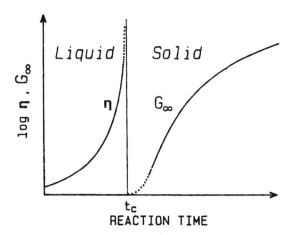

FIGURE 15.9 Schematic of steady-shear viscosity and equilibrium modulus of a crosslinking polymer. (From Ref. 32.)

lation theories that have been developed in recent years [33 and references therein]. The process, however, involves making simplifying assumptions that may not hold for real systems. Additionally, gelation can also result from physical clustering of particles [33] or from the presence of secondary bonds [34], where a theoretical approach may be difficult to adopt. For all these reasons, it is desirable to determine the point of liquid-to-solid transition in an experimental manner. Since it is difficult to measure either an infinite-shear viscosity or the onset of a nonzero equilibrium modulus, the technique of choice is small amplitude oscillatory shearing. Here, as shown in Fig. 15.10 for a PDMS [32], the storage modulus of the uncured polymer (at a fixed frequency) is much smaller than the loss modulus. With increasing time or extent of curing, though, both moduli increase, but the storage modulus increases at a faster rate and eventually equals the loss modulus. This is thought to happen in the vicinity of the gel point [35]. Eventually both moduli level off, but the storage modulus of the viscoelastic solid is much larger than the corresponding loss modulus. Note that in these experiments, crosslinking was stopped at different extents of curing prior to making the dynamic mechanical measurements.

Storage and loss moduli, as a function of frequency and corresponding to the data shown in Fig. 15.10, are displayed in Fig. 15.11 at different extents of cure. Samples at times less than t_c behave like viscoelastic liquids in that both moduli vary as expected with frequency [see Eq. (6.14) and (6.15)]. However, samples at times exceeding t_c tend to show a plateau in G' at low frequencies; this is characteristic of a solid. Thus, these data bound the solid-to-liquid transition. At

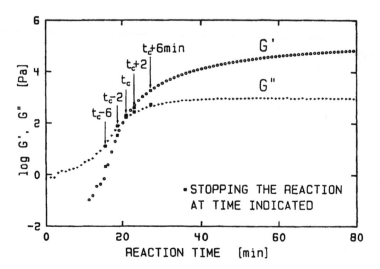

FIGURE 15.10 Evolution of storage and loss moduli during crosslinking of a silicone fluid in an oscillatory-shear experiment at constant frequency. (From Ref. 32.)

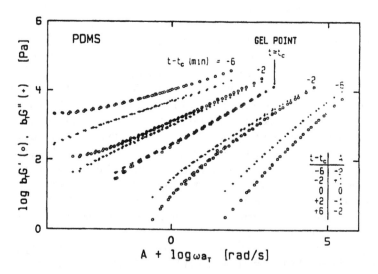

FIGURE 15.11 Reduced storage and loss moduli of PDMS samples for which the reaction has been stopped at intermediate states of conversion. t_c is the instant of intersection (see Fig. 15.10) of G' and G''. The curves have been shifted sideways to avoid overlap. (From Ref. 32.)

t, itself, G' is found to superpose with G'' at all frequencies, and the plot of G' versus frequency is a straight line of slope 0.5 on logarithmic coordinates. As a consequence, for this polymer at the gel point,

$$G' = G'' = C\omega^{0.5} \tag{15.13}$$

where C is a constant independent of temperature. That the congruence of G' and G'' does, indeed, represent the gel point is revealed by the theory of linear viscoelasticity. From Eq. (6.19), the zero-shear viscosity tends to infinity. Simultaneously, the equilibrium modulus $G(\infty)$, which is G' as the frequency approaches zero, is zero from Eq. (15.13); this fact can also be demonstrated by relating the dynamic data in Fig. 15.11 to $G(t)$ using the method of Sec. IV in Chap. 6 [33]. Consequently, the twin requirements for the onset of gelation are met. Furthermore, from Eq. (6.12) and (6.13), Eq. 15.13 can hold only if the stress relaxation modulus is given by [32]:

$$G(t) = St^{-1/2} \tag{15.14}$$

in which the parameter S is known as the gel strength.

Later work of Chambon and Winter [36] showed that Eq. (15.13) and (15.14) were special cases of a more general result. In general, the exponent in Eq. (15.14) is n, which takes the value 0.5 only for a balanced stoichiometric ratio of the reactants. Also, at the gel point, G' and G'' when plotted against frequency on logarithmic coordinates are parallel to each other instead of being congruent. Thus, G' and G'' is each proportional to ω^n at the gel point. As a consequence, $\tan \delta$, which is the ratio of G'' to G', is still independent of frequency. This, then, is the correct criterion for the onset of gelation, instead of the crossover point between G' and G'' on a plot of moduli versus time or extent of cure at a fixed frequency. Representative $\tan \delta$ data as a function of reaction time at three different frequencies for the PDMS system are displayed in Fig. 15.12 for unbalanced stoichiometry [37]; the point of onset of gelation is very easy to distinguish.

Since the stress relaxation modulus of the crosslinking polymer at the gel point is known, the rheology of this "critical gel" is known for small deformations. Thus, according to Eq. (6.8):

$$\tau(t) = S \int_{-\infty}^{t} (t - s)^{-n} \dot{\gamma}(s) \, ds \tag{15.15}$$

which is called the gel equation [32,36]. Winter and coworkers extended this equation to finite strains by replacing the infinitesimal-strain measure in the integrand by the Finger measure of strain; predicted stresses for the startup of shear flow at a constant shear rate and predicted strains during creep flow under a constant shear stress agreed with experimental data on PDMS as long as the total

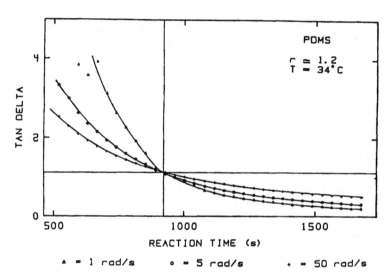

Figure 15.12 Loss tangent measured at three different frequencies on a cross-linking PDMS system. At the gel point, tan δ is independent of frequency and the curves pass through a single point. (From Ref. 37, with permission from Elsevier Science.)

shear strain was less than two units [38]. When larger strains were imposed, mechanical rupture of the gel took place.

Beyond the point of gelation, the polymer is a viscoelastic solid for which the crosslink density increases as the extent of conversion increases. This results in an increase in the shear modulus. All the rheological properties can now be explained within the framework of the theory of rubber elasticity [1,23,39].

VI. POLYMER VITRIFICATION

As mentioned earlier in this chapter, there is a one-to-one relationship between the degree of cure for a thermosetting polymer and the polymer glass transition temperature; T_g is a monotonically increasing function of conversion. The quantitative relationship can be determined experimentally with the help of dynamic mechanical thermal analysis (see Fig. 6.2), and Gillham and coworkers have pioneered the use of a freely oscillating torsional pendulum for this purpose [2]. If the temperature of cure, T_{cure}, is less than $T_{g\infty}$, the glass transition temperature of the fully cured polymer, the polymer T_g will rise and reach T_{cure} sometime during the course of the reaction. When this happens, the reaction mass will vitrify, or become glassy, and the reaction will either be extinguished or will slow down very appreciably. This can happen either before gelation or after gelation. In the

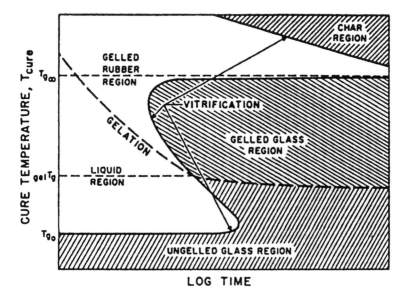

FIGURE 15.13 Generalized time–temperature–transformation (TTT) cure diagram. (From Ref. 40. Copyright © 1983 by John Wiley & Sons, Inc. Reprinted by permission of John Wiley & Sons, Inc.)

former case, the material is liquid prior to gelation, in the latter case, it is a rubber. Note that while gelation is a chemical process, vitrification is a physical process. The reaction resumes if the temperature of cure is raised. These various events can be represented on a time–temperature–transformation (TTT) cure diagram, shown schematically in Fig. 15.13 [40]. Because gelation occurs at a fixed conversion, the time required for gelation decreases with increasing T_{cure}, since reaction rates increase with increasing temperature; this behavior is sketched in Fig. 15.13, and it can be calculated from a knowledge of curing kinetics. For the same reason, gelation cannot occur at temperatures below the temperature at which the times for gelation and vitrification are the same; this is because $T_g = T_{cure}$ at a degree of conversion that is lower than the conversion needed for gelation. The application of TTT diagrams to composites processing has been illustrated by Simon and Gillham [41].

REFERENCES

1. P.J. Flory. Principles of Polymer Chemistry. Cornell University Press, Ithaca, NY, 1953.
2. J.K. Gillham, J.B. Enns. On the cure and properties of thermosetting polymers using torsional braid analysis. Trends Polym. Sci. 2:406–419 (1994).

3. Introduction to Composites. SPI Composites Institute, Washington, DC, 1992.
4. P.K. Mallick. Fiber-Reinforced Composites. 2nd ed. Marcel Dekker, New York, 1993.
5. G.L. Batch, C.W. Macosko. Heat transfer and cure in pultrusion: model and experimental verification. AIChE J. 39:1228–1241 (1993).
6. C.W. Macosko. RIM Fundamentals of Reaction Injection Molding. Hanser, Munich, 1989.
7. S.G. Advani, ed. Flow and Rheology in Polymer Composites Manufacturing. Elsevier, Amsterdam, 1994.
8. G.C. Howard, C.R. Fast. Hydraulic Fracturing. Soc. Pet. Eng. of AIME, Dallas, 1970.
9. M.J. Economides, K.G. Nolte, eds. Reservoir Stimulation. 2nd ed. Prentice Hall, Englewood Cliffs, NJ, 1989.
10. A.V. Tungare, G.C. Martin, J.T. Gotro. Chemorheological characterization of thermoset cure. Polym. Eng. Sci. 28:1071–1075 (1998).
11. P.J. Halley, M.E. Mackay. Chemorheology of thermosets—an overview. Polym. Eng. Sci. 36:593–609 (1996).
12. Y. Otsubo, T. Amari, K. Watanabe, T. Nakamichi. Rheological behavior of high-solid coatings during thermal curing. J. Rheol. 31:251–269 (1987).
13. R.B. Prime. Thermosets. In: E.A. Turi, ed. Thermal Characterization of Polymeric Materials. 2nd ed. Vol 2. Academic Press, San Diego, 1997, pp. 1397–1766.
14. K.W. Lem, C.D. Han. Thermokinetics of unsaturated polyester and vinyl ester resins. Polym. Eng. Sci. 24:175–184 (1984).
15. M.R. Kamal, S. Sourour. Kinetics and thermal characterization of thermoset cure. Polym. Eng. Sci. 13:59–64 (1973).
16. M.R. Kamal. Thermoset characterization for moldability analysis. Polym. Eng. Sci. 14:231–239 (1974).
17. A. Dutta, M.E. Ryan. The relationship between isothermal and non-isothermal kinetics for thermoset characterization. Thermochim. Acta 33:87–92 (1979).
18. A. Hale, C.W. Macosko, H.E. Bair. Glass transition temperature as a function of conversion in thermosetting polymers. Macromolecules 24:2610–2621 (1991).
19. R.A. Fava. Differential scanning calorimetry of epoxy resins. Polymer 9:137–151 (1968).
20. L.J. Lee. Curing of compression molded sheet molding compound. Polym. Eng. Sci. 21:483–492 (1981).
21. J.F. Stevenson. Free radical polymerization models for simulating reactive processing. Polym. Eng. Sci. 26:746–759 (1986).
22. C.D. Han, D.S. Lee. Analysis of the curing behavior of unsaturated polyester resins using the approach of free radical polymerization. J. Appl. Polym. Sci. 33:2859–2876 (1987).
23. A. Kumar, R.K. Gupta. Fundamentals of Polymers. McGraw-Hill, New York, 1998.
24. R.P. White Jr. Time-temperature superpositioning of viscosity–time profiles of three high temperature epoxy resins. Polym. Eng. Sci. 14:50–57 (1974).
25. M.B. Roller. Characterization of the time–temperature–viscosity behavior of curing B-staged epoxy resin. Polym. Eng. Sci. 15:406–414 (1975).

26. Y.A. Tajima, D. Crozier. Thermokinetic modeling of an epoxy resin. I. Chemoviscosity. Polym. Eng. Sci. 23:186–190 (1983).

27. C.D. Han, K.W. Lem. Chemorheology of thermosetting resins. I. The chemorheology and curing kinetics of unsaturated polyester resin. J. Appl. Polym. Sci. 28:3155–3183 (1983).

28. J.M. Dealy, K.F. Wissbrun. Melt Rheology and Its Role in Plastics Processing. Chapman and Hall, London, 1995, pp. 410–423.

29. W.I. Lee, A.C. Loos, G.S. Springer. Heat of reaction, degree of cure, and viscosity of Hercules 3501-6 resin. J. Comp. Mater. 16:510–520 (1982).

30. A. Letton, P.L. Chiou. Development of a cure model for determining optimum cure cycles in kinetically complex systems. In: A.A. Collyer, L.A. Utracki, eds. Polymer rheology and processing. Elsevier, London, 1990, pp. 431–458.

31. D.S. Lee, C.D. Han. A chemorheological model for the cure of unsaturated polyester resin. Polym. Eng. Sci. 27:955–963 (1987).

32. H.H. Winter, F. Chambon. Analysis of linear viscoelasticity of a crosslinking polymer at the gel point. J. Rheol. 30:367–382 (1986).

33. H.H. Winter, M. Mours. Rheology of polymers near liquid–solid transitions. Adv. Polym. Sci. 134:165–234 (1997).

34. A.G. Ward, P.R. Saunders. The rheology of gelatin. In: F.R. Eirich, ed. Rheology. Vol. II. Academic Press, New York, 1958, pp. 313–362.

35. C.-Y. M. Tung, P.J. Dynes. Relationship between viscoelastic properties and Gelation in thermosetting systems. J. Appl. Polym. Sci. 27:569–574 (1982).

36. F. Chambon, H.H. Winter. Linear viscoelasticity at the gel point of a crosslinking PDMS with imbalanced stoichiometry. J. Rheol. 31:683–697 (1987).

37. E.E. Holly, S.K. Venkatraman, F. Chambon, H.H. Winter. Fourier transform mechanical spectroscopy of viscoelastic materials with transient structure. J. Non-Newt. Fluid Mech. 27:17–26 (1988).

38. S.K. Venkatraman, H.H. Winter. Finite shear strain behavior of a crosslinking polydimethylsiloxane near its gel point. Rheol. Acta 29:423–432 (1990).

39. F.G. Mussatti, C.W. Macosko. Rheology of network forming systems. Polym. Eng. Sci. 13:236–240 (1973).

40. J.B. Enns, J.K. Gillham. Time-temperature–transformation (TTT) cure diagram: modeling the cure behavior of thermosets. J. Appl. Polym. Sci. 28:2567–2591 (1983).

41. S.L. Simon, J.K. Gillham. Thermosetting cure diagrams: calculation and application. J. Appl. Polym. Sci. 53:709–727 (1994).

16

Flow Through Porous Media

I. INTRODUCTION

The inertialess flow of a fluid through a porous medium is encountered extremely frequently in engineering applications. Common examples include filtration, packed-bed reactors, and enhanced oil recovery by water or polymer flooding. In the field of polymer processing, as part of the process of synthetic fiber formation, thermoplastic melts are pumped through a filter pack that may consist of sand, wire mesh, or steel spheres and whose purpose is to homogenize the temperature, degrade any polymer gels, and remove foreign matter that may plug the capillaries used to shape the fibers. In many polymer composites manufacturing operations, the proper flow of thermosetting polymeric fluids through fiber preforms is central to the success of the entire operation.

For example, in the resin transfer molding (RTM) process, shown schematically in Fig. 16.1, a cold prepolymer is injected into a heated mold, where it impregnates the fiber reinforcement contained in the mold; mold-filling times are of the order of minutes [1]. Curing of the polymer is initiated by heat transfer from the mold walls, and this results in the formation of the composite part. Since curing reactions are typically exothermic, resion viscosity initally decreases due to an increase in temperature, but the viscosity increases again as the polymer molecular weight increases. Consequently, curing should not begin much before the mold is filled or else premature polymer gelation can cause the resin to bypass some of the interstitial regions in the fiber preform and lead to the creation of voids and dry spots, which are undesirable.

Voids can be eliminated from the cured composite if the fiber preform is allowed to consolidate during processing. This is done in the autoclave bag mold-

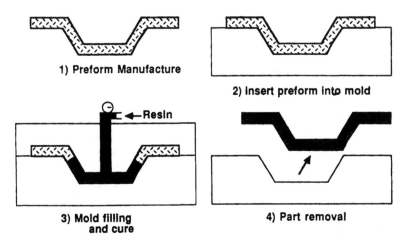

1) Preform Manufacture

2) Insert preform into mold

←—Resin

3) Mold filling and cure

4) Part removal

FIGURE 16.1   RTM process steps. (From Ref. 1, with permission from Elsevier Science.)

ing process, shown schematically in Fig. 16.2 [2]. Here the mold is filled with prepregs that are sheets containing continuous fibers embedded in a partially cured polymer matrix, typically an epoxy. A porous material called a bleeder is placed on top of the pregregs, and restraints are arranged around the prepregs to prevent lateral motion of the fibers and to generate essentially one-dimensional flow of the resin in the z direction. A plastic sheet or vacuum bag is placed around the entire assembly which is then subjected to an elevated time-varying (curing) temperature. At some time during the curing process a downward pressure is

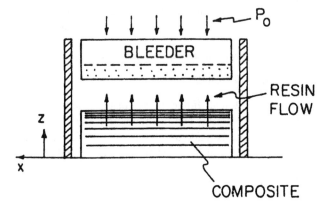

FIGURE 16.2   Schematic of the autoclave bag molding process. (Reprinted from Ref. 2, with permission from Technomic Publishing Co., Inc., copyright 1982.)

applied to squeeze excess resin out of the composite through the bleeder. This has the effect of consolidating the individual plies, eliminating air bubbles, and filling all voids with resin [2,3].

In both RTM and bag molding and, for that matter, in other, similar operations, such as compression molding, we wish to determine processing conditions that lead to uniform and complete curing of the polymer without the entrapment of gases or the occurrence of hot spots or voids, and we want to do this with as short a cycle time as possible. These objectives can be achieved only with the help of process models. A key input to these models is the mathematical relation between the average fluid velocity in the porous medium and the imposed pressure gradient causing the flow [1–3]. The output of the process models includes the mold filling pattern in RTM and the temperature, pressure, and extent of cure as a function of time and position in both RTM and bag molding; the resin viscosity is related to temperature and degree of cure, using the equations developed in Chap. 15.

The flow of fluids through porous materials has been considered by a very large number of authors in the past [4–8, for example, and references therein]. However, reliable answers have not always been forthcoming to the conceptually simple problem of predicting the relationship of flow rate to pressure drop for a given porous medium [9]. This is because the results often depend quite sensitively on both the pore structure and the fluid rheology, especially for polymeric fluids, for these are viscoelastic. Indeed, the observed pressure drop for a given flow rate of a polymeric fluid through a packed bed can, under appropriate conditions, be significantly greater than that of a purely viscous liquid having the same viscosity function [5,10–12]. Since neither the details of the pore structure of the medium nor the fluid constitutive equation are always known with any great certainty, we usually couple a specific model of the pore structure with a specific model of the fluid rheology in order to analyze the problem at hand. Before presenting details, though, it is useful to seek guidance from dimensional analysis.

II. DIMENSIONAL CONSIDERATIONS

Let us consider the flow of a Newtonian liquid of viscosity η and density ρ through an unconsolidated, isotropic porous bed of thickness L under the influence of a pressure drop Δp. If the bed is made up of spherical particles, each of diameter D, the variables that are likely to be relevant (and these are fluid, geometrical, and process variables) are [13]:

$$\eta, \rho, D, \varepsilon, \frac{|\Delta p|}{L} \text{ and } v$$

in which ε is the void fraction, or porosity, defined as the ratio of the void volume in the bed to the total volume; all the pores are assumed to be interconnected, allowing liquid to pass from one side to the other. Also, v is the superficial velocity, taken to be the ratio of the volumetric flow rate to the bed cross-sectional area. If the particles making up the bed are equiaxed but nonspherical, we can use the hydraulic radius concept to define an equivalent diameter; if there is a distribution of sizes, we can use an average diameter.

Since our list of variables has six quantities that have three dimensions among them, a total of three dimensionless groups can be formed, and it is readily determined that [13]:

$$\left(\frac{|\Delta p|}{\rho v^2}\right)\left(\frac{D}{L}\right) = \text{function of} \left(\frac{Dv\rho}{\eta}, \varepsilon\right) \tag{16.1}$$

where the three groups are identified to be a friction factor or the dimensionless pressure gradient, the Reynolds number, and the porosity.

If the fluid is non-Newtonian, other variables that characterize the flow behavior will also enter the problem. For a purely viscous, shear-thinning fluid, for example, we would include the consistency index and the power-law index in place of the viscosity in the list of variables. If the fluid, in addition, happens to be elastic, a fluid characteristic time or relaxation time, θ, will also need to be included. This last quantity can be made dimensionless with respect to D/v, which yields the Deborah number ($\theta v/D$). It is to be noted that in the literature most of the discussions of the influence of fluid elasticity on the pressure drop across a packed bed have been done with reference to the Deborah number.

It is well known that dimensional analysis can help only in identifying the relevant dimensionless groups but not in providing the functional relationship among them. This relationship can be determined either experimentally or analytically if one postulates a specific model. Even so, we would want to verify the results of the theoretical model by conducting experiments.

III. CAPILLARY MODEL

The simplest idealization of a porous medium is shown in Fig. 16.3: a solid containing cylindrical passages, each of diameter D_{eff}. The resistance to flow is taken to be the same as for the actual porous medium, and this implies that, for a given pressure drop, the volumetric flow rate is the same in the two cases. Consequently, the average velocity v_{eff} through each capillary is v/ε, where v is again the superficial velocity. Now, the velocity of a Newtonian liquid flowing through a tube of circular cross section is related to the pressure drop by means of the Hagen–Poiseuille equation as (see, for example, Ref. 13):

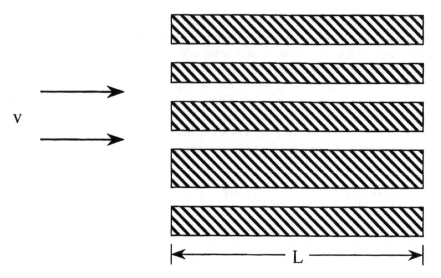

V

$$\longleftarrow \qquad\qquad L \qquad\qquad \longrightarrow$$

FIGURE 16.3 Capillary model of a porous bed.

$$v_{\text{eff}} = \frac{D_{\text{eff}}^{2}|\Delta p|/L}{32\eta L} \qquad\qquad (16.2)$$

and we need to relate D_{eff} to the geometry of the actual porous bed. This is done by requiring that the hydraulic diameter be the same in both cases. The hydraulic diameter is four times the volume available for flow divided by the total wetted surface area. For a porous bed composed of spheres of diameter D, the result is [13]:

$$D_{\text{eff}} = \frac{2\varepsilon D}{3(1 - \varepsilon)} \qquad\qquad (16.3)$$

Combining all of the foregoing, we obtain an expression for the superficial velocity in terms of known quantities:

$$v = \frac{D^{2}\varepsilon^{3}|\Delta p|/L}{72\eta(1 - \varepsilon^{2})} \qquad\qquad (16.4)$$

When both sides of Eq. (16.4) are multiplied by the bed cross-sectional area A, we obtain the following simple expression, called Darcy's law:

$$Q = \frac{kA|\Delta p|}{\eta L} \qquad\qquad (16.5)$$

in which Q is the volumetric flow rate and k is the permeability which is given by:

$$k = \frac{D^2\varepsilon^3}{72(1 - \varepsilon)^2} \tag{16.6}$$

The permeability is generally measured in units of darcy; if the permeability is 1 darcy, a flow rate of 1 cm^3/sec results for a 1-cP viscosity liquid on applying a pressure drop of 1 atm across a cube having sides of 1 cm. Since the surface area per unit volume, s, for a sphere is $6/D$, the permeability is often written as $\varepsilon^3/[s^2\kappa(1 - \varepsilon)^2]$, where κ is called the Kozeny constant; according to Eq. (16.6), $\kappa = 2$. It is clear that the permeability depends only on the properties of the porous medium and not on the properties of the fluid. Also note that Eq. (16.4) can be rewritten as:

$$D\varepsilon^3 \frac{|\Delta p|/L}{\rho v^2(1 - \varepsilon)} = \frac{72\eta(1 - \varepsilon)}{D\rho v} \tag{16.7}$$

where the left side of f_p, a modified friction factor called the packed-bed friction factor, and the right side is 72 divided by a modified Reynolds number Re$_p$ called the packed-bed Reynolds number. Thus,

$$f_p \text{Re}_p = \text{constant} \tag{16.8}$$

Clearly, Eq. (16.8) is a specific form of Eq. (16.1); it is known as the Blake–Kozeny equation. The predictions of this equation are compared in Fig. 16.4 with experimental data over wide ranges of both dimensionless groups. Data correlate well with the Ergun equation [14]:

$$f_p = \frac{150}{\text{Re}_p} + 1.75 \tag{16.9}$$

and the second term on the right can be neglected at low values of the packed-bed Reynolds number. Even so, the constant in Eq. (16.8) turns out to be 150 instead of the expected value of 72. This is explained by saying that the path of the typical fluid element is not straight but tortuous, because real pores are not isolated circular cylinders but a complicated network of interconnected passages. As a consequence, L in Eq. (16.7) has to be replaced by an equivalent length that exceeds L by a factor of 25/12. Some time ago, Kemblowski and Michniewicz [15] pointed out that data often correlate better with Eq. (16.8) if the constant in that equation has the value 180 rather than 150. They also provided theoretical arguments in support of the higher value. Indeed, data for molten polyethylene terephthalate in a shear rate range in which the viscosity is constant do tend to agree with the use of the value 180 [16]. A more detailed discussion of the strengths and weaknesses of the capillary model and a description of other mod-

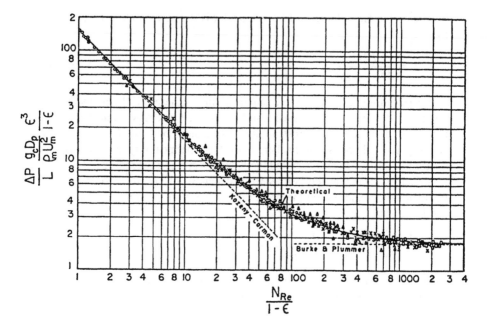

FIGURE 16.4 A comprehensive plot of pressure drop in fixed beds. (From Ref. 14. Reproduced with permission of the American Institute of Chemical Engineers.)

els, such as the submerged-object model, where flow through a packed bed is considered to be equivalent to flow around an assemblage of submerged objects, may be found in the book by Chhabra [17].

IV.   EFFECT OF A SHEAR-THINNING VISCOSITY

The steady-shear viscosity of most polymer melts decreases as the shear rate increases. This can be taken into account in the capillary model of a porous medium if we use the power-law model (see Chap. 3) to describe the fluid rheology:

$$\tau = K\dot{\gamma}^n \tag{16.10}$$

in which n has a value less than unity. The equivalent form of Eq. (16.2) now is [13]:

$$v_{\text{eff}} = \frac{n}{3n+1}\left[\left(\frac{D_{\text{eff}}}{2}\right)^{n+1}\frac{|\Delta p|}{2KL}\right]^{1/n} \tag{16.11}$$

which when combined with the previous definitions of v_{eff}, D_{eff}, and the permeability k yields [18]:

$$v = \left(\frac{k|\Delta p|}{HL}\right)^{1/n} \tag{16.12}$$

The quantity H appearing in Eq. (16.12) accounts for the additional dependence of v on k and ε due to shear thinning; it is given as:

$$H = \left(\frac{K}{12}\right)\left[\frac{(9n + 3)}{n}\right]^n (150k\varepsilon)^{(1-n)/2} \tag{16.13}$$

In terms of the packed-bed friction factor, Eq. (16.12) takes the form:

$$f_p \text{Re}' = 150 \tag{16.14}$$

where the modified Reynolds number Re$'$ is

$$\text{Re}' = \frac{Dv^{2-n}\rho}{H(1 - \varepsilon)} \tag{16.15}$$

and it is clear that we recover Eq. (16.7) from Eq. (16.14) if we set n equal to unity in Eq. (16.13).

Eq. (16.14) has been found to adequately describe the flow behavior of several polymer solutions through beds filled with materials such as sand and glass beads [18,19]. Additionally, it predicts the correct results for the flow of molten polymers such as polyethylene [20,21] and polypropylene [22] as well. It must be remembered, however, that Eq. (16.14) has been derived by assuming power-law behavior for the polymer viscosity. Since the power law is a good approximation over a limited range of shear rates only, it is necessary to check that the shear rates in the bed do lie in this range. For the flow of a power-law fluid through a long circular tube, the shear rate at the wall is given by Eq. (2.3) and is:

$$\dot{\gamma}_w = \frac{2(3n + 1)v_{eff}}{nD_{eff}} \tag{16.16}$$

in which v_{eff} has to be replaced by v/ε and D_{eff} has to be replaced by the expression in Equation (16.3). Thus,

$$\dot{\gamma}_w = \frac{3(3n + 1)v(1 - \varepsilon)}{n\varepsilon^2 D} \tag{16.17}$$

For a bed packed with material not in the shape of spheres, it is more convenient to replace D in Eq. (16.17) in terms of the permeability via Eq. (16.6). The result is

$$\dot{\gamma}_w = \frac{3(3n + 1)v}{n(150k\varepsilon)^{1/2}} \tag{16.18}$$

Since the permeability can be determined from results obtained using a Newtonian liquid, the shear rate range in the packed bed can be estimated with the help of Eq. (16.18).

In closing this section, we mention that Kemblowski and Michniewicz [23] have extended the foregoing analysis to the flow of fluids that obey the Carreau equation (see Chap. 3); this equation is more representative of the rheological behavior of polymer melts and solutions, in that it allows for a constant viscosity at low shear rates. Still, Duda et al. [24] have argued that capillary models are inadequate for the description of the flow of nonlinear, purely viscous solutions in porous media because these models do not account for the additional pressure drop that must arise in expansion and contraction regions of the flow field. A comparison between the various theoretical derivations and experimental data for shear-thinning fluids may be found in Ref. 25.

V. PERMEABILITY MEASUREMENT

In order to measure the permeability of a porous medium, and assuming the validity of Darcy's law, we need to carry out experiments on pressure drop versus flow rate using Newtonian liquids of known viscosity. These experiments may be conducted in either a transient or a steady manner; possible experimental schemes are shown in Fig. 16.5 [26]. Here it is useful to employ transparent flow cells so that the mold-filling pattern may also be observed. In Fig. 16.5a, liquid is injected at constant pressure or at constant flow rate into the center of the cell, and the flow is radially outwards. This flow field is meant to simulate mold filling during RTM, as illustrated earlier in Fig. 16.1; the mold is filled with several layers of fiber mat, each having a circular hole centered over the injection gate. In Fig. 16.5b, liquid flows at a fixed flow rate through a cylindrical porous bed in the axial direction; this geometry simulates autoclave bag molding or one-dimensional flow through an oil reservoir. Here one has to be careful to ensure that fluid does not channel between the cell wall and the edge of the fibrous reinforcement, or else the measured permeability will be larger than the true value.

The radial-flow experiment cannot be run at steady state. If the flow rate is held constant, the pressure at the injection point must increase with time, because the distance over which flow occurs increases progressively. Conversely, if the injection pressure is held fixed, the flow rate into the cell decreases monotonically, for the same reason. Thus, one has to measure the time dependence of either the pressure or the flow rate. Additionally, it is desirable to record the progress of the flow front as a function of time as well, and data analysis for

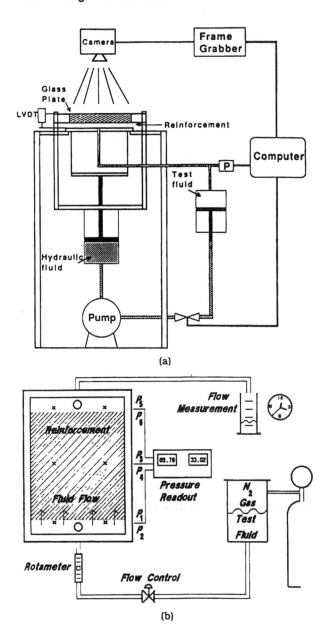

(a)

(b)

FIGURE 16.5 Schematic diagram of permeability measurement equipment: (a) radial (transient) flow; (b) one-dimensional (steady) flow (From Ref. 26.)

obtaining the permeability requires the simultaneous solution of Eq. (16.5) (in cylindrical coordinates) with the overall mass balance [1,27]. In the axial-flow experiment, of course, the pressure drop and the flow rate can be kept constant at the same time, and data analysis to give the permeability is trivial and follows directly from Eq. (16.5).

For an isotropic, homogeneous porous medium, such as a packed bed composed of particulates, both techniques of permeability measurement should give the same permeability value. Differences can, however, sometimes arise, and these are related to the fact that the porous medium in the transient experiment is unsaturated while that in the steady-state experiment is saturated; in the former case, pore impregnation and wetting effects in the vicinity of flow front can change the flow behavior and influence the results [1,26]. More seriously, though, a porous medium may be homogeneous but nonisotropic. In this instance, the permeability may be different in different directions. A flow front whose shape is elliptical rather than circular during radial flow is a clear indication of a nonisotropic porous medium [26]. Such a situation is easy to visualize for fibrous porous media that are relevant to composites processing—as discussed in the next section, the permeability of a fiber bundle in the direction of the fibers can be quite different from the permeability perpendicular to the fibers.

VI. FIBROUS POROUS MEDIA

When we consider resin flow through the fiber bundle making up a prepreg in Fig. 16.2, we find that the permeability for transverse flow, i.e., flow perpendicular to the fibers, approaches zero much before the porosity approaches zero. This is contrary to the predictions of Eq. (16.6), and it happens at low porosities when the solids volume fraction reaches its maximum packing fraction. Nonetheless, permeability for flow along the aligned fibers is finite since parallel flow is clearly possible. At the other extreme of high porosities ($\varepsilon > 0.8$), the capillary model is again geometrically incorrect, and we really have flow around dispersed solid particles [28]. As a consequence, and in order to make the permeability directional, it is necessary to write the permeability as a matrix K and to determine the numerical value of the different elements by solving the Navier–Stokes equation for flow through a periodic or random array of cylinders. Eq. (16.5) is now rewritten as

$$v = -K \frac{\nabla p}{\eta} \qquad (16.19)$$

in which v is the superficial velocity and ∇p is the pressure gradient. If the porous medium is homogeneous, i.e., if the structure does not vary from position to position, the K matrix can be made diagonal by picking one axis of the coordinate system to lie in the fiber direction; in this situation, the permeability is called *orthotropic* [1]. In practical terms, the fluid flow problem has to be solved for flow along the fibers and for flow perpendicular to the fibers. It needs to be emphasized, though, that while the permeability can no longer be obtained with the help of Eq. (16.6), Darcy's law remains valid for the flow of Newtonian liquids through fibrous porous media; this has been verified by experiment [29–30, for example].

A number of authors have solved for the flow resistance offered by an array of cylinders to the axial or transverse flow of a Newtonian liquid [1,31–34 and references therein]. The calculations have been done as a function of both porosity and the packing geometry and in general have employed numerical techniques. At low porosity values, though, analytical solutions have been obtained using a lubrication approach. A typical flow geometry is illustrated in Fig. 16.6; here, transverse flow through a square array of cylinders is considered. The corresponding results, both theoretical and experimental, for the Kozeny constant are shown in Figure 16.7a [33]; there is good agreement between data (filled circles and unfilled squares) and calculations. Results for axial flow are given in Fig. 16.7b; the single datum point in the figure is for a square array, and it agrees with the theoretical results. Since the permeability depends inversely on the Kozeny constant, it is clear that axial flow is preferred over transverse flow in a fibrous porous medium.

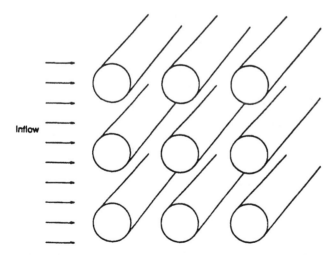

FIGURE 16.6 Flow across a regular array of cylinders (square packing arrangement). (From Ref. 34a.)

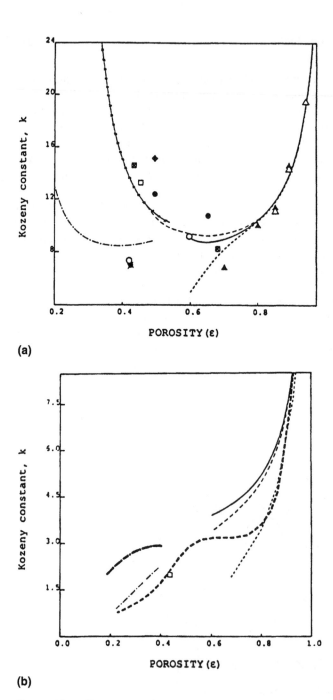

FIGURE 16.7 Experimental and theoretical results for the flow of a Newtonian liquid through aligned cylinder arrangements. (a) transverse flow; (b) axial flow. (From Ref. 33.)

Once the components of the permeability tensor (matrix) have been determined in a coordinate system aligned with the fibers, the permeability for flow in any other direction can be computed by matrix transform methods [1]. In polymer composites processing, though, we frequently stack together a number of fiber mats, each having a different fiber orientation. The permeability tensor of such a combination of layers, which may even be composed of different materials, can be obtained by proper averaging over the various orientations and permeabilities of the individual layers, as shown by Advani et al. [1]. These same authors have also provided results for the permeability of power-law fluids flowing across a regular array of cylinders [34a]. Corresponding results for Carreau model fluids are also available [34b].

VII. PORE GEOMETRY EFFECTS

Although the capillary model used for the particulate porous media and flow across cylinders for fibrous media capture the essential physics of the respective situations, both models can be refined to even better represent reality. In the case of the capillary model, a popular modification has been to use periodically constricted tubes; as shown in Fig. 16.8, the capillary cross section remains circular, but the diameter of the tube or tube bundle varies periodically with axial position. By employing this geometry it is hoped to account for the two-dimensional nature of the true velocity profile—the presence of a nonzero radial velocity resulting from the converging-diverging nature of actual pores. While this modification makes the assumed flow geometry more realistic, it also implies that solutions have to be obtained numerically, even for Newtonian liquids [35,36]. As far as experimental results are concerned, Dullien and Azzam found that, for creeping flow of Newtonian liquids, Darcy's law was valid even for flow through tubes consisting of short, alternating segments of two different diameters [37]. Deiber and Schowalter came to the same conclusion based on data obtained using a tube with sinusoidal axial variations in diameter [38]. The only change in going from a straight to a wavy tube is a change in the numerical value of the permeability. That this should be so becomes evident from a rearrangement of Eq. (16.2).

$$-\frac{dp}{dx} \propto \frac{Q}{D^4} \tag{16.20}$$

where dp/dx is the pressure gradient and Q is the volumetric flow rate. It is reasonable to expect that Eq. (16.20) will hold locally even for a tube having axial diameter variations, provided that the diameter changes gradually. In such a case, the pressure drop over a length L is [9]:

$$\Delta p \propto Q \int_0^L \frac{dx}{D^4} \tag{16.21}$$

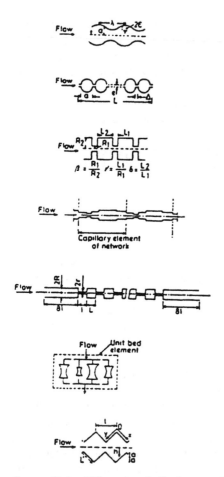

FIGURE 16.8 Different periodically constricted tubes employed to model the geometry of flow through a packed bed. (From Ref. 17. Copyright CRC Press, Boca Raton, Florida.)

and this depends strongly on the diameter–distance relationship. By comparing Eq. (16.5) and (16.20) we find that the flow rate through both the straight and wavy tubes is proportional to the pressure drop, but the constant of proportionality, or equivalently the permeability, depends on the D–x relationship in the case of the periodically constricted tube. This is exactly what is revealed by the experimental data.

Turning to models of fibrous porous media, we have seen that the flow across cylinders is a good representation of a homogeneous porous medium. Many porous media of practical interest, however, are heterogeneous—the me-

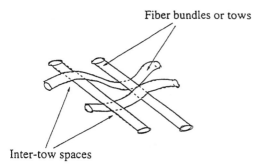

Fiber bundles or tows

Inter-tow spaces

FIGURE 16.9   Arrangement of fiber bundles in a bidirectional fiber mat. (From Ref. 39.)

dium porosity varies with position in a random or periodic manner. An example of this is the preform used in resin transfer molding and illustrated in Fig. 16.9. The preform is made by weaving or stitching together a fiber bundle, or *tow*. Thus, not only is the preform porous, but the fiber bundle from which it is constructed is also porous. However, the bundle porosity is significantly less compared to the preform porosity. This leads to a dual-scale porous medium whose permeability cannot be predicted using Fig. 16.7. Conceptually, a dual-scale porous medium can be obtained from a single-scale fibrous porous medium by gathering together some of the fibers into bundles [39]. This opens up large passages for fluid to flow through; and the net result is that, at the same overall porosity, the permeability of the dual-scale medium can be much larger than the permeability of the single-scale medium, especially at low porosities. Permeability calculations for idealized dual-scale porous media may be found in Ref. 39. Sometimes though, when working with fiber reinforcements that do not resemble model porous media, the only recourse may be to actually measure the preform permeability.

VIII.   VISCOELASTIC EFFECTS

As was seen in Chap. 7, polymer solutions and polymer melts exhibit a resistance to extension that can be orders of magnitude greater than the resistance to shearing. For the flow of polymeric fluids through a porous medium, this can translate into a pressure drop that is larger than the value expected based on Eq. (16.14) when it is realized that a porous medium consists of alternate converging and diverging sections in which the deformation field is a combination of shear and extension. As a result of the extensional-flow component, the product of the packed-bed friction factor with the packed-bed Reynolds number can increase quite significantly with increasing volumetric flow rate. This is shown in Fig. 16.10, taken from the work of Marshall and Metzner [11]; the Reynolds number here, N_{RE}, is the quantity defined in Eq. (16.5) divided by 150. The porous medium employed was a sintered bronze disk; the three polymeric fluids were chosen to obtain a wide variation in the degree of viscoelasticity. The elastic nature of the fluids was quantified with the help of the Deborah number De, defined as $\theta v_{eff}/D$, or the ratio of the fluid relaxation time to the characteristic process time. Had Eq. (16.14) been valid, the ordinate in Fig. 16.10 would have had a value of unity regardless of the value of the flow rate or the Deborah number. Instead, at the highest value of the Deborah number examined, the friction factor is as much as 20 times larger than the purely viscous value. This is an immense effect, and it has been repeatedly verified for the flow of high-molecular-weight polymer solutions through granular media, even at extremely low polymer concentrations [10–12, 40–41 and reference therein]. For polymer melts, on the other hand, viscoelastic effects are either negligible [15–16, 20–23] or not as dramatic [22] as in the case of polymer solutions. Consequently, polymer viscoelasticity may not be much of a concern during composites processing operations such as RTM and compression molding.

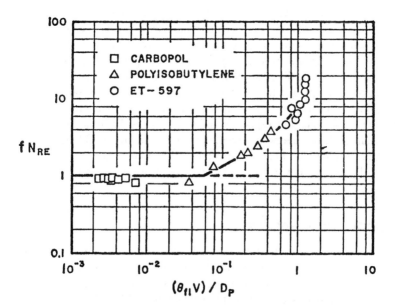

FIGURE 16.10 Dependence of viscoelastic effects on the Deborah number of the flow process. (Reprinted with permission from Ref. 11. Copyright 1967 American Chemical Society.)

In terms of correlating data, f_pN_{RE} is usually plotted as a function of the Deborah number, and the graph often follows the relation proposed by Wissler [42]:

$$f_pN_{RE} = 1 + ADe^2 \qquad (16.22)$$

However, widely different values of the constant A are observed in practice [22, 43–44]. In addition, as the Deborah number increases, f_pN_{RE} does not increase without bound as implied by Eq. (16.22); in fact, maxima are frequently observed [12,45–46]. This has been found to be true even when data have been generated using well-characterized polymer solutions and well-characterized porous media. Although the lack of data superposition in some cases [obtaining different values of A in Eq. (16.22)] may be the result of how the fluid relaxation time is measured or how the characteristic process time is defined, there is an important interaction between the pore geometry and fluid viscoelasticity that is not accommodated by the use of the Deborah number alone [47]. The essential argument is that flow through a porous medium is a transient one in which steady-state stress levels are never attained. Stress levels can be increased by increasing the deformation rate, which can come about by increasing the flow rate, but this reduces the residence time of a fluid element in the system. Since a polymeric fluid has to be deformed for some time to allow the stresses to rise, we have two competing, coupled effects. At low flow rates, the deformation rate is too low for significant stress growth and no viscoelastic effects are observed. At high flow rates, although the deformation rate is high, the residence time is too short to allow for stress buildup. It is only in an intermediate range of flow rates that significant viscoelastic effects occur, and this explains the observed maxima [47]. Finally, the height of the maximum depends on the geometry, because the geometry can govern whether the deformation rate effect will overshadow the residence time effect or vice versa.

In order to better understand the phenomenon of viscoelasticity in particulate porous media, several investigators have used idealized porous media, for which experimental results could be compared with theoretical predictions computed using reasonable fluid constitutive equations. The flow geometries employed include converging channels, tubes consisting of short, alternating segments of two different diameters, and tubes having sinusoidal axial variations in diameter. From these studies, a number of clear results have emerged. As with real porous media, "shear thickening" is observed for the flow of dilute polymer solutions, and distinct maxima are obtained [48–50]. Also, the extent of flow resistance depends on the geometry—for wedge-shaped channels there is no shear thickening [50], while for conical channels the apparent shear viscosity is several times the true value [48–50]; this surprising difference can be explained, at least qualitatively, by calculations done using the upper convected Maxwell rheological equation [51]. There is also quantitative agreement between computations done with the Oldroyd B model for flow through a corrugated tube [52–54] and corresponding experiments utilizing constant-viscosity elastic fluids [54–55], but in this case shear thickening is neither predicted nor observed over a shear rate range in which the experimental fluids behave like Oldroyd B liquids. Very large elastic effects, though, can be measured in this flow geometry for highly viscoelastic (but shear-thinning) liquids [55,56], and such effects can also be calculated using appropriate constitutive equations [53].

The conclusions reached about viscoelastic effects in granular porous media also carry over to fibrous porous media. Transverse permeabilities measured for constant-viscosity elastic polymer solutions flowing across ordered arrays of cylinders are substantially smaller than those of Newtonian liquids of the same viscosity [29,30]. This is equivalent to saying that these solutions shear-thicken or show a flow resistance that is greater than that of comparable Newtonian fluids. This is demonstrated in Fig. 16.11 for the flow of three different constant-viscosity, ideal elastic liquids through a square array of cylinders [30]. It is seen that deviations from Newtonian behavior arise at the same value of the Deborah number for all the fluids; but the larger the polymer molecular weight, the larger is the maximum in the flow resistance. This is related to the fact that the ratio

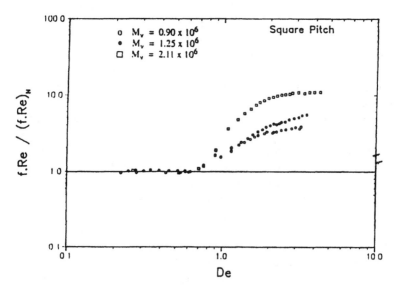

FIGURE 16.11 Normalized fRe product versus Deborah number for the flow of dilute polyisobutylene-in-polybutene solutions flowing across a square array of cylinders. (From Ref. 30.)

of extensional viscosity to shear viscosity of polymer solutions increases with increasing molecular weight. It is also found that the flow resistance is greater when a triangular arrangement is used instead of a square arrangement; the triangular geometry results in greater polymer chain extension [29]. Similar results are obtained for the flow of shear-thinning viscoelastic polymer solutions [30,57].

IX. CONCLUDING REMARKS

For a given flow rate of a polymeric fluid and as a result of elastic effects, the pressure drop across a porous medium can differ significantly from that calculated on the basis of either Eq. (16.5) or Eq. (16.14) and their suitable modifications for fibrous media. However, this deviation is important mainly for the flow of high-molecular-weight polymer solutions and not for the creeping flow of polymer melts. Thus, unless viscoelastic effects are suspected to be important, Eq. 16.5 and 16.14 may be the best design equations to use at the present time. Note that we have considered only the flow of fluids through saturated porous media in this chapter. Effects related to pore wettability can be important when one fluid is used to displace another fluid from the porous medium. We have also not considered other complicating factors, such as blind or deadend pores, polymer degradation, and polymer adsorption. These are considered elsewhere [7].

REFERENCES

1. S.G. Advani, M.V. Bruschke, R.S. Parnas. Resin transfer molding flow phenomena in polymeric composites. In: S.G. Advani, ed. Flow and Rheology in Polymer Composites Manufacturing. Elsevier, Amsterdam, 1994, pp. 465–515.
2. G.S. Springer. Resin flow during the cure of fiber reinforced composites. J. Composite Materials 16:400–410 (1982).
3. A.C. Loos, G.S. Springer. Curing of epoxy matrix composites. J. Composite Materials 17:135–169 (1983).
4. R.E. Collins. Flow of Fluids Through Porous Materials. Van Nostrand-Reinhold, Princeton, NJ, 1961.
5. J.G. Savins. Non-Newtonian flow through porous media. Ind. Eng. Chem. 61:18–47 (1969).
6. A.E. Schiedegger. Physics of Flow Through Porous Media. 3rd ed. University of Toronto Press, Toronto, 1974.
7. F.A.L. Dullien. Porous Media: Fluid Transport and Pore Structure. Academic Press, New York, 1979.
8. R.A. Greenkorn. Flow Phenomena in Porous Media. Marcel Dekker, New York, 1983.
9. A.B. Metzner. Flows of polymeric solutions and emulsions through porous media—current status. In: D.O. Shah, R.S. Schechter, eds. Improved Oil Recovery by Surfactant and Polymer Flooding. Academic Press, New York, 1977, pp. 439–451.

10. D.L. Dauben, D.E. Menzie. Flow of polymer solutions through porous media. J. Pet. Tech. 19:1065–1073 (1967).

11. R.J. Marshall, A.B. Metzner. Flow of viscoelastic fluids through porous media. Ind. Eng. Chem. Fundam. 6:393–400 (1967).

12. D.F. James, D.R. McLaren. The laminar flow of dilute polymer solutions through porous media. J. Fluid Mech. 70:733–752 (1975).

13. M.M. Denn. Process Fluid Mechanics. Prentice-Hall, Englewood Cliffs, NJ, 1980.

14. S. Ergun. Fluid flow through packed columns. Chem. Eng. Progr. 48:89–94 (1952).

15. Z. Kemblowski, M. Michniewicz. A new look at the laminar flow of power law fluids through granular beds. Rheol. Acta 18:730–739 (1979).

16. F.C. Wampler, D.R. Gregory. Flow of molten poly(ethylene terephthalate) through packed beds of glass beads. AIChE J. 18:443–445 (1972).

17. R.P. Chhabra. Bubbles, Drops, and Particles in Non-Newtonian Fluids. CRC Press, Boca Raton, FL, 1993, pp. 217–297.

18. R.H. Christopher, S. Middleman. Power-law flow through a packed tube. Ind. Eng. Chem. Fundam. 4:422–426 (1965).

19. N.Y. Gaitonde, S. Middleman. Flow of viscoelastic fluids through porous media. Ind. Eng. Chem. Fundam. 6:145–147 (1967).

20. D.R. Gregory, R.G. Griskey. Flow of molten polymers through porous media, AIChE J. 13:122–125 (1967).

21. N. Siskovic, D.R. Gregory, R.G. Griskey. Viscoelastic behavior of molten polymers in porous media. AIChE J. 17:281–285 (1971).

22. Z. Kemblowski, M. Dziubinski. Resistance to flow of molten polymers through granular beds. Rheol. Acta 17:176–187 (1978).

23. Z. Kemblowski, M. Michniewicz. Correlation of data concerning resistance to flow of generalized Newtonian fluids through granular beds. Rheol. Acta 20:352–359 (1981).

24. J.L. Duda, S.A. Hong, E.E. Klaus. Flow of polymer solutions in porous media: Inadequacy of the capillary model. Ind. Eng. Chem. Fundam. 22:299–305 (1983).

25. W. Kozicki, C. Tiu. A unified model for non-Newtonian flow in packed beds and porous media. Rheol. Acta 27:31–38 (1988).

26. R.S. Parnas, A.J. Salem. A comparison of the unidirectional and radial in-plane flow of fluids through woven composite reinforcements. Polym. Compos. 14:383–394 (1993).

27. K.L. Adams, W.B. Russel, L. Rebenfeld. Radial penetration of a viscous liquid into a planar anisotropic porous medium. Int. J. Multiphase Flow 14:203–215 (1988).

28. L. Skartsis, J.L. Kardos, B. Khomami. Resin flow through fiber beds during composite manufacturing processes. Part 1: Review of Newtonian flow through fiber beds. Polym. Eng. Sci. 32:221–230 (1992).

29. C. Chmielewski, C.A. Petty, K. Jayaraman. Crossflow of elastic liquids through arrays of clinders. J. Non-Newt. Fluid Mech. 35:309–325 (1990).

30. C. Chmielewski, K. Jayaraman. The effect of polymer extensibility on crossflow of polymer solutions through cylinder arrays. J. Rheol. 36:1105–1126 (1992).

31. A.S. Sangani, A. Acrivos. Slow flow past periodic arrays of cylinders with application to heat transfer. Int. J. Multiphase Flow 8:193–206 (1982).

32. A.S. Sangani, C. Yao. Transport processes in random arrays of cylinders. II. Viscous flow. Phys. Fluids 31:2435–2444 (1988).
33. L. Skartsis, B. Khomami, J.L. Kardos. Resin flow through fiber beds during composite manufacturing processes. Part II. Numerical and experimental studies of Newtonian flow through ideal and actual fiber beds. Polym. Eng. Sci. 32:231–239 (1992).
34a. M.V. Bruschke, S.G. Advani. Flow of generalized Newtonian fluids across a periodic array of cylinders. J. Rheol. 37:479–498 (1993).
34b. A. Tripathi, R.P. Chhabra. Transverse laminar flow of non-Newtonian fluids over a bank of long cylinders. Chem. Eng. Commun. 147:197–212 (1996).
35. A.C. Payatakes, C. Tien, R.M. Turian. A new model for granular porous media: Part I. Model formulation. AIChE J. 19:58–67 (1973).
36. A.C. Payatakes, C. Tien, R.M. Turian. Part II. Numerical solution of steady incompressible Newtonian flow through periodically constricted tubes. AIChE J. 19:67–76 (1973).
37. F.A.L. Dullien, M.I.S. Azzam. Flow rate–pressure gradient measurements in periodically nonuniform capillary tubes. AIChE J. 19:222–229 (1973).
38. J.A. Deiber, W.R. Schowalter. Flow through tubes with sinusoidal axial variations in diameter. AIChE J. 25:638–645 (1979).
39. K.M. Pillai. Flow modeling in dual scale porous media. Ph.D. dissertation, University of Delaware, Newark, DE, 1997.
40. G. Chauveteau. Fundamental criteria in polymer flow through porous media. In: J.E. Glass, ed. Water-Soluble Polymers. Adv. Chem. Series 213. ACS, Washington, DC, 1986, pp. 227–267.
41. S. Flew, R.H.J. Sellin. Non-Newtonian flow in porous media—a laboratory study of polyacrylamide solutions. J. Non-Newt. Fluid Mech. 47:169–210 (1993).
42. E.H. Wissler. Viscoelastic effects in the flow of non-Newtonian fluids through a porous medium. Ind. Eng. Chem. Fundam. 10:411–417 (1971).
43. S. Vossoughi, F.A. Seyer. Pressure drop for flow of polymer solution in a model porous medium. Can. J. Chem. Eng. 52:666–669 (1974).
44. I. Machac, V. Dolejs. Flow of viscoelastic liquids through fixed bed of particles. Chem. Eng. Commun. 18:29–37 (1982).
45. M. Barboza, C. Rangel, B. Mena. Viscoelastic effects in flow through porous media. J. Rheol. 23:281–299 (1979).
46. W.M. Kulicke, R. Haas. Flow behavior of dilute polyacrylamide solutions through porous media. 1. Influence of chain length, concentration, and thermodynamic quality of the solvent. Ind. Eng. Chem. Fundam. 23:308–315 (1984).
47. R.K. Gupta, T. Sridhar. Viscoelastic effects in non-Newtonian flows through porous media. Rheol. Acta 24:148–151 (1985).
48. D.F. James, J.H. Saringer. Extensional flow of dilute polymer solutions. J. Fluid Mech. 97:655–671 (1980).
49. D.F. James, J.H. Saringer. Flow of dilute polymer solutions through converging channels. J. Non-Newt. Fluid Mech. 11:317–339 (1982).
50. D.F. James, J.H. Saringer. Planar sink flow of a dilute polymer solution. J. Rheol. 26:321–325 (1982).
51. A.K. Chakraborty, A.B. Metzner. Sink flows of viscoelastic fluids. J. Rheol. 30:29–41 (1986).

52. S. Pilitsis, A.N. Beris. Calculations of steady-state viscoelastic flow in an undulating
     tube. J. Non-Newt. Fluid Mech. 31:231–287 (1989).
53. R. Zheng, N. Phan-Thien, R.I. Tanner, M.B. Bush. Numerical analysis of viscoelas-
     tic flow through a sinusoidally corrugated tube using a boundary element method.
     J. Rheol. 34:79–102 (1990).
54. D.F. James, N. Phan-Thien, M.M.K. Khan, A.N. Beris, S. Pilitsis. Flow of test fluid
     M1 in corrugated tubes. J. non-Newt. Fluid Mech. 35:405–412 (1990).
55. S. Huzarewicz, R.K. Gupta, R.P. Chhabra. Elastic effects in flow through sinuous
     tubes. J. Rheol. 35:221–235 (1991).
56. J.A. Deiber, W.R. Schowalter. Modeling the flow of viscoelastic fluids through po-
     rous media. AIChE J. 27:912–920 (1981).
57. L. Skartsis, B. Khomami, J.L. Kardos. Polymeric flow through fibrous media, J.
     Rheol. 36:589–620 (1992).

17

Melt Fracture

I. INTRODUCTION

In this book, for the most part, we have considered those situations that involve steady flow. When we did consider time-dependent flow, the time dependence was deliberately imposed, as during the measurement of dynamic mechanical properties in Chap. 6. Often, however, the time dependence is both unintentional and unwelcome. In particular, a great many instabilities have been found to arise during polymer processing, and the topic has been reviewed on a number of occasions [1-3, for example]. While some of these instabilities are observed with Newtonian and other purely viscous liquids, others are unique to elastic liquids. Some occur in confined flows, others are free surface instabilities. Collectively, the different instabilities can serve to limit production rates in industrial operations and to render unsuitable viscometric measurements made for the purpose of fluid characterization [4]. Examples of unstable Newtonian fluid behavior include (1) the phenomenon of draw resonance, which arises during fiber or film manufacture by extrusion and subsequent drawdown in cross-sectional area by the use of high-velocity rollers, (2) the formation of ribs on rotating rollers during coating operations, and (3) the occurrence of secondary flows, such as Taylor vortices in rotational Couette flow. In viscoelastic fluids, the addition of fluid elasticity can modify either the point of onset of the instability or the nature of the instability itself. Concerning purely elastic instabilities, one finds, for example, that upon extruding polymer melts or solutions through a circular die at slow to moderate extrusion speeds, the polymer emerges as a smooth extrudate of uniform diameter. As the extrusion speed or shear rate is increased, a critical value is reached above which the extrudate appears distorted in a periodic or random manner. This behavior is generally undesirable, and it is known as *melt fracture*; it may also be accompanied by surges in the flow rate. The phenomenon, though, is very general and is independent of the shape of the die, for it has been observed with slits and tubes of noncircular cross section as well. In addition, a major unifying feature of all these observations is that fracture is an elastic phenomenon, since it has not been reported for Newtonian liquids. This points to the role of fluid rheology in giving rise to secondary flows and unstable flows.

The mathematical process of determining when a particular flow is likely to become unstable, at least to small disturbances, is conceptually straightforward [5]. Indeed, there is good agreement between observations and theoretical hydro-dynamic instability predictions when the analysis is carried out assuming isother-mal, Newtonian fluid behavior, as in the case of draw resonance during fiber spinning [6,7]. The difficulty arises in extending these analyses to non-Newtonian fluids, mostly because polymeric liquids do not faithfully follow any one postu-lated rheological constitutive equation; in addition there are questions concerning the boundary condition that should be employed [8]. This is nowhere more true than in the case of melt fracture, and in this chapter we focus attention on melt fracture alone. This is in recognition both of the importance of the phenomenon to extrusion that has resulted in an enormous literature on the topic and of recent progress aimed at explaining the origins of this instability. What has emerged is a fascinating interplay between fluid rheology and fluid adhesion that depends on the surface characteristics of the die. The result has been the opening up of the possibility of delaying the onset of melt fracture to significantly higher flow rates than was hitherto thought possible.

II. HISTORICAL PERSPECTIVE

The literature on melt fracture dates back more than 50 years, and observations made during the extrusion of polymeric liquids through a circular die or capillary have been described succinctly by Middleman [2]; some of the different types of extrudate distortion that can occur are sketched in Fig. 17.1. (Very instructive photographs of actual melt fracture and other instabilities may be found in the book by Boger and Walters [9].) Note that melt fracture is observed for both polymer melts and polymer solutions [10], and it is often, but not always, pre-ceded by noticeable surface roughness, known as matte, orange peel, or shark-skin. Also, in some cases there may be a second stable region at flow rates higher than those that trigger the initial onset of melt fracture [11,12]. In general, two broadly different trends have been noted [1]. For one group of polymers, which incudes high-density polyethylene (HDPE), the onset of melt fracture is charac-terized by a discontinuity in the flow curve, and this suggests the possibility of a stick-slip phenomenon leading to flow rate or pressure oscillations. Further, the severity of melt fracture is increased by increasing the length of the capillary.

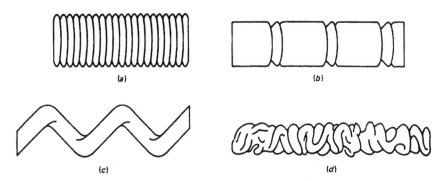

FIGURE 17.1 Examples of melt fracture. (From Ref. 2, with permission of The McGraw-Hill Companies.)

Additionally, these polymers fill the entire entry angle in the reservoir upstream of the capillary. As opposed to this, another group of polymers, whose unstable behavior can be represented by low-density polyethylene (LDPE), shows no discontinuity in the flow curve. Here, melt fracture severity is increased by shortening the capillary. Also, these materials flow into the capillary in a "wine glass" manner; flow birefringence measurements reveal that the corners contain fluid that recirculates in an unsymmetrical manner once the instability sets in. Furthermore, the amplitude of surface distortions can be reduced by using a tapered entry. All this suggests that the instability is initiated in the die entry and the disturbance decays during its passage to the die exit.

Since HDPE is a linear polymer and LDPE is branched, it would seem that the differences between the two sets of behavior are the result of chain branching. However, there are enough polymers, such as polystyrene and polypropylene, whose fracture characteristics are not in accord with such a conclusion. Nonetheless, it is accepted that melt fracture is a die-entry or a die-land effect, and attention must be focused on these two areas if we want to understand the phenomenon further. By contrast, matte or sharkskin is considered to arise from the presence of excessive local tensile stresses at the die exit, where the polymer separates from the die [13]. Kurtz has also observed that a change in the slope of the curve of shear stress versus shear rate correlates with the onset of sharkskin, especially in linear low-density polyethylene [14].

There are fewer studies involving polymer solutions as compared to those involving polymer melts. However, the fracture behavior of a wide variety of polymers has been investigated, and this has typically been done by observing flow patterns upstream of the die. It has been found that elastic fracture of polymer solutions is similar to that of LDPE melts [11], and the onset of fracture can be delayed by decreasing the polymer concentration, decreasing the polymer molecular weight at a fixed mass concentration, increasing the temperature, or decreasing the entry angle. The strong influence of the entry angle suggests that, at least for some polymers, elastic fracture is related to the build-up of tensile stresses in the converging region at the entrance to the die [15]. Fracture occurs in a manner similar to the breaking of a rubber band when this stress exceeds the rupture stress. When fluid fractures, the oriented molecules cause the polymer to snap back to an unoriented state. The orientation must then build up before fracture can take place again. This periodic fracture thus produces periodic variations in the appearance of the extruded strands.

This overly simplified picture enables us to understand many of the effects observed when the experimental conditions are changed. For example, tensile stresses result when the size of the flow channel goes from a larger to a smaller cross section. The change in cross section need not be abrupt but can be a gradual taper. The less drastic the change in cross section or the smaller the taper, the smaller the stretch rate and the less is the tensile stress. Similarly, increasing the temperature raises the onset of melt fracture to higher values of the deformation rate because higher temperatures require higher deformation rates in order to reach a critical fracture stress. The influence of a molecular weight and concentration in the case of polymer solutions can be understood in the same manner.

An observation that has frequently been made with polymer melts is that fracture occurs when the shear stress at the wall reaches a value of about 10^5 Pa [1,2]. This, however, does not carry over to polymer solutions. Some polystyrene-in-benzene solutions, for example, fracture at wall shear stress values that are two orders of magnitude smaller than 10^5 Pa [16], while other solutions fail to show fracture even when the wall-shear stress exceeds this value [10]. It would seem that the appropriate parameter for correlating the onset of melt fracture should be a dimensionless group that includes a measure of fluid elasticity. Since all viscoelastic fluids possess a relaxation time, θ, a logical dimensionless group is the Weissenber number, $U\theta/D$, where U and D are, respectively, the characteristic velocity and characteristic length. Another possible group is the recoverable shear, which is defined in a viscometric flow to be half the ratio of the primary normal stress difference to the shear stress. If one uses Eq. (5.5) and identifies U/D with the shear rate, one can show that these two groups are identical to each other [10]. A theoretical stability analysis utilizing the upper-convected Maxwell equation predicts that the pipe flow of a viscoelastic liquid becomes unstable to two-dimensional disturbances at a recoverable shear of 2.6 [17]. Also, most published experimental studies place the critical value of the recoverable shear for the onset of melt fracture at between 1 and 10 [1]; this criterion appears to work for both polymer melts and polymer solutions [1,10]. Note that the critical value of the recoverable shear usually reduces as the polymer becomes a more monodisperse [8]. Also note that the existence of this correlation sheds no light on the reason why a polymeric fluid fractures; it merely says that fluid elasticity is implicated in the phenomenon.

III. TYING MELT FRACTURE TO WALL SLIP

Motivated by the need to eliminate melt fracture that occurred at relatively low deformation rates in the commercially important, narrow-molecular-weight-distribution, linear low-density polyethylenes, Ramamurthy conducted an extensive series of capillary viscometer experiments directed at establishing the presence or absence of slip at the polymer/metal interface [18]. Volumetric flow rate measurements made with capillaries of different radii but under conditions of constant wall-shear stress and plotted according to Eq. (9.5) are shown in Fig. 17.2. It is clear that beyond a critical value of the wall-shear stress, wall slip is present since the slope of plot of the apparent wall-shear rate versus the reciprocal radius is nonzero; as indicated in the figure, the slip velocity increases with increasing wall-shear stress beyond this critical value. Ramamurthy found that the onset of wall slip correlated fairly well with the onset of melt fracture showing that the two phenomena are inextricably linked to each other. Since wall slip, or lack of adhesion to the wall, should depend on the surface characteristics of the die wall, Ramamurthy also investigated the influence of materials of construction on the onset of melt fracture. However, little effect was noticed when experiments were conducted using capillaries made from a number of ferrous and nonferrous metals. Surprisingly, though, when different metal inserts were employed for the die land region in a tubular blown film die (see Fig. 17.3), the onset of melt fracture in this geometry could be promoted, significantly delayed, or prevented altogether! The use of titanium nitride coatings resulted in the most severe melt fracture, while alloys of copper and zinc containing up to 40% zinc exhibited no melt fracture at all after the passage of an induction time period. During this induction time, zinc was lost from the surface of the brass, which turned noticeably red.

Based on the preceding results, Ramamurthy concluded that melt fracture in linear polyethylenes was a consequence of the breakdown of polymer/metal adhesion in the die land region. He also demonstrated that the use of additives such as fluoroelastomers could eliminate melt fracture in a blown film die made from conventional chrome-plated steel; the additives migrate to the die wall, coat the metal surface, and act in the same way as inserts. Since the publication of these results, several researchers have verified that molten polymers do slip at viscometer walls. Hatzikiriakos and Dealy, for example, used a piston-driven capillary rheometer to extrude a high-density polyethylene melt through a stainless steel die [19]. When slip velocities were computed, two distinct regions of behavior were observed; the slip velocity increased discontinuously at a particular

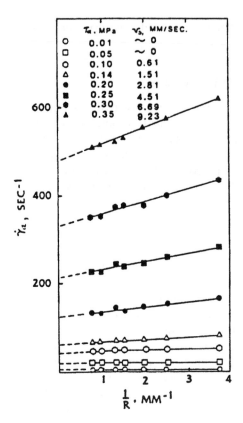

FIGURE 17.2 Wall slip measurements for a linear low-density polyethylene at 220°C. (From Ref. 18.)

wall-shear stress. The lower-stress region corresponded to the occurrence of sharkskin; the higher-stress region was related to gross melt fracture. In the higher-stress region, the true shear rate in the polymer was close to zero, and the material slipped through the capillary like a rubbery solid.

In a different study, Hatzikiriakos and Dealy empoyed a sliding-plate vis-cometer (see Sec. VI of Chap. 2) to measure slip of high-density polyethylene (HDPE) in the absence of entrance effects and with no pressure gradient present [20]; Fig. 17.4 shows that beyond a critical shear stress of 0.09 MPa, the flow curve of HDPE depends on the gap spacing, demonstrating the presence of wall slip. If it is assumed that the slip velocity, u_s, is the same at each plate surface, the true shear rate $\dot{\gamma}$ will be

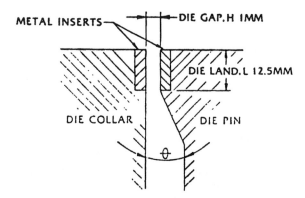

FIGURE 17.3 Schematic die land insert design for screening materials of construction. (From Ref. 18.)

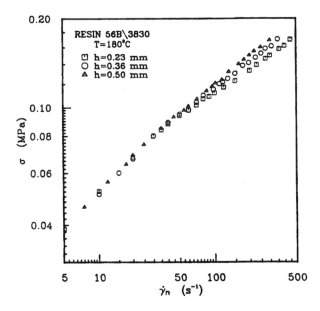

FIGURE 17.4 The effect of gap spacing on the flow curve of HDPE. (From Ref. 20.)

$$\dot{\gamma} = \frac{V - 2u_s}{h} \tag{17.1}$$

where V is the velocity of the moving plate and h is the gap spacing. Rearranging Equation 1:

$$\frac{V}{h} = \dot{\gamma} + \frac{2u_s}{h} \tag{17.2}$$

So a straight line should result if V/h, or the nominal shear rate $\dot{\gamma}_n$, is plotted against the reciprocal of h while keeping the shear stress unchanged. This is because a constant value of the shear stress implies a constant value of the true shear rate. The slope of this straight-line plot should equal $2u_s$. Hatzikiriakos and Dealy found that HDPE slip velocity determined in this manner was a unique function of the wall-shear stress but depended on the temperature of measurement. Their results for the slip velocity are displayed in Fig. 17.5, and these reveal a power-law relationship between slip velocity and the wall-shear stress. These and other authors have also proposed models to relate the slip velocity, under both steady-and transient-stress conditions, to temperature, pressure, flow geometry, and bulk rheology of the polymers examined [21–24].

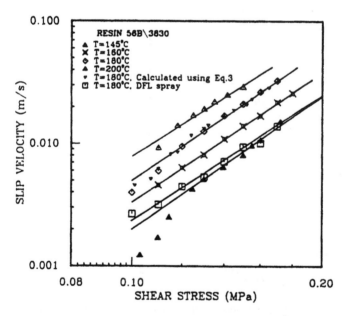

FIGURE 17.5 Slip velocity of HDPE as a function of the wall shear stress. (From Ref. 20.)

IV. ADHESION PROMOTION OR SLIP PROMOTION?

Since melt fracture appears to be the result of loss of adhesion between molten polymer and the die wall, it is logical to conclude that additives or die wall materials of construction that promote adhesion should lead to smooth extrudates. This, indeed, was the opinion of Ramamurthy [18], and it was partially confirmed by Hatzikiriakos and Dealy [20,25]. The latter authors coated the surfaces of a sliding-plate rheometer with a fluorocarbon commonly used as a mold release agent in injection molding. The result was a significant *reduction* in the slip velocity of HDPE and a concomitant improvement in extrudate quality. On the other hand, the use of a coating of a fluoroelastomer, which also elimated melt fracture, was accompanied by the reduction in the critical-shear stress for the onset of slip and an increase in the slip velocity [19,20,25]! Kazatchkov et al. found that with polypropylenes also a fluoropolymer coating acted as a slip promoter [26].

Why both adhesion promoters and slip promoters should prevent melt fracture is not entirely clear. If it is assumed though that for each polymer a critical wall-shear stress exists for the onset of melt fracture, the presence of wall slip will reduce the true shear rate at a given volumetric flow rate. (That the true shear rate does decrease in the presence of a fluoropolymer coating is established by the fact that the extent of die swell reduces [27].) Thus, with wall slip, the flow rate can be boosted and melt fracture delayed till the true shear rate at the wall again reaches a level at which the wall-shear equals the critical-shear stress [28]. This may explain why slip promoters reduce melt fracture.

V. CONCLUDING REMARKS

The cause and cure of melt fracture have remained an enigma for a long time. Although significant progress has been made in the past several years, during which wall slip has been identified as being an important element in melt fracture, a complete understanding of the phenomenon is still not at hand. A point that is currently being debated is whether slip occurs due to failure of adhesion at the wall (adhesive failure) or due to failure within the polymer but in the proximity of the wall (cohesive failure). While measurements of the critical-shear stress for the onset of slip appear to depend on the work of adhesion of the interface, suggesting adhesive failure [29,30], other measurements made on LLDPE using a sliding-plate rheometer show the presence of cohesive failure at a clean interface [31]. Clearly, rheologists will continue to be busy for the foreseeable future!

REFERENCES

1. C.J.S. Petrie, M.M. Denn. Instabilities in polymer processing. AIChE J. 22:209–236 (1976).

2. S. Middleman. Fundamentals of Polymer Processing. McGraw-Hill, New York, 1977.
3. R.G. Larson. Instabilities in viscoelastic flows. Rheol. Acta 31:213–263 (1992).
4. E.S.G. Shaqfeh. Purely elastic instabilities in viscometric flows. Annu. Rev. Fluid Mech. 28:129–185 (1996).
5. M.M. Denn. Stability of Reaction and Transport Processes. Prentice Hall, Englewood Cliffs, NJ, 1975.
6. S. Kase. Studies on melt spinning. IV. On the stability of melt spinning. J. Appl. Polym. Sci. 18:3279–3304 (1974).
7. R.K. Gupta, R.L. Ballman. A study of spinline dynamics using orthogonal collocation. Chem. Eng. Commun. 14:23–33 (1982).
8. M.M. Denn. Issues in viscoelastic fluid mechanics. Annu. Rev. Fluid Mech. 22:13–34 (1990).
9. D.V. Boger, K. Walters. Rheological Phenomena in Focus. Elsevier, Amsterdam, 1993.
10. R.C. Chan, R.K. Gupta, T. Sridhar. Elastic fracture of polymeric fluids. Chem. Eng. Commun. 53:85–96 (1987).
11. A.V. Ramamurthy. Flow instabilities in a capillary rheometer for an elastic solution. Trans. Soc. Rheol. 18:431–452 (1974).
12. J.H. Southern, R.L. Ballman. Solution fracture barrier in wet spinning. Textile Res. J. 53:230–235 (1983).
13. F.N. Cogswell. Stretching flow instabilities at the exits of extrusion dies. J. Non-Newt. Fluid Mech. 2:37–47 (1977).
14. S.J. Kurtz. Die geometry solutions to sharkskin melt fracture. Proc. IX Intl. Congress on Rheol., Acapulco, Mexico, 1984, Vol. 3, pp. 399–407.
15. F.N. Cogswell, P. Lamb. The mechanism of melt distortion. Trans. J. Plast. Inst. 35:809–813 (1967).
16. J.H. Southern, D.R. Paul. Elastic fracture of polystyrene solutions. Polym. Eng. Sci. 14:560–566 (1974).
17. R. Rothenberger, D.H. McCoy, M.M. Denn. Flow instability in polymer melt extrusion. Trans. Soc. Rheol. 17:259–269 (1973).
18. A.V. Ramamurthy. Wall slip in viscous fluids and influence of materials of construction. J. Rheol. 30:337–357 (1986).
19. S. Hatzikiriakos, J.M. Dealy. The effect of interface conditions on wall slip and melt fracture of high density polyethylene. SPE ANTEC 37:2311–2314 (1991).
20. S.G. Hatzikiriakos, J.M. Dealy. Wall slip of molten high density polyethylenes. I. Sliding plate rheometer studies. J. Rheol. 35:497–523 (1991).
21. D.A. Hill, T. Hasegawa, M.M. Denn. On the apparent relation between adhesive failure and melt fracture. J. Rheol. 34:891–918 (1990).
22. S.G. Hatzikiriakos, J.M. Dealy. Wall slip of molten high density polyethylenes. II. Capillary rheometer studies. J. Rheol. 36:703–741 (1992).
23. M.M. Denn. Surface-induced effects in polymer melt flow. Proc. XI Intl. Congress on Rheol., Brussels, 1992, pp. 45–49.
24. S.G. Hatzikiriakos, N. Kalogerakis. A dynamic slip velocity model for molten polymers based on a network kinetic theory. Rheol. Acta 33:38–47 (1994).

25. S.G. Hatzikiriakos, J.M. Dealy. Effect of interfacial conditions on wall slip and sharkskin melt fracture of HDPE. Intern. Polym. Process. 8:36–43 (1993).

26. I.B. Kazatchkov, S.G. Hatzikiriakos, C.W. Stewart. Extrudate distortion in the capillary/slit extrusion of a molten polypropylene. Polym. Eng. Sci. 35:1864–1871 (1995).

27. S.G. Hatzikiriakos, P.Hong, W. Ho, C. Stewart. The effect of Teflon™ coatings in polyethylene capillary extrusion. J. Appl. Polym. Sci. 55:595–603 (1995).

28. E.E. Rosenbaum, S.G. Hatzikiriakos, C.W. Stewart. Flow implications in the processing of tetrafluoroethylene/hexafluoropropylene copolymers. Intern. Polym. Process. 10:204–212 (1995).

29. S.G. Hatzikiriakos, C.W. Stewart, J.M. Dealy. Effect of surface coatings on wall slip of LLDPE. Intern. Polym. Process. 8:30–35 (1993).

30. S.H. Anastasiadis, S.G. Hatzikiriakos. The work of adhesion of polymer/wall interface and its association with the onset of wall slip. J. Rheol. 42:795–812 (1998).

31. R.S. Jeyaseelan, J.M. Dealy, A.D. Rey. Wall slip of molten polymers: recent observations. Proc. XII Intl. Congress on Rheol., Quebec City, Canada, 1996, pp. 103–104.

Index

387